*Ecology and Evolution
of Darwin's Finches*

Hermanos IV, the most isolated of a group of four islands near Isabela, has the outline of a giant tortoise, for which the Galápagos have been named.

Ecology and Evolution of Darwin's Finches

PETER R. GRANT

PRINCETON UNIVERSITY PRESS

Princeton, New Jersey

Copyright © 1986 by Princeton University Press
Published by Princeton University Press,
41 William Street, Princeton, New Jersey 08540
In the United Kingdom: Princeton University Press,
Chichester, West Sussex

LIBRARY OF CONGRESS CATALOGING-IN-PUBLICATION DATA

Grant, Peter R. 1936-
Ecology and evolution of Darwin's finches.

Bibliography: p.   Includes index.
1. Finches—Evolution. 2. Finches—Ecology. 3. Birds—Galapagos Islands—Evolution.
4. Birds—Galapagos Islands—Ecology. I. Title. II. Title: Darwin's Finches.
QL696.P246G73 1986   598.´83   86-9352
ISBN 0-691-08427-0 (alk. paper)   ISBN 0-691-08428-9 (pbk. : alk. paper)

This book has been composed in Linotron Times Roman

Princeton University Press books are printed on acid-free paper
and meet the guidelines for permanence and durability of
the Committee on Production Guidelines for Book Longevity of the
Council on Library Resources

Printed in the United States of America

10 9 8 7 6 5 4 3 2

For
Rosemary
Nicola and Thalia

Contents

Preface   *xi*

ONE   Introduction   *3*

*Charles Darwin, 6. After Darwin, 9. The first synthesis,
9. Evolutionary inference, 10. Plan of the book, 12.*

TWO   Characteristics of the Islands   *19*

*Origins and ages, 19. Distribution and sizes, 21. Climate,
21. Plants, 27. Vegetation, 28. Changes in the past, 29.
Changes in recent times, 30. Cocos Island, 31.
Summary, 31.*

THREE   General Characteristics and Distributions of Finches   *45*

*The main groups, 45. Genera, 51. Species, 51.
Subspecies, 60. Distributions, 60. Patterns among the
islands, 62. Extinctions, 64. Other land birds, 64.
Summary, 65.*

FOUR   Patterns of Morphological Variation   *77*

*Introduction, 77. The major simple patterns, 77. The mi-
nor simple patterns, 79. Correlations between traits, 79.
Size, 80. Allometry, 82. Shape, 82. Multivariate
shape variation, 89. Geographical variation in size, 92.
Summary, 95.*

FIVE   Growth and Development   *100*

*Introduction, 100. Variation in egg size, 100. Absolute
growth, 102. Relative growth, 106. Summary, 111.*

SIX   Beak Sizes, Beak Shapes, and Diets   *113*

*Introduction, 113. Feeding mechanics, 113. Feeding
types, 116. Ecological significance of beak differences between species, 117. Dietary differences between species,
118. Dietary differences between populations of the same*

species, 128. Dietary differences among individuals in a variable population, 132. Summary, 138.

SEVEN   The Importance of Food to Finch Populations   *147*

Introduction, 147. Plant phenology in the arid zone, 147. Finch phenology, 148. Finch populations in relation to food supply, 152. Extreme conditions, 152. Food limitation of population sizes, 154. The frequency of food limitation, 168. Other factors limiting finch populations, 171. Interspecific competition for food, 173. Summary, 173.

EIGHT   Population Variation and Natural Selection   *175*

Introduction, 175. Relative variation, 175. Theoretical background, 177. Field studies, 180. Genetic variation, 180. Natural selection, 183. Sexual selection, 192. Countervailing selection, 193. A summary of selection pressures, 195. Sexual dimorphism, 196. Genetic drift, 197. Enhancement of genetic variation, 199. Variation in relation to abundance, 207. Other species, 208. Summary, 219.

NINE   Species-Recognition and Mate Choice   *222*

Introduction, 222. The possible cues used in species-recognition, 222. Morphological cues, 224. Song, 251. Song and bill morphology as species cues, 241. Imprinting, 242. The learning of heterotypic song, 244. Misimprinting, 246. Beyond species-recognition: mate choice, 249. Summary, 251.

TEN   Evolution and Speciation   *253*

Evolution, 253. Origins, 253. The number of species, 256. The pattern of speciation, 257. The time framework, 260. Allopatric speciation, 263. Alternative models of speciation, 273. Parapatric speciation, 274. Sympatric speciation, 275. Alternatives to gradual genetic change, 280. Conclusions and summary, 283.

ELEVEN   Ecological Interactions during Speciation   *285*

Introduction, 285. Ecological isolation, 285. Causes of initial differentiation, 286. An alternative view, 288. Differentiation entirely in allopatry, 289. The food supply

hypothesis, 291. Lack's evidence for competition, 294.
Tests of the competition hypothesis, 300. Different explan-
ations reconciled, 310. Conclusions and summary, 312.

TWELVE Competition and Finch Communities   *314*

Introduction, 314. Combinations of species, 315. Structure
determined by competition, 317. Minimum differences be-
tween coexisting species, 321. Greater than minimum dif-
ferences, 323. A digression on methods of analysis, and on
bias, 328. Predictive models, 331. The classical case of
character release, 340. Conclusions and summary, 346.

THIRTEEN The Evolution of Reproductive Isolation   *348*

Introduction, 348. Experimental tests, 348. Implications of
the experimental results, 350. Reinforcement? 353.
Absence of species from islands, 354. Summary, 355.

FOURTEEN Adaptation: Body Size, Plumage Coloration, and
Other Traits   *357*

Introduction, 357. Historical survey, 358. Body size, 359.
Plumage, 364. Other features, 371. Summary, 373.

FIFTEEN Reconstruction of Phylogeny   *375*

Introduction, 375. Reconstructing the process of morpho-
logical divergence, 375. Comparison with contemporary
selection, 379. Further evolution, 380. Ontogeny, 381.
Phylogeny, 383. Summary, 387.

SIXTEEN Recapitulation and Generalization   *389*

Introduction, 389. Patterns and processes among modern
finches, 389. Evolution, 397. Generalizations, 401.

APPENDIX: Spanish and English Names of the Major
Galápagos Islands   *413*

References   *415*

Author Index   *437*

Subject Index   *442*

Preface

FIFTEEN years ago a curator of birds at a prominent museum asked me what my next research would be. I told him I was thinking of studying Darwin's Finches, to which he replied "But I thought they had been done!" By this he meant that all the species had been well characterized taxonomically, and a comprehensive interpretation of their evolution had been published by David Lack in 1947, extended by Robert Bowman in 1961, and incorporated into the fabric of modern, general, evolutionary theory (Mayr 1963).

In fact Lack's admirable book had not solved all problems posed by the finches (Wynne-Edwards 1947, Ratcliffe and Boag 1983), nor had Bowman's monograph. A commentary on Lack's book by Richards (1948, p. 84) had, by 1971, lost none of its freshness:

> . . . Mr. Lack's theories come out reasonably well, but there seems to be a permanent dichotomy between those who find their zest in the natural selection theory and those who cannot. The problem, for instance, of whether certain specific characters (e.g. beaks of *Geospizinae*) are or are not adaptive (vague term) is one involving studies of field populations, genetics, and variation. It is a problem which might be largely solved by teams of workers dealing for a number of years with a particularly favourable example. It is not one, I think, which can be settled in a four months stay, even if supplemented by the examination of skins in museums. It is not, of course, that the beaks may not be adapted to different uses, but that it is extremely difficult to prove that this is so to an extent which affects the level of the population.

This book meets the challenge thrown down by Richards. It provides an account, from long-term field studies, of ecological and evolutionary processes occurring in populations of Darwin's Finches over a period of more than a decade. It integrates these processes with the evolutionary patterns so well described by Lack (1947) and Bowman (1961) to produce an explanation for the evolution of the group as a whole. It then compares the adaptive radiation of the finches with several other diverse groups of animals to see how far the finches can be treated as a model of evolutionary diversification.

It would be false of me, however, to imply that I studied Darwin's Finches specifically in order to meet Richards' challenge. The book, and the research that gave rise to it, had different origins. I chose to study the finches for two, quite different, reasons. The first arose from a confusion about the significance of population variation. Leigh Van Valen had proposed in 1965 that population variation in the size of such traits as the beaks of birds is adaptive, in the sense that different variants in a population occupy slightly different ecological niches and experience slightly different selective forces. He tested this idea with several comparisons involving, among others, Canary Island, Azores Island, and European mainland samples of chaffinches (*Fringilla coelebs*) and blue tits (*Parus caeruleus*). In general he found strong empirical support for the idea. His comparisons were not the most logical, however, and when I performed the proper comparisons I found little evidence to support his views (Grant 1979a, 1979b). Little or no evidence is not the same as contrary evidence, so to investigate a more promising situation in which natural selection could be studied, I asked what are the most variable populations of birds that might be amenable to ecological and behavioral field study? From David Lack's book I knew the answer: Darwin's Finches.

The second reason sprang from a similar confusion concerning interspecific competition. On the one hand, I had just completed a series of field experiments which had demonstrated competition between species of rodents (Grant 1972a). On the other hand, I had taken a critical look at the evidence for character displacement, that is, the tendency for differences between ecologically similar species to be enhanced where they occur together as a result of natural selection having minimized competition between them (Brown and Wilson 1956); the evidence, including that for the so-called classical case of character displacement, looked very weak (Grant 1972b). Therefore I was confronted with this dilemma: competition was a demonstrable process in nature, now, but some of the best evidence for competition as a historical process had crumbled; hence its importance in contributing to the structure of communities was called into question. Since the classical case of character displacement was invalid (Grant 1972b, 1975a), it was logical to turn attention to the classical case of character release. Character release refers to changes in the morphology and ecology of a species apparently resulting from the absence of restraints from a competitor species (Grant 1972b). The classical case involves two species of Darwin's Finches on the Galápagos Islands (Brown and Wilson 1956).

Despite having two good reasons for studying the finches, I might never have begun research on them without the stimulus of a proposal from a prospective postdoctoral Fellow, Ian Abbott. He had developed a plan for detecting the effects of interspecific competition among Darwin's Finches, having been struck by the fact that Lack and Bowman had come to different views about competition from examining the same material. We prepared a research proposal and sought financial support. And here we ran into a problem that confronts many research workers today. Funds are possible to obtain, providing the project is good, if preliminary data can be marshaled to show that the proposed project has a good chance of being successful. But without funds, how could we get the preliminary data, 6000 miles from home, to show this? We were very fortunate to be funded by the Faculty of Graduate Studies and Research of McGill University, in what must have been partly an act of faith. Had we not received the money we would probably have given up, as alternative prospects looked bleak. I am grateful beyond measure for that initial support.

The first visit to Galápagos in 1973 was an end in itself, and was dominated by the fieldwork of Ian and Lynette Abbott. That I have returned in most years since is due to conversations with Tjitte de Vries, then a member of the staff of the Charles Darwin Research Station on Isla Santa Cruz. At the end of the first visit he pointed out that our work, conducted in the late wet season and early dry season, should be repeated at the end of the dry season when environmental conditions are so different. After I had followed his advice, he persuaded me that no two years are alike, and our research has continued ever since. The book grew out of this continuing research.

I have had the good fortune to collaborate with several enthusiastic and talented people: Ian Abbott, Lynette Abbott, Jamie Smith, Peter Boag, Laurene Ratcliffe, Trevor Price, Dolph Schluter, Lisle Gibbs, Stephen Millington, and my wife, Rosemary. They and I, variously, constitute the "we" used frequently in this book. We have been ably assisted by David Anderson, Chris Chappell, Robert Curry, Phillip deMaynadier, James Gibbs, Erick Greene, Sylvia Harcourt, Margaret Kinnaird, William Johnson, Steven Latta, Douglas Nakashima, Scott Stoleson, Jon Weiland, Michael Wells, David Wiggins, Tom Will, my daughters Nicola and Thalia, and many others to whom I can record my gratitude only collectively and anonymously.

Our research has been carried out with the permission and support of the Dirección General de Desarrollo Forestal, Quito, Ecuador, the Servicio Parque Nacional Galápagos, the Charles Darwin Foundation, and the Charles Darwin Research Station. I must single out for special acknowledgment of their considerable help, first Miguel Cifuentes, Intendente of S.P.N.G., and second the directors of the Charles Darwin Research Station since 1973: Peter Kramer, Craig MacFarland, Hendrick Hoeck, Friedemann Köster, and Günther Reck. The research has been funded by the National Research Council (Canada), the National Science Foundation (U.S.) and grants from the Chapman Fund of the American Museum of Natural History, from McGill University, and from the University of Michigan. As part of the terms for receiving support from the National Science Foundation, I declare that "Any opinions, findings, and conclusions or recommendations expressed in this publication are those of the author and do not necessarily reflect the views of the National Science Foundation.''

I thank three groups for giving me the impetus to put my thoughts together on Darwin's Finch evolution by inviting me to give series of lectures: to Staffan Ulfstrand, Per Brinck, and the Nordic Ecology Council for the opportunity to give seminars at the Universities of Uppsala and Lund in Sweden in 1981; to Henry Horn and Robert May for the invitation to give the MacArthur lectures at Princeton University in 1982; and to Chris Simon and Ken Kaneshiro for the invitation to give the inaugural lectures of the Hawaiian Evolutionary Biology Program at the University of Hawaii in 1984. The book was written while I was at the University of Michigan. While preparing it I have been supported by a Fellowship from the J. S. Guggenheim Foundation. I was able to improve the book in a multitude of ways, thanks to the many suggestions offered by Ian Abbott, Peter Boag, Bob Curry, Lisle Gibbs, Trevor Price, Laurene Ratcliffe, Dolph Schluter, Jamie Smith, and my wife, who read the manuscript. Bob Storer, Tom Sherry, and Tracey Werner gave me help with certain sections. The photographs are my own except where otherwise credited, and I am very grateful to the many people who kindly supplied their photographs: Peter Boag, Bill Clark, Eberhard Curio, Bob Curry, David Day, Tjitte de Vries, Nicola Grant, Michael Harris, Cleveland Hickman Jr., Hendrick Hoeck, Friedemann Köster, Yael Lubin, Doug Nakashima, Åke and Ulla Norberg, Roger Perry, Fritz Pölking, Alan Root, Dolph Schluter, Tom Sherry, Alan Simon, Jamie Smith, Heidi Snell, and Tracey Werner. Thalia Grant helped by drawing Figures 2, 6, 14, 23, 30, 31, 74, 100, and 101, Bob Bowman and Laurene Ratcliffe kindly supplied some sonagrams, and Margaret Madouse and Tina Dzienis typed the manuscript. Finally, the research would never have taken the direction it did, nor would the book have been written, if I had not had the enthusiastic companionship and support of my family.

Princeton University
September 1985

*Ecology and Evolution
of Darwin's Finches*

CHAPTER ONE

Introduction

A MAJOR TASK for biologists is to explain organic diversity in terms of evolutionary principles: to explain why so many different types of species have arisen, and why they vary so much in form, function, and behavior. The way to answer these questions is to establish the relationships among species, with the help of fossils wherever this is possible, and to interpret past evolutionary processes in terms of ecological and evolutionary forces acting on modern species. Some groups of organisms are especially suitable for study, for example the *Drosophila* flies, *Cepaea* snails and *Heliconius* butterflies. Among the favorable groups must be numbered Darwin's Finches on the Galápagos and Cocos Island (Fig. 1). They are a textbook example of adaptive radiation: the evolutionary diversification of a single lineage into a variety of species with different adaptive properties.

The diversity of form and function within this single subfamily of passerine birds (Geospizinae) is remarkable, especially in beak structure and associated feeding habits (Plates 1–6, Fig. 2). For example, one species, the woodpecker finch (*Cactospiza pallida*), uses cactus spines, leaf petioles, or twigs (Plate 2) to pry insect larvae and pupae and termites out of cavities in the dead branches of trees in arid zone woodland (Gifford 1919, Lack 1947), and a related species, the mangrove finch (*C. heliobates*), performs similar maneuvers in mangroves (Curio and Kramer 1964). On the islands of Wolf and Darwin, sharp-beaked ground finches (*Geospiza difficilis*) perch on large seabirds (boobies, *Sula* species), peck at the developing feathers of the tail and wing (Plate 5), draw blood, and drink it (Gifford 1919, Bowman and Billeb 1965, Schluter and Grant 1984a). They also crack the eggs of seabirds by kicking or pushing them against rocks, then drink the contents (Schluter and Grant 1984a). Other finches (*G. fortis*, *G. fuliginosa*, and *G. difficilis*) remove ticks from tortoises, marine iguanas (Plate 6), and land iguanas (Plate 11), responding in some instances to ritualized soliciting behavior of the reptiles (MacFarland and Reeder 1974, Eibl-Eibesfeldt 1984). In addition to these unique feeding habits, Darwin's Finches show a very broad range of habits and skills, from the cactus-exploiting habits of the cactus finch, *G. scandens* (Plates 4 and 12), to the bark-stripping habits of tree finches such as the vegetarian finch, *Platyspiza*

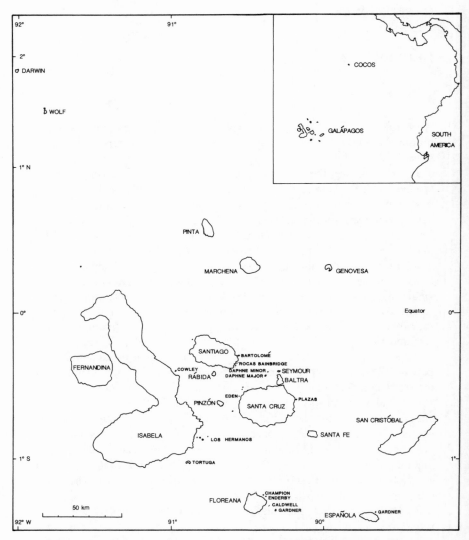

FIG. 1. Galápagos Islands. Cocos Island is approximately 600 km to the northeast (see also Fig. 5). See Appendix for English names of major islands.

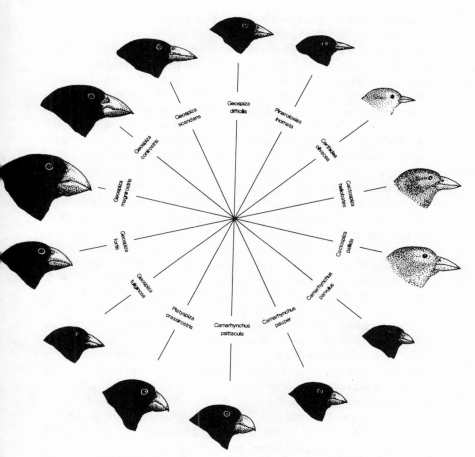

FIG. 2. The adaptive radiation of fourteen species of Darwin's Finches. Based in part on drawings in Swarth (1931) and a diagram in Bowman (1961).

crassirostris (Plates 4 and 13). They rip open rotting cactus pads, kick over stones, probe flowers and search for arthropods in leaf litter and on the exposed rocks of the shoreline at low tide. They consume nectar, pollen, leaves, buds, a host of arthropods, and seeds and fruits of various sizes.

Their diversity is but one aspect of their fascination. It is complemented by the extreme similarity of some of the species, which is highlighted by the following three quotations: "The series before me includes specimens that can almost as well be referred to either of two contiguous species, so that their position can only be determined by assigning to each species what must be called arbitrary standards of measurements of the bill alone" (Salvin 1876, p. 469). This is not just a problem that arises from inadequate samples, for nearly one hundred years later Bowman (1961, pp. 262–263) wrote: "Even with the exceptionally fine material I have had available in the form of skeletons, skins, and entire specimens in alcohol, I would not guarantee the correct species designation of all birds examined, even from one island such as Indefatigable [Santa Cruz], so similar are certain individuals of two different species. This applies particularly to very large individuals of *Geospiza fortis* and certain small individuals of *G. magnirostris*." With regard to those same species, Lack (1947, p. 87) summed it up as follows: "[Some] individuals are so intermediate in appearance that they cannot safely be identified—a truly remarkable state of affairs. In no other birds are the differences between species so ill-defined."

Thus there are species that are so similar it would appear they have only recently formed from a common ancestor, and others so different in feeding habits and associated beak structure as to suggest an ancient origin: ". . . the extraordinary variants that crop up in many of the series [of specimens in museums] give an impression of change and experiment going on" (Swarth 1934, p. 231). All of this is fruitful variation for biologists who attempt to understand the patterns and processes of evolution.

The attempt began with Darwin: "The most curious fact is the perfect gradation in the size of the beaks of the different species of Geospiza— Seeing this gradation and diversity of structure in one small, intimately related group of birds, one might really fancy that, from an original paucity of birds in this archipelago, one species had been taken and modified for different ends" (Darwin 1842, p. 458; see also Fig. 3). It is my purpose in this book to explore and interpret those *modifications* and *ends*.

CHARLES DARWIN

Collectively these species are known as Darwin's Finches (Lowe 1936) in recognition of Darwin's part in bringing them fame. Darwin was not the first person to see them or to write about them, however; those honors belong to some member of the entourage of the Inca Tupac Yupanqui (Beebe 1924),

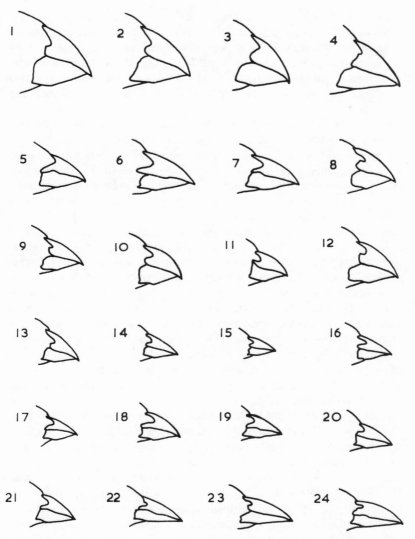

FIG. 3. The intergradation of sizes and shapes among the six species of ground finches is illustrated with 24 male specimens from twelve islands. The species are *G. magnirostris* (1–3), *G. conirostris* (4–7), *G. fortis* (8–13), *G. fuliginosa* (14–15), *G. difficilis* (16–21), and *G. scandens* (22–24). From Abbott et al. (1977).

and to Colnett (1798), respectively. But Darwin, on his visit to the Galá-pagos in 1835, was the first person to record their habits carefully, and, to-gether with Captain FitzRoy and companions on the H.M.S. *Beagle*, to col-lect specimens which were later described by Gould (1837). He wondered about their origin, their evolution, and their variation, and wrote about them in several works (Darwin 1839, 1841, 1845). Scientifically he placed them on the map.

But Darwin fell far short of providing a detailed explanation for the evo-lution of the finches. There are several reasons for this. For one thing, he did not see all the species, and collected specimens of only nine of them. More importantly, he was thoroughly confused by them. The small differ-ences between some species on the same island, and between populations on different islands, made it difficult for him to define the limits to variation of individual species (e.g. see Sulloway 1982a, p. 6), and hence to deter-mine the number of species, given the small number of specimens he had. In fact, while he was on the Galápagos he only distinguished, and recog-nized as finches, six of the species. As mentioned just above, he observed an almost perfect gradation in the size of beaks among the different species of *Geospiza*. Problems of identification, which are experienced by every visitor to Galápagos today (Harris 1974), were unwittingly compounded by Darwin's failure to label separately the specimens he collected on different islands. To some extent these problems were resolved after his return to England by John Gould, who also pointed out that the finches constituted a single group, and not a heterogeneous collection of finches, tanagers, and warblers as Darwin, impressed by the diversity of beaks, had originally sup-posed (Sulloway 1982a, 1982b).

Darwin was confronted by an additional difficulty, once the problems of classification had been removed. He lacked clear evidence of geographical replacement of finch species: that is, the occurrence of similar species on different islands, with one species "replacing" another. Geographical re-placement was a requirement for his argument that species were mutable. Mockingbirds, tortoises, and plants on the Galápagos provided this evi-dence (Sulloway 1984), but the distributions of finches were too imperfectly known to allow more than the tentative conclusion: ". . . hence we may feel almost sure that these islands possess their representative species of these two sub-groups [*Cactornis* and *Camarhynchus*]" (Darwin 1842, p. 475). Even in this case his facts were probably wrong (Sulloway 1982a). So although he was able to devise an explanation for evolution and adaptive radiation on islands in general, he did not apply it specifically to the finches.

Just how important the finches were to Darwin's thinking and theorizing about evolution is difficult to know, and even historical studies of all of his writings leave room for guesswork. It seems clear to me that his reflections on them, after he returned to England (Lack 1947, Sulloway 1984), contributed to his recognition of that great organizing principle in biology, the principle of natural selection. But, owing to the complexities of the evidence, he did not mention the finches in any of his four notebooks on transmutation of species, or even in the *Origin of Species* (Sulloway 1982a).

AFTER DARWIN

In the hundred years that followed Darwin's visit to the Galápagos, problems arising from uncertain taxonomy and distribution, and from inadequate description of subspecific variation, were resolved. A great deal of collecting activity for museums took place at the turn of the last century, culminating in the year-long expedition of the California Academy of Sciences in 1905–1906. It is fortunate this expedition was not mounted a year earlier, for while its members were assembling thousands of specimens in the Galápagos, the San Francisco earthquake destroyed the collections in the parent institution. Many years later Swarth (1931) used these specimens and others from museums around the world to produce the first modern taxonomic treatment of the finches. It dealt with some species that were discovered only at the end of the nineteenth century, and hence unknown to Darwin: notably the large cactus finch *Geospiza conirostris* (Ridgway 1890), the medium tree finch *Camarhynchus pauper* (Ridgway 1890), and the mangrove finch *Cactospiza heliobates* (Snodgrass and Heller 1901). David Lack (1945) then simplified the taxonomic results, and rendered the judgment that has persisted until today; there are fourteen species, thirteen on the Galápagos and one on Cocos Island.

Swarth's monograph did more than just classify the species, it laid the foundation for the first comprehensive model of the diversification of the finches, developed and published by David Lack in 1945. Often referred to as the allopatric model of speciation, it incorporated and elaborated the ideas sketched out and applied to Darwin's Finches a little earlier by Stresemann (1936), and was remarkably faithful to Darwin's (1859) general reasoning on the causes of adaptive radiation. Lack acknowledged these sources for his ideas, indeed every one of the sixteen chapters of his book is preceded by a quotation from Darwin's works, but Stresemann's important contributions were eclipsed by Lack's more comprehensive treatment (Lack 1945, 1947).

THE FIRST SYNTHESIS

All but a few specialists who had examined the morphology and anatomy of the finches, from John Gould onwards, agreed that they formed a monophyletic group, the members of which were more similar to each other than to any living finch species in Central or South America. There was general, although not unanimous, agreement that the Galápagos had never been connected to the mainland by a land bridge. The problem, then, which Lack confronted, was to account for the generation of fourteen species from a single and unknown ancestral species that had colonized the islands by overwater flight from the continent: to Cocos and then to Galápagos, to Galápagos first, or to both locations separately and independently (Steadman 1982).

Lack marshaled all the information available on morphological variation, distribution, and feeding and breeding habits, and placed it in the newly developed framework of the Huxley-Mayr modern synthesis of evolutionary thought (Huxley 1940, 1942; Mayr 1942). The most explicit statement of Lack's views on the evolution of Darwin's Finches is worth quoting in full (Lack 1947, p. 159):

> It may be supposed that the ancestral finch first became differentiated into various forms in geographical isolation on different islands. After a significantly long period of isolation, some of these forms had become so different that, when by chance they met on the same island, either they were already intersterile, or hybrids between them were at a selective disadvantage so that intersterility was evolved. Thus new species originated. But when they met, such new species must have tended to compete with each other; both could persist together only if they had diverged sufficiently for each to prove better adapted to one part of the original food supply or habitat, which was then divided between them. After such restriction in ecology, each became increasingly specialized. Each then spread to other islands, so that it became differentiated into new geographical forms, some of which in their turn later met on the same island and kept distinct, thus giving rise to further new species, and resulting in further ecological restriction and structural specialization. In this way there appeared the adaptive radiation whose end-forms are alive today.

Those end-forms show different degrees of morphological similarity, and hence of presumed relationship. On this basis Lack (1947) constructed a tentative evolutionary tree which is reproduced in Figure 4. The ecologically restricted and structurally most specialized species are shown at the top of the tree. These include one that resembles a woodpecker (*Cactospiza pallida*) and another that resembles a warbler (*Certhidea olivacea*). The supposedly least modified species, *Geospiza difficilis*, is closest to the basal stock.

EVOLUTIONARY INFERENCE

Darwin, Swarth, Stresemann, and Lack all attempted to explain current patterns of Darwin's Finch species in terms of past evolutionary processes. The

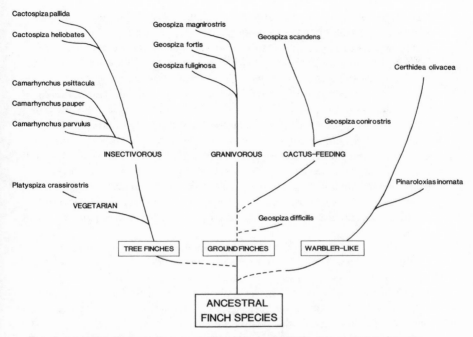

FIG. 4. A tentative phylogenetic tree, constructed by Lack (1947). An ancestral finch species gave rise to three main groups, each of which differentiated into two or more species. The uncertain nature of the origin of the groups is indicated by broken lines. Redrawn from Lack (1947).

patterns were observed, dimly by Darwin and more clearly by the others; the processes were inferred. For example, Lack observed the following pattern: there are small morphological differences between some allopatric populations (subspecies), larger differences between other allopatric populations (species), and yet larger differences between sympatric populations of different species. He inferred a process of differentiation that began in allopatry and continued in sympatry. He hypothesized the causes and characteristics of the process.

Inferences about historical events are inevitably uncertain (Mayr 1961). Some ecologists go so far as to assert that processes cannot be inferred from patterns (e.g. Connor and Simberloff 1983, Wiens 1984). I agree, but only providing the phrase "with certainty" is added. Faced with uncertainty, an evolutionary biologist behaves somewhat like a detective called in to solve a crime. He combines all available clues about past events, including those provided by fossils, with a study of current processes, postulates the most likely courses of past events given this information, and then tests the postulates with new data whenever possible to see which has the most explanatory power. For organisms like Darwin's Finches that are not suitable for some types of laboratory investigation, this means applying the techniques of population biology to the study of phenotypic variation.

The studies of Darwin, Swarth, Stresemann, and Lack were restricted by ignorance of current processes. Darwin's visit to the Galápagos lasted for only five weeks, and Lack stayed for less than four months. Neither returned. Stresemann never visited the islands and Swarth only did so, briefly, after publishing his taxonomic monograph. Therefore, their goals were to offer coherent evolutionary explanations for the patterns known to them, not to test those explanations by subsequent observation and experimentation. Darwin (1842, p. 603) wrote candidly of the drawbacks of short studies: "... as the traveller stays but a short time in each place, his descriptions must generally consist of mere sketches, instead of detailed observations. Hence arises, as I have found to my cost, a constant tendency to fill up the wide gaps of knowledge by inaccurate and superficial hypotheses."

In the last forty years travel to the Galápagos islands and working conditions there have become much easier. The islands now form part of the National Parks system of Ecuador, and research has been helped by the establishment in 1964 of the Charles Darwin Research Station on Isla Santa Cruz. It should no longer be necessary for anyone to say, as Lack (1947, pp. 5–6) did about his working environment, that it is "sometimes curious and nearly always unpleasant." Scientists can now make repeated visits for long periods, check on their previous findings, correct past errors, embark on ecological studies that were simply beyond the reach of earlier visitors, and test hypotheses of past and present processes.

PLAN OF THE BOOK

This book integrates the results of modern studies with those of earlier ones to arrive at a comprehensive theory of the evolution of Darwin's Finches. It is organized in four sections. The first introduces the general characteristics of islands and finches. The second describes morphological patterns of finches, and shows how differences in the growth of individuals give rise to differences among adults of different species. The third deals with the ecology of finches; the major topics are the relationship between beak shape and diets, the importance of a fluctuating food supply to individuals and to populations, and the breeding ecology and behavior of finches, especially the ways in which mates are chosen and hybridization is usually avoided. The final section discusses evolution, the central problem being to explain the diversification of the group, which is accomplished by using information on morphology, biochemistry, genetics, distribution, ecology, and behavior.

Birds elsewhere in the world have undergone similar radiations, as have numerous types of other organisms. I shall refer to these relatively sparingly until the final chapter of the book. The final chapter first summarizes and synthesizes the major points of the preceding chapters, and then concludes with a discussion of the evolution of other groups of organisms in the light of current understanding of Darwin's Finches.

PLATE 1. Extremes in the diversity of beak shape and size. Upper: Large ground finch, *Geospiza magnirostris*. Lower: Warbler finch, *Certhidea olivacea (photos by A. Root)*.

PLATE 2. Woodpecker finch, *Cactospiza pallida*, using a tool to extract insect larvae from a dead branch of *Bursera graveolens*; the tool is a cactus spine (upper) or twig (lower). Santa Cruz (*photos by A. Root, R. Perry*).

PLATE 11. Medium ground finch, *Geospiza fortis*, searching for ticks on the back of a land iguana, *Conolophus subcristatus*. The erect posture and arched

PLATE 3. Upper: Large ground finch, *Geospiza magnirostris*. Genovesa (*photo by D. Schluter*). Lower: Sharp-beaked ground finch, *Geospiza difficilis*, foraging for arthropods in epiphytic-covered *Zanthoxylum fagara*. Pinta (*photo by D. Nakashima*).

PLATE 4. Upper: Vegetarian finch, *Platyspiza crassirostris*. Santa Cruz (*photo by R. I. Bowman*). Lower: Cactus finch, *Geospiza scandens*. The bill of all species is pale in the nonbreeding season, as here, and black in the breeding season (see Plate 3). Santa Cruz (*photo by H. Snell*).

PLATE 5. Sharp-beaked ground finch, *Geospiza difficilis*, feeding on the blood at the base of the tail of boobies. Wolf. Upper: On a red-footed booby, *Sula sula (photo by D. Nakashima).* Lower: On a masked booby, *Sula dactylatra (photo by F. Köster).*

PLATE 6. Medium ground finch, *Geospiza fortis*, searching for ticks on the back of a marine iguana, *Amblyrhynchus cristatus*. Fernandina. Sequence is from upper left to lower right (*photos by F. Pölking*).

PLATE 7. Color polymorphism in the beaks of young finches: large ground finch, *Geospiza magniros-tris*, on Genovesa. Upper: Yellow morph. Lower: Pink morph.

PLATE 8. Daphne Major, in years of contrasting rainfall. Upper: March 1976, a normal year. Lower: March 1977, a drought year (*photos by P. T. Boag*).

PLATE 9. The effects of El Niño, 1982–1983, on the vegetation of Daphne Major. Upper: March 1976, a normal wet season. Blue-footed boobies, *Sula nebouxii*, are nesting or resting on the crater floor. Lower: August 1983, shortly after the rains had ceased. The wall of the crater is covered with vegetation, principally *Cacabus miersii* (light patches) and *Heliotropium angiospermum* (dark patches). In some cases the cactus (see above) is smothered by these plants. Eighteen species of plants now grow on the crater floor, and boobies are absent altogether (*photo by N. Grant*).

PLATE 10. Foods for finches. Upper left: Flowers of *Opuntia helleri*. Genovesa. Upper right: Flowers and fruits of *Cordia lutea*. Genovesa (*photo by J.N.M. Smith*). Middle: Flowers of a vine, *Sarcostemma angustissima*. Santa Cruz (*photo by J.N.M. Smith*). Lower: Fruit of *Bursera graveolens*, showing red aril around a black-tipped stone. Genovesa.

PLATE 12. Cactus finch, *Geospiza scandens*. Santa Cruz (*photo by A. Root*).

PLATE 13. Upper: Vegetarian finch, *Platyspiza crassirostris*, stripping the bark of a twig of *Croton scouleri*. Marchena. Lower: Close-up of a stripped *Croton* twig (*photos by D. Day, D. Schluter*).

CHAPTER TWO

Characteristics of the Islands

ORIGIN AND AGES

The Galápagos archipelago straddles the equator in the Pacific Ocean about 1000 km west of the Ecuadorean part of continental South America (Fig. 1). The islands are volcanic, being the product of outpourings of mantle material from a Galápagos hotspot about 150 km wide (Cox 1983), and comprising magmas, lava flows, and cratered tuff cones formed by the compaction of volcanic ash (Plates 14–17). Some of the tuff cones are small islands (Plates 15 and 16), while others may have been once but are now hills surrounded by lava on the large islands (Plate 14). Volcanic activity continues today (Simkin 1984).

According to the hotspot hypothesis (e.g. see Morgan 1971), the hotspot has a fixed position. Crustal plates ride over the hotspot and are occasionally perforated by molten rock arising from deep in the earth's mantle. In the case of the Galápagos, not one but two plates have moved over the hotspot, the Cocos and Nazca plates. It has been calculated from bathymetric and paleomagnetic data that the relative positions and motions of the plates have changed during the last 30 million years (Hey 1977). About 5 million years ago the Nazca plate was, for the first time, positioned over the hotspot. Since all of the islands lie on the Nazca plate (Fig. 5), 5 million years before present is the upper limit to the origin of the existing islands. Evidence of older, now submerged, islands (guyots), which would suggest greater antiquity of the archipelago, has not been found on either plate (Cox 1983).

Alternative estimates of the ages of the islands are consistent with the maximum figure of 5 million years. The oldest known rocks, from the Plazas (see Fig. 1), have been dated by the potassium/argon method at 4.2 ± 1.8 (standard deviation) million years (Cox 1983, Simkin 1984). Marine fossiliferous deposits on the islands are younger (Hickman and Lipps 1985). There has been uncertainty as to the exact ages of the islands as islands, rather than as submarine peaks or ridges. For example McBirney and Williams (1969) believed the smooth and dissected lava flows on Española were submarine in origin and later uplifted, but Hall et al. (1983) have since found evidence of subaerial (above water) formation, which indicates that islands existed at least 3.3 million years ago, the oldest potassium/argon date for Española (Cox 1983, Simkin 1984). In general the southeastern is-

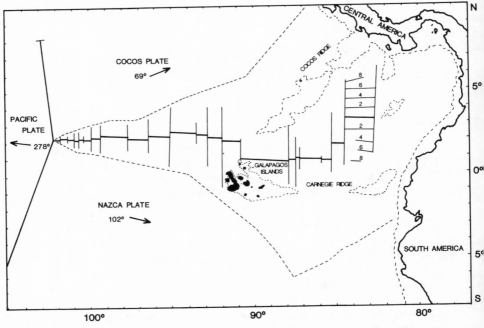

FIG. 5. Reconstruction of the sea floor of the eastern Pacific to show the placement of the Galápagos Islands on the Nazca plate, and of Cocos Island on the Cocos plate. Broken lines indicate the boundary of the wedge-shaped portion of the ocean floor as the Nazca and Cocos plates have moved apart during the past 25 million years. Plates are currently moving apart at 71 mm/year along vectors indicated by arrows. New ocean floor is being generated along the faulted spreading boundary shown by the heavy, staggered, lines running east-west. The light, parallel, lines, drawn only at the eastern end of the spreading boundary, are isochrons of constant age in millions of years; they increase in distance from the spreading boundary, and show the rate and direction of spreading of ocean floor. Broken contour lines mark 1000 fathoms. Redrawn from Hey (1977), Cox (1983), and Simkin (1984).

lands appear to be the oldest, and the northern and western islands appear to be the youngest. Most if not all of the current major islands had been formed by about 0.5 to 1.0 million years ago.

The submarine Cocos and Carnegie ridges lie between the Galápagos and the continent (Fig. 5). They are probably hotspot traces, having moved towards the continent, on the Cocos and Nazca plates respectively, in the last 20 to 25 million years. The Galápagos are the youngest part of the Carnegie ridge, and they are estimated to be moving towards the continent at the rate of 71 mm/year. There is no geological evidence that the islands were ever linked to the mainland by either an isthmus or a chain of islands (Cox 1983).

DISTRIBUTION AND SIZES

The islands are heterogeneous in three features that are important to the biota: area, elevation, and isolation. The largest and highest island, Isabela, is more than 100 km long, 467,000 ha in area, and 1700 m high, while some islets are less than one ha in area and only a few meters above sea level. Isabela is unusual in comprising six major volcanoes separated by low-lying land, chiefly lava flows. The other high islands are dominated by one volcano (e.g. Pinta), or have a ridged topography. The closest islands, such as the Plazas and Santa Cruz, are separated by less than one km, while the greatest gap (~120 km) separates the two northern islands of Wolf and Darwin from Pinta. Isolation distances were not altered very much by the lowering of sea level in recent Ice Age times; some of the small islands were connected to nearby large islands, but the large islands are separated by deep channels and remained well isolated from each other at that time (Simpson 1974).

This simple summary of the main physical features of the islands glosses over another important factor of consequence to the finches; not all the islands are entirely vegetated (Adsersen 1976). Santiago, Fernandina, Pinta, and Marchena, as well as Isabela, have extensive areas of fairly recently formed lava that have been scarcely colonized by plants. On Marchena (Fig. 6) the vegetated parts of the island are patchily distributed in little islands within the "sea" of lava (see also Plate 14), and combined together these patches would cover less than half of the island's area. The vegetation on Fernandina is similarly highly restricted. This means that island area is not always a simple index to the amount of habitat that is suitable for finches.

CLIMATE

Seasonality. For islands lying astride the equator, the Galápagos have a remarkably seasonal climate. The major influences on the climate are cool water masses originating off the coast of Peru and flowing northwestwards, and warm water masses originating to the north and flowing southeastwards (Colinvaux 1984, Houvenaghel 1984).

Seasonality takes the form of a hot and wet period, from approximately January to May, and a cooler and drier period for the rest of the year. The warm waters in the intertropical convergence zone to the north of the islands move south in December and January and begin to influence the islands. Air temperatures increase, moisture-laden warm air rises and forms large cumulus clouds, especially over the high islands, and rain then falls, occasionally heavily. Cumulus clouds rarely develop under the cooler conditions that prevail after May following the recession of the warm water influence,

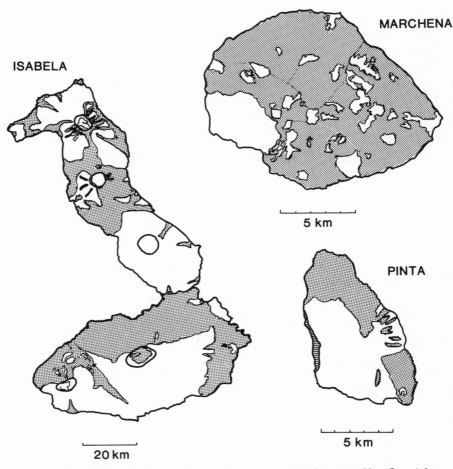

FIG. 6. Three islands that illustrate the extent of recent and sparsely vegetated lava flows (stippled) which have dissected the older, more thickly vegetated, areas. Redrawn from Hamann (1981) and Snell et al. (1983).

and precipitation then more often takes the form of light, misty rains known locally as *garúa*.

Seasonality in air temperature is illustrated in Figure 7, and seasonality in amount of precipitation is shown in Figures 7 and 8. Even though rain falls heavily in the months January to May it does so on relatively few days. For instance, in the period 1965–1975 precipitation was recorded on 10 to 40% of the days in the hot-wet season at the Charles Darwin Research Sta-

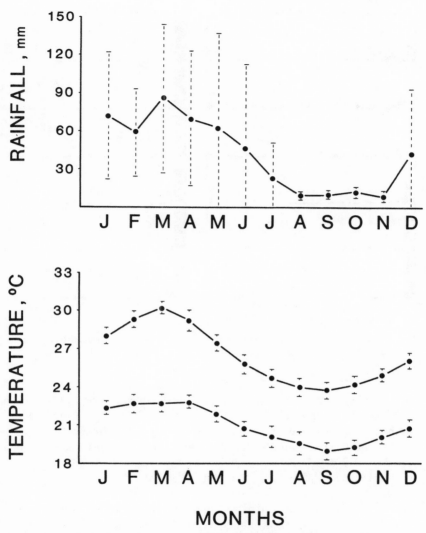

FIG. 7. Seasonality on the Galápagos: mean monthly precipitation, and mean monthly temperature minima and maxima, at the Charles Darwin Research Station, south Santa Cruz, at an altitude of 6 m. Ninety-five percent confidence limits for the estimates of each mean are shown by broken lines; they reflect the degree to which conditions vary from year to year. The data were collected in the period 1965-1984. Note the warm and wet season January to April or May, and the cool, dry, season in the remainder of the year. The heavy rain in the El Niño event of 1982-1983 (see text) significantly elevated the monthly averages for May, June, July, and December, and broadened the confidence intervals for those months. Temperatures were unusually high that year, but they scarcely affected the 20-year mean values.

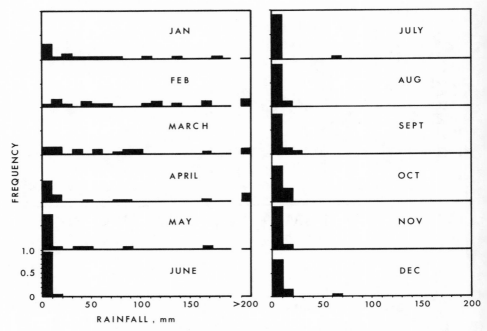

FIG. 8. The monthly frequencies of rainfall in units of 10 mm at Puerto Bacquerizo on San Cristóbal, altitude 6 m. From Grant and Boag (1980).

tion (CDRS) on the south coast of Santa Cruz (Grant and Boag 1980): April had rain least often, on an average of only 5.8 ± 1.7 days (SE). Conversely, in the remainder of the year precipitation occurred more frequently, on 25 to 60% of the days, with September experiencing the most frequent rain (16.3 ± 1.7 days). Total accumulation, however, is relatively trivial at coastal locations during this season (Figs. 7 and 8).

The amount of rain varies from place to place. More rain falls on high islands than on low ones, and more falls at high altitudes than at lower ones. The Research Station at 6 m altitude received an average of 406 mm of precipitation a year in the period 1965–1975, whereas Bella Vista, at an altitude of 194 m on the same side of the island, received three times that amount (Grant and Boag 1980). Northern sides of high islands receive less rain at a given altitude than do southern sides of the same island, because they are in the rain-shadow of the prevailing southeasterly winds.

As will be discussed later, precipitation has a more profound effect on Darwin's Finches than does temperature, and for this reason I will describe the seasons as wet and dry, although the terms hot and cool would be equally correct (e.g. Harris 1974).

Variation in precipitation. As is characteristic of arid regions in general, rainfall is erratic in occurrence and quantity. Precipitation is much more variable in the wet season than in the dry season (Fig. 8), a feature that is reflected in the broader confidence limits on the estimates of mean monthly rainfall in the wet season in Figure 7. There are two elements to this variation. Precipitation in a given month, say March, may vary from 0 mm in one year to more than 200 mm in the next; and successive months in the wet season of a given year may differ by the same amount.

Such variation may be regular, even though large, or irregular and unpredictable. To measure the predictability of precipitation we used a technique adopted from information theory by Colwell (1974), and found predictability to be very low (Grant and Boag 1980). It is lower in the wet season than in the dry season, low between successive wet seasons, and low between successive months within a wet season. These remarks apply to precipitation at coastal stations. At higher elevations rainfall is more evenly distributed throughout the year; therefore monthly rain varies less and is more predictable than at lower elevations.

The most striking feature of the Galápagos climate is the extraordinary year-to-year variation in rainfall. Figure 9 shows records of annual rainfall at coastal sites on two islands spanning 35 years. Since more than 90% of the annual rain falls in the wet season, the variation among years is an expression of how wet the wet seasons are. In some years heavy rains begin in late December or early January, and the last one occurs in May. In other years there are a few light showers but no heavy rains. These are drought years (see Plate 8). Therefore the length of the wet season, as well as the amount of rain that falls, varies considerably among years, and there is no obvious pattern to the variation (Grant and Boag 1980).

The heaviest and most extensive rains are associated with El Niño events: the occurrence of unusually warm surface water of low salinity along the coasts of Peru and Ecuador which usually first appears in about December and extends westward to the Galápagos (Halpern et al. 1983). The combined atmospheric and oceanographic causes of the El Niño phenomenon are still only partly known (e.g. see Philander 1983, Cane and Zebiak 1985), they are undoubtedly complex, and they vary from one event to the next. The regular oscillation of atmospheric pressure across the Pacific, known as the Southern Oscillation, is strongly coupled to fluctuating oceanographic conditions, and in recognition of this fact some modern treatments of the El Niño phenomenon refer to El Niño–Southern Oscillation, or ENSO, events (Philander 1983, Rasmusson 1985).

El Niño events recur irregularly in Galápagos at intervals of 2 to 11 years, with a rough average frequency of one in 7 years (Grant 1985a). There is an element of arbitrariness in classifying them because they are simply magnifications of the regular, seasonal, ocean warming that begins in the latter

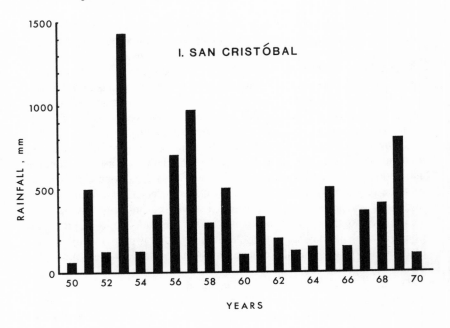

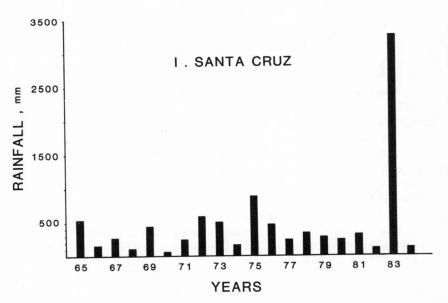

FIG. 9. Annual variation in rainfall at Puerto Bacquerizo, San Cristóbal, and the Charles Darwin Research Station on Santa Cruz. Partly from Grant and Boag (1980).

part of each calendar year. Since 1950 there have clearly been six such events that influenced the Galápagos and gave rise to extensive rains: 1953, 1957–1958, 1965, 1972–1973, 1975, and 1982–1983 (Plate 9). In contrast to these, the event of 1976–1977 recorded off the coast of Peru had no pronounced effects on Galápagos. Rainfall records were not kept before 1950, with one small exception (Alpert 1961), but there is circumstantial evidence of exceptionally heavy and extensive rains on the Galápagos in two years in the first half of the century: 1925 and 1939 (Grant 1984a). Nineteen thirty-nine was the year of David Lack's visit.

These extremes in 1925 and 1939 were apparently surpassed (Grant 1984a) in 1982–1983 (Fig. 9) with an extraordinary El Niño event that has been described as the strongest oceanographic warming trend off the coast of South America this century (Cane 1983). It had an unusual origin in the mid-Pacific. Heavy rains began on Galápagos in early November 1982 and ceased more than eight months later in the second half of July (Grant and Grant 1985). During this period a total of 3258 mm of rain was recorded on Santa Cruz (Fig. 9) and 3648 mm of rain were recorded at the Puerto Bacquerizo site on San Cristóbal. For comparison, the largest amount of rain previously recorded over the same period of months, during the El Niño event of 1952–1953, was 1399 mm (see Fig. 9).

PLANTS

New species and subspecies of plants are still being discovered and described. At the latest count of species, subspecies, and varieties (Porter 1984), Galápagos have 736 vascular plants, 195 of which are weeds and garden escapees introduced by humans, the remainder (541) being indigenous. Considering just species, 170 of the 497 indigenous species (34%) are endemic, i.e. found only on Galápagos. According to Porter (1983), they were derived from a minimum of 101 colonizations.

From where, and how, did they reach Galápagos? Darwin (1839) and Hooker (1847) identified the location of the main relatives of Galápagos plants as the cooler regions of South America. A large fraction of the Galápagos flora has an affinity with the flora of the humid tropics of South and Central America, and, surprisingly, some Galápagos plants are most similar to others now living in the West Indies. To answer the question of how all these plants reached the remote islands, Porter (1976) took the properties of their seeds and fruits into account in assigning species to probable agents of transport. He estimated that approximately 60% of plant taxa were carried to the islands by birds, another 30% were borne on air currents, and the remainder reached the islands by sea.

Some plant taxa have become differentiated into several species on the islands. The best known of these is *Scalesia*, a genus in the Compositae.

Fourteen species are currently recognized (Wiggins and Porter 1971), putting them on a par with Darwin's Finches. They vary in growth form, from small shrubs to trees as much as 10 m tall. A parallel differentiation took place in prickly pear cactus (*Opuntia*), all derived from one or possibly two invasions. There are fewer species than in *Scalesia*, only six, but subspecies are highly distinctive, and variation in growth form (Plates 18 and 19) is of comparable magnitude to that of *Scalesia*. On Marchena, for example, *O. helleri* grows to no more than about a meter high, whereas on Santa Cruz *O. echios* reaches 10 m or more in height (Racine and Downhower 1974) and a full meter in trunk circumference. Most of the endemic taxa are found in arid habitats, as are most of the species that originated through speciation on the Galápagos (Porter 1984).

VEGETATION

Plant communities change gradually in composition along altitudinal gradients on the islands (Reeder and Reichert 1975, Eliasson 1984, Hamann 1981, 1984). Fairly rapid change in the composition over a relatively short distance of the gradient allows us to categorize the vegetation into four basic types or zones (Plates 20–23). From low to high elevation they are (a) arid zone, (b) transitional forest, (c) moist forest, and (d) a fern-sedge-grass zone (Stewart 1911). Characterization of the transitional zone is somewhat arbitrary; nevertheless it occurs on all high islands that have a moist forest, and on some medium elevation islands, like Pinzón, that do not. Structural and compositional features of these habitats have been described in the literature in detail, most recently by Eliasson (1984) and Hamann (1984). Such details permit finer distinctions to be made and more comprehensive classifications to be constructed than I have used here; for example, Hamann (1981) has recognized as many as nine zones or formations.

All small islands and the extensive lowlands of the large islands are covered by arid zone, dry forest, habitat, although the smallest islets lack trees and should properly be described as supporting arid zone vegetation. This zone is floristically rich. The conspicuous components are torchwood trees of the genus *Bursera*, cactus of either a branching (*Opuntia*) or columnar (*Jasminocereus*) growth form, and several types of shrubs and small trees including *Croton scouleri* and *Cordia lutea*. Less conspicuous but just as important to finches are the many species of grasses and herbs. Sustained plant growth requires continued rainfall because the soil is shallow and poorly consolidated, moisture retention is low, and there is no permanent reservoir of water. Furthermore, evaporation is high owing to intense sunlight, directional winds, and convection air currents. Many plants are dry-deciduous, but there are several evergreen shrubs including *Waltheria ovata*, *Castela galapageia*, and *Scutia pauciflora*.

The moist, humid, forest of *Scalesia, Zanthoxylum, Psidium, Pisonia,* and other tree species has a higher and closed canopy (Plate 23). Epiphytes are abundant. The soil is richly developed, and may be half a meter or more deep. On some islands, notably Santa Cruz and Floreana, the forest is so dominated by *Scalesia pedunculata* that it is described as *Scalesia* forest rather than more generally as moist forest (Hamann 1981, 1984). The transition zone is aptly named, for it is intermediate in composition and structure between arid and humid forest vegetation. Dominant tree species are *Pisonia* (Plate 22), *Piscidia*, and *Psidium*. The grassy zone is a mixture of ferns, sedges, grasses, and small herbs.

The number of vegetation zones on an island is largely a function of its elevation and associated climate. Moist forest is present only on the high islands of Santa Cruz, Santiago, Isabela, Floreana, Pinta, and Fernandina; it may have been present on San Cristóbal before human settlement (see below). It occurs at lower elevations on the south side than on the north side of the islands, presumably as a result of greater precipitation from the prevailing southeasterly winds. Itow (1975) recorded this rain shadow effect on Santa Cruz; the upper limit to the arid zone vegetation occurs at an altitude of 40 m on the south side but at 430 m on the north side.

The preceding descriptions apply to all islands, except that the two western ones, Fernandina and Isabela, differ from the rest in supporting xerophytic vegetation at the highest altitudes. On the southeastern slope of Fernandina the vegetation just below the caldera rim is more xerophytic than at slightly lower elevations. The drought-tolerant vegetation coincides with the position of the upper limit of the cloud belt. Similar conditions prevail at high elevations on Volcan Cerro Azul at the south end of Isabela and on V. Darwin at the north end (Hamann 1981).

Other vegetation zones, such as the littoral zone dominated by mangroves (Plate 24), are more restricted. Santa Cruz has a habitat found nowhere else in the archipelago, although formerly present on San Cristóbal. On the south side an endemic melostome, *Miconia robinsonia*, occurs in dense concentrations of 2–3 m tall shrubs below the grass zone. On this island too the moist forest is patchy; Bowman (1961) has classified as separate habitats one component dominated by *Scalesia pedunculata*, and another characterized by *Psidium galapageum* and *Zanthoxylum fagara* which are draped with epiphytic mosses and ferns. The second of these two zones has now largely disappeared due to human (farming) activity.

CHANGES IN THE PAST

The climate of Galápagos has not remained stable over the last 50,000 years. This is known from an analysis of particles and plant products in cores taken from the sediment of El Junco lake on the summit of San Cristóbal (Colinvaux 1972, 1984). Inferences can be made about changes in water level, cloud cover, and heat budget from the composition of the cores at different levels.

The present climate has persisted for the last 3,000 years, and it also prevailed between about 6,200 and 8,000 years ago. In the intervening period of 3,200 years it was drier, and possibly hotter, than now. Going back further, it was drier before 8,000 years ago. The most different climatic regime from the present one occurred from about 10,000 to 34,000 years ago; this was a time of little precipitation or evaporation. Further back than this the data are scanty but hint at a climate not very dissimilar to the current one from 34,000 to 48,000 years ago. The preceding years are climatically unknown. Thus climatic conditions have fluctuated, and we know of drier conditions but not of wetter conditions than those experienced by Galápagos at present.

The chief implication from this perspective of a fluctuating climate is that the vegetation zones shifted altitudinally, with the arid zone vegetation being more extensive altitudinally than at present for long periods of time. It was most extensive between 10,000 and 34,000 years ago, partly because of dry conditions, and partly because more land was exposed at certain times in this period, associated with an advance of glaciers in the north and a general lowering of the sea level (Quinn 1971). Mesic habitat on some high elevation islands, such as San Cristóbal, was reduced or absent during this period (Colinvaux and Schofield 1976a, 1976b, Colinvaux 1984). As mentioned earlier, the moist forest at high elevation presently contains a low number of endemic plant species, except among woody plants (Hamann 1981, 1984), probably because it is a relatively young habitat (Johnson and Raven 1973).

CHANGES IN RECENT TIMES

Crater lakes occur on San Cristóbal, Isabela, Fernandina, and Genovesa, but the mineral-rich waters are not suitable for human consumption. A year-round supply of drinkable water exists on San Cristóbal, Floreana, Santa Cruz, Isabela, and Santiago. All but Santiago have permanent human settlements, and of these all but Santa Cruz were settled in the last century. The vegetation has been considerably altered on the settled islands by the clearance of forest, the planting of citrus trees, avocados, sugarcane, and other crop plants, and the introduction of house plants that have escaped. These destructive influences have been magnified by the effects of mammals introduced to these islands and, in some instances, to uninhabited islands as well; the nine culprits are cattle, horses, donkeys, goats, pigs, dogs, cats, rats, and mice. Only Fernandina, Genovesa, Darwin, Wolf, and the smaller islands have remained free from destructive effects mediated directly or indirectly by humans. At great effort and expense the Galápagos National Park Service has eradicated goats from Española, Rábida, Santa Fe, South Plaza, Pinta, and Marchena. Vegetation is making a slow but perceptible recovery. Santiago and the settled islands remain regrettably far from their pristine state (Hamann 1984).

Additional information on various aspects of the Galápagos environment and inhabitants can be found in three recent books edited by Bowman et al. (1983), Perry (1984), and Berry (1985).

Cocos Island

Cocos is a small (47 km^2), single, volcanic island (Fig. 5) approximately 630 km to the northeast of Galápagos and 500 km from Central America (Costa Rica). Like the Galápagos, it has never been connected to the continent, and it is of approximately comparable age (Dalrymple and Cox 1968). The temperature is about the same as at low elevations on Galápagos islands, but much more rain falls annually. Sherry (1985) gives a figure of 8 m, which is more than twice the amount recorded at Galápagos coastal stations during the El Niño event of 1982–1983, but is possibly comparable to the maximum at higher elevations. Rainfall on Cocos is more evenly distributed throughout the year, so there is no marked seasonality. The vegetation comprises a single zone from the coast to the summit (573 m) of thick, humid, forest, which is richly endowed with epiphytes, vines, and lianas (Plate 25).

Summary

The Galápagos islands are volcanic. They originated through hotspot activity no earlier than 5 million years ago. Most if not all of the major islands had been formed by a half to a million years ago. They vary greatly in size, elevation, and degree of isolation, and have never been connected to the mainland.

The predominant vegetation is dry, mainly deciduous, forest at low elevations, transitional forest and moist forest at high elevations, and a grassy habitat at the highest elevations on a few islands. Zonation is a function of precipitation and temperature. The climate, especially precipitation, is strongly seasonal at low elevations, less so at high elevations. Rainfall is heaviest, and air temperatures are highest, in the period January to May. In the remainder of the year precipitation may be lacking altogether at coastal sites. Variation in annual rainfall is extreme and irregular; for example, at a coastal site on San Cristóbal 37 mm were recorded in the whole of 1950, but during the extraordinary El Niño event of 1982–1983 a total of 3648 mm was registered in nine months. In the last 50,000 years climatic conditions have fluctuated between present and drier conditions, suggesting that arid zone vegetation was sometimes more extensive in the past and that mesic vegetation was either more restricted or entirely eliminated from some islands then.

In climate and vegetation, Cocos island is more similar to the humid tropics of South and Central America. It is similar to the Galápagos in origin and age.

PLATE 14. Upper: Crescent-shaped Cerro Ballena surrounded by a recent lava flow, Isabela. Hermanos II and III are just offshore, and Hermanos IV is more distant and just to the right of C. Ballena. Lower: Seaward slope of Cerro Ballena, dissected by erosion channels. Principal vegatation is *Bursera graveolens* and *Castela galapageia*.

PLATE 15. Upper: Hermanos III. Lower: Hermanos III, an eroded tuff cone now reduced to a crescent like Cerro Ballena (Plate 14). The four Hermanos were previously called the Crossmans.

PLATE 16. Upper: Daphne Major, an intact tuff cone. Lower: Rocas Bainbridge, almost devoid of trees, off the Southeast coast of Santiago.

PLATE 17. Upper: Wolf. *Croton scouleri* covers the left flank. Lower: Fernandina. Slabs of lava from a recent flow form a fringing apron to this high island.

PLATE 18. Upper: *Opuntia megasperma*, Española. The tree is approximately 5 m high, and the trunk has a diameter of close to 1 m. The surrounding trees are mainly *Bursera graveolens*. Lower: *Opuntia helleri*, Darwin. As on Wolf, Marchena, and Genovesa, this cactus has a shrubby growth (1–2 m high). A stand of *Croton scouleri* trees is in the background (*photo by R. I. Bowman*).

PLATE 19. *Opuntia echios*, Santa Cruz, with a height of 10 m and a trunk diameter of 0.5 m.

PLATE 20. Seasonal difference in the appearance of vegetation. Upper: Dry season view of arid zone vegetation on the north side of Santa Cruz, characterized by *Bursera graveolens* trees, *Scalesia pedunculata* shrubs, and an occasional cactus bush (*Opuntia echios*). The lava forms irregularly shaped boulders, quite different from the plate-like lava in Plate 18. Lower: Wet season view at the same location.

PLATE 21. Difference between islands in arid zone vegetation. Upper: Santa Cruz, south side and close to the Charles Darwin Research Station. Dominant trees are *Opuntia echios, Jasminocereus thouarsii, Bursera graveolens, Croton scouleri, Cordia leucophlyctis,* and *Tournefortia psilostachya*. Lower: Genovesa. Dominant trees are *Bursera graveolens, Croton scouleri,* and *Cordia lutea*. Cactus growth (see Plate 18) is patchily distributed.

PLATE 22. Transition zone vegetation. Upper: Epiphytic growth on *Bursera graveolens*. Pinta. Middle: *Pisonia-Zanthoxylum* thicket. Santa Cruz. Lower: *Pisonia floribunda*, a typical member of transition zone vegetation. Santa Cruz.

PLATE 23. Upper: Forest of *Scalesia pedunculata* on Santa Cruz. Lower: sedge-grass-fern habitat above *Scalesia* forest on Santa Cruz.

PLATE 24. Mangroves, fringing the southeastern coastline of Isabela. They are islands of vegetation surrounded by sea and recent flows of lava. Tortuga is in the distance.

PLATE 25. Cocos Island. Upper: Wafer Bay. The moist evergreen forest extends to the shoreline. Lower: Epiphytic bromeliads in the highlands (*photos by N. Grant*).

CHAPTER THREE

General Characteristics and Distributions of Finches

THE MAIN GROUPS

Darwin's Finches entered the literature inauspiciously as "resembling the Java sparrow, in shape and size, but of black plumage" (Colnett, 1798). Relatively drab with short tails, they are nowhere near as colorful as the equally famous honeycreeper finches of Hawaii, the predominant colors being black, brown, and grey. They vary most conspicuously in beak size and shape (Plates 1, 3, 26–37) but also in body size, ranging from less than 10 grams *(Certhidea olivacea)* to more than 40 g for some individuals of *Geospiza magnirostris*. An unusual feature of the group is the sharply angled edge of the mandibles near their base, especially in the largest species *G. magnirostris* (Snodgrass 1903, e.g. see Fig. 10); beneath the surface of the bill the tomium of the upper mandible forms an abrupt angle with the zygomatic bar. This gives the impression that the bill is pointing downwards, whereas in other finches elsewhere in the world it projects straight out of the head.

The fourteen species are listed in Table 1 in four groups: a ground finch group and a tree finch group, with the warbler finch and the Cocos finch treated as two further groups. Names of the two main groups reflect their habits. All species feed on the ground and in the vegetation, but all the ground finches spend much more time on the ground than do the tree finches. The warbler finch and the Cocos finch also feed mainly in vegetation, but differ between themselves and from the others in their behavior.

Appearances of the finches, as well as their habits, set the groups apart. I will start with the last two species, which differ as much from each other as do any of the remainder. Adult males of the Cocos finch are black; adult females are brown and streaked. The warbler finch is a greenish color; males and females are alike. It is unique in possessing an orange-tawny throat patch. This is almost exclusively a feature of males (Rothschild and Hartert 1902, Lack 1945), and occurs most frequently in the population on Santiago but rarely if at all in other populations.

All species of ground finches resemble the Cocos finch in adult plumage, but differ from it in the way in which adult male plumage is acquired. Young male ground finches in immature plumage resemble females, but with suc-

FIG. 10. Plumages of *Geospiza fortis*. All birds are initially brown. Females retain the brown plumage in successive molts, whereas males gradually acquire fully black coloration through stages 1 to 5. The classification of plumages is similar to one introduced by Snodgrass and Heller (1904). From Price (1984a).

cessive molts they gradually acquire a fully black plumage (Fig. 10). Black plumage is acquired by Cocos finch males in a mosaic fashion over the head and body (Lack 1945, 1947), and not in a systematic progression from the head posteriorly (Fig. 10).

The remaining group, the tree finches, is heterogeneous, but in none of them are the males as black as ground finch males. Adult males of *Platyspiza* (Plates 4 and 27) and *Camarhynchus* species (Plate 29) have about as much black color as do ground finches at stages 2 or 3 (Fig. 10). Adult males of the *Cactospiza* species have no black color; instead they have a greenish plumage (*C. pallida*), like the warbler finch, or a somewhat browner plumage (*C. heliobates*) resembling that of female ground finches. Females of

TABLE 1. Darwin's Finch species

Scientific Name	English Name	Approximate Weight (grams)
Geospiza fuliginosa	Small Ground Finch	14
Geospiza fortis	Medium Ground Finch	20
Geospiza magnirostris	Large Ground Finch	35
Geospiza difficilis	Sharp-beaked Ground Finch	20
Geospiza scandens	Cactus Ground Finch	21
Geospiza conirostris	Large Cactus Ground Finch	28
Camarhynchus parvulus	Small Tree Finch	13
Camarhynchus pauper	Medium Tree Finch	16
Camarhynchus psittacula	Large Tree Finch	18
Platyspiza crassirostris	Vegetarian Finch	34
Cactospiza pallida	Woodpecker Finch	20
Cactospiza heliobates	Mangrove Finch	18
Certhidea olivacea	Warbler Finch	8
Pinaroloxias inornata	Cocos Finch	16

NOTE: Sources for this table are Lack (1947), Grant (1984b) and Grant et al. (1985); see Lack (1947) for the authors of the species names, and for subspecific names. I have simplified the English names by omitting hyphens from, for example, ground-finch. The word "ground" is often dropped from the names of the two cactus finches. Bowman (1983) suggests "Honeycreeper Finch" for *Pinaroloxias inornata*. One problem with scientific names needs to be mentioned. The Sharp-beaked Ground Finch was described by Gould (1837) and given the name *Geospiza nebulosa*. Owing to subsequent confusion about the identity of the specimens and their source island, the name *G. difficilis* Sharpe came to replace *G. nebulosa*. Sulloway (1982a, 1982b), through historical researches, has recently identified the specimens and island (Floreana), and Steadman (1985) has confirmed with fossils the existence of a population of these finches on Floreana. Nevertheless, I have used the name *difficilis* throughout because *nebulosa*, although not forgotten (e.g. Lack 1945, 1947, 1969), was not used without qualification for more than fifty years until resurrected by Sulloway (1982a, 1982b) and Steadman (1982). I agree with Lack (1969) that replacing the well known name of *difficilis* with *nebulosa* is likely to cause confusion.

PLATE 26. Cocos finch, *Pinaroloxias inornata*: variation in feeding habits. Upper: Bark-stripping for arthropods. Lower: Feeding from *Terminalia catappa* fruit (*photos by T. K. Werner/T. W. Sherry*).

Upper: Gleaning from grass leaf blades (*photo by T.K. Werner/T.W. Sherry*). Lower: Cracking rice grains, like seeds, in camp (*photo by A. C. Graham*).

PLATE 27. Upper: Male vegetarian finch, *Platyspiza crassirostris*. The head is black, the back and wings are brown, and the ventral surface is creamy-white. Marchena. Lower: Female large tree finch, *Camarhynchus psittacula*. Santiago (*photos by D. Schluter, D. Day*).

the tree finch group are similarly heterogeneous. Females of the *Camarhyn-chus* species and *Cactospiza pallida* have a greenish cast to the plumage and are scarcely streaked, those of *Platyspiza* are brown and streaked like the ground finch females, and female *C. heliobates* are intermediate.

GENERA

While there is agreement on the number of species, there is no agreement on the number of genera. It is the tree finches that are the problematical species. I have followed Swarth (1931) and Bowman (1961) in recognizing three genera in this group. Lack (1947, p. 14) would have followed suit had the finches been continental passerine birds, but he combined the tree finches in the single genus *Camarhynchus* because he believed that fourteen species should not be split into so many genera. He later revised this opinion (Lack 1969; also Harris 1974) and placed the tree finches in two genera, *Platyspiza* and *Camarhynchus* (including *C. pallida* and *C. heliobates*). Steadman (1982) has recommended returning to the earlier practice (Gould 1837, Salvin 1876, Rothschild and Hartert 1899, 1902, Snodgrass 1902) of treating all species as members of a single genus, *Geospiza*, in order to emphasize their similarities. The similarities are overemphasized, in my opinion, by ignoring the fact that four specimens, which link the genera in an apparently unbroken chain of variation, are probably hybrids (Stresemann 1936; see also next chapter).

Genera are to some extent arbitrary categories. I have followed Swarth (1931) and Bowman (1961) in recognition of the distinctive appearance of clusters of finches. These clusters differ biochemically, but weakly (Yang and Patton 1981). Although Lack (1947) adopted fewer genera, those he recognized (*Geospiza*, *Camarhynchus*, *Certhidea*, and *Pinaroloxias*) correspond to the four groups set out in Table 1.

SPECIES

Agreement on the occurrence of fourteen species should not be taken to mean the absence of taxonomic problems. Sympatric populations that do not interbreed, or interbreed rarely, are clearly separate species. They can be recognized as such by their different appearance. Some populations on different islands are so similar in appearance to each other that they clearly belong to the same species, and in fact most allopatric populations can be confidently grouped into species. Problems arise with moderately to strongly differentiated, allopatric, populations. Would their members interbreed if given the opportunity, and hence do they belong to one, two, or more species?

These difficulties are pronounced with some ground finch populations. I shall start with *Geospiza difficilis*. As currently recognized this species comprises six extant populations and at least two extinct ones (Table 2). They are more differentiated than are those of any other species (Fig. 11). The average weight of adult males on Pinta and Fernandina is 19–20 g, but males on Genovesa average only 12 g, while the figures for Wolf, Darwin, and Santiago are 21 g, 25 g, and 27 g, respectively (Schluter and Grant 1984a, Grant et al. 1985); one remarkably fat bird on Santiago was found to weigh 45 g (D. Schluter, pers. comm.)! Are these really all one species? Since the populations do not occur sympatrically, and immigrants from one population have never been recorded on an island supporting another population, we do not know the outcome of a natural interbreeding test. Instead, a taxonomic judgment has to be based on affinity, or resemblance. The best guide would be a measure of genetic affinity. Analyses of biochemical variation that can be linked to genetic variation have been performed with Darwin's Finch material by using the technique of electrophoresis (Ford et al. 1974, Yang and Patton 1981), but their resolving power is insufficient to throw light on this particular problem. We are therefore forced to rely on the traditional mode of comparative morphology. Results of detailed examination of individual character variation (Swarth 1931, Lack 1945) and of multivariate morphological analyses (Grant and Schluter 1984, Grant et al. 1985) are consistent with the taxonomic judgment of a single well-differentiated species. There are similarities also among the songs sung by males of the different populations (Chapter 9).

Variation among populations of *G. magnirostris* presents a similar problem, exacerbated by the extinction of the populations of questionable taxonomic status on San Cristóbal and Floreana (see later). All other populations of these species resemble each other closely, but the five males and three females collected on the two southern islands by Darwin, FitzRoy, and two assistants in 1835 (Sulloway 1982a) are distinctly larger than all other specimens from all other islands. They probably weighed about 45 g on average, to judge from the allometric relationships between either wing length or tarsus length and weight among other *G. magnirostris* populations (Grant et al. 1985). I follow previous authors in considering them conspecific with *G. magnirostris* (e.g. Steadman 1984) because they are no more different from *G. magnirostris* elsewhere, in either size or shape, than are some populations of *G. difficilis* from each other. But there is an intriguing possibility that these large forms lived together with typical members of the species (Grant et al. 1985). Some specimens of typical members may have been collected on the same islands, but this is far from certain (Sulloway 1982a, Grant et al. 1985). If the large and typical forms were sympatric they were

TABLE 2. Species of Darwin's Finches on the seventeen major islands (>1.5 km²) of the Galápagos. B = breeding, (B) = probably breeding, E = extinct, (E) = probably present as a breeding population formerly, and now extinct.

	G. magnirostris	G. fortis	G. fuliginosa	G. difficilis	G. scandens	G. conirostris	C. psittacula	C. pauper	C. parvulus	P. crassirostris	C. pallida	C. heliobates	C. olivacea
Seymour		B	B		B								B
Baltra		B	B		B								B
Isabela	B	B	B	(E)	B		B		B	B	B	B	B
Fernandina	B	B	B	B			B		B	B	(B)	B	B
Santiago	B	B	B	B	B		B		B	B	B		B
Rábida	B	B	B		B		B		B	B			B
Pinzón	B	B	B		(B)		E		B	E	B		B
Santa Cruz	B	B	B	E	B		B		B	B	B		B
Santa Fe	B	B	B		B		B		B	(E)			B
San Cristóbal	E	B	B	(E)	B				B		B		B
Española			B			B							B
Floreana	E	B	B	E	B		B	B	B	B			B
Genovesa	B			B		B							B
Marchena	B	B	B		B		B			B			B
Pinta	B	B	B	B	B		B		(B)	B			B
Darwin	B			B									B
Wolf	B			B									B

NOTE: Sources for this table are Harris (1973), Grant and Schluter (1984), and unpublished observations. Some species occur as occasional vagrants on islands where they do not breed. Harris (1973) gives a partial list of known occurrences. See the Appendix for the English names of the islands which are no longer, or scarcely, used.

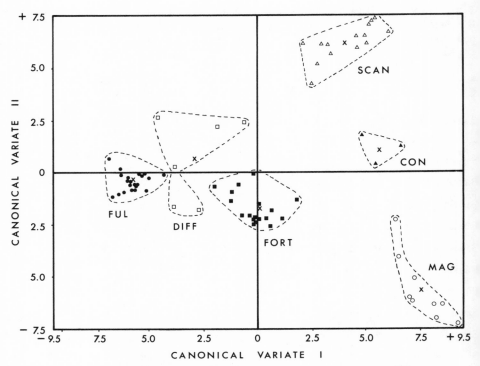

FIG. 11. Differences in beak morphology among the six species of ground finches summarized on two canonical axes, the first representing beak size increasing from left to right, the second representing increasing beak pointedness from bottom to top. The analysis treats variation in six beak dimensions simultaneously, and maximizes the separation of the species. The mean position of each species is shown by a cross, and all other symbols refer to individual populations. Names of species have been abbreviated. The *G. difficilis* points comprise five extant populations and one extinct on Santa Cruz. Not included are one extant (Fernandina) and one extinct (Floreana) population of this species. The point for the former, known from measurements of live specimens (Schluter and Grant 1984a), would fall within the polyhedron, whereas the point for the Floreana population would be outside and close to *G. fortis* (see. Fig. 22). From Grant (1983c).

probably two species, because the range of variation in size shown by the combined material is far greater than would be expected for a single species. Perhaps an analysis of fossil material from Floreana and San Cristóbal (Steadman 1985) will resolve this problem. No modern study can hope to do so, as the populations became extinct, probably as a result of human activity, some time after Darwin's visit (Sulloway 1982a, 1982b, Steadman 1984).

A different type of problem is presented by *G. conirostris*. Breeding populations occur on Genovesa and on Española and its satellite Gardner (see Fig. 1). Here the problem is to decide whether they belong to the same species, *G. conirostris*, or to one or two other species.

The distinctive feature of *G. conirostris* is a lateral flattening of the bill (Plate 32). In particular, the base of the lower mandible is flatter and less convex in profile than it is in all other *Geospiza* species (e.g. see Plates 32 and 35). However, some individuals of *G. scandens* from Marchena resemble some individuals of *G. conirostris* on the adjacent island of Genovesa (Grant et al. 1985; see also Plate 35). On the islands of Española and Gardner (Plate 32) *G. conirostris* resembles, in bill proportions, two species occurring elsewhere in the archipelago but not on those two islands: *G. magnirostris* and *G. fortis* (Plates 3 and 36). It does not resemble *G. scandens* (Plates 4 and 12). The question, then, is whether *G. conirostris* is a distinct species, or whether its populations are really highly differentiated forms of *G. scandens* on the one hand and either *G. magnirostris* or *G. fortis* on the other. Again I have followed traditional practice, supported by the results of multivariate analysis (Grant and Schluter 1984, Grant et al. 1985; see also Fig. 12) and biochemical analysis (J. L. Patton, pers. comm.), of retaining the name *G. conirostris* for three differentiated populations of a single species.

Perhaps the most enigmatic population of all is the large ground finch population on the remotest and least accessible island, Darwin. Thirty-four specimens were collected within a period of ten years at the turn of the last century. The specimens at each collecting time, and in combination, are extraordinarily variable (Chapter 8); furthermore, frequency distributions of traits show a slight bimodal tendency suggesting the possibility of not one population but two. They have been classified as a distinct species *G. darwini* (Rothschild and Hartert 1899), a subspecies of *G. conirostris* (Rothschild and Hartert 1902, Lack 1945, 1947) and as two species, namely *G. conirostris* and *G. magnirostris* (Swarth 1931, Bowman 1961). In body and beak size, and shape, some specimens resemble typical *G. magnirostris* most clearly while others resemble typical *G. conirostris* most clearly (Fig. 13). However, in the curvature of the sides of the beaks, especially the lateral base of the lower mandibles, they all resemble *G. magnirostris*. On this basis they have been tentatively classified as *G. magnirostris* (Grant et al. 1985) and included in Figures 11 and 12 as such (their positions in these figures are the closest to *G. conirostris*). But this is no ordinary population, with the extreme variation in bill dimensions tending towards bimodality (Fig. 13) and suggesting genetic heterogeneity due to hybridization.

Further study of the finches on Darwin is needed, yet the physical difficulties are formidable as the slopes of the island are too steep to be climbed.

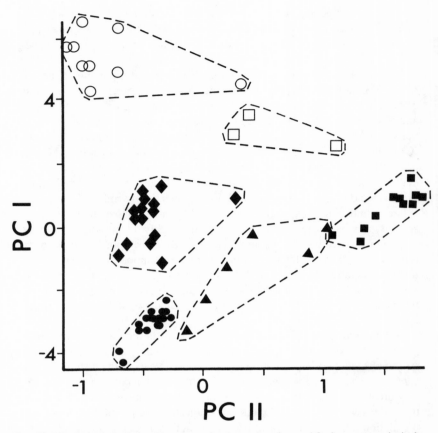

FIG. 12. Morphological relationships among the six species of ground finches on two principal component axes. Nine dimensions were used in the analysis, based on a correlation matrix of untransformed data; six beak dimensions, wing, tarsus, and hallux. PC I represents overall size, increasing from bottom to top, and PC II represents beak pointedness, increasing from left to right. Unlike canonical variates analysis (Fig. 11), principal components analysis does not effect a maximal separation of groups. Note the extreme similarity of some populations of *G. difficilis* and *G. scandens*, and *G. magnirostris* (I. Darwin) and *G. conirostris*. Symbols: *G. magnirostris* (○), *G. conirostris* (□), *G. scandens* (■), *G. fortis* (♦), *G. difficilis* (▲), and *G. fuliginosa* (●). From Grant and Schluter (1984).

The only visit to the top of the island (Plate 18) was made by helicopter in 1964 when, in one afternoon, no large finches were seen (R. I. Bowman, pers. comm.). It seemed at this time as if the taxonomic problem had been a transitory one which might have disappeared with the finches. But Dolph Schluter (pers. comm.) saw two individuals on a visit by sea (in 1980) to the rocky slope where specimens had been collected earlier. One was captured

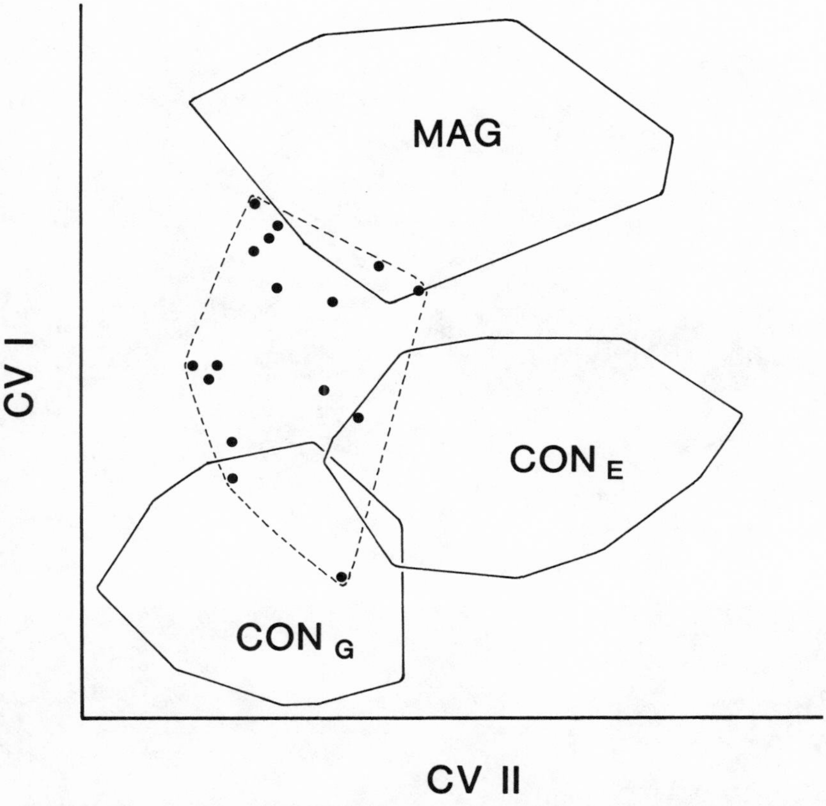

FIG. 13. Morphological relationships among populations of *G. magnirostris* and *G. conirostris* summarized on two canonical axes derived from multiple discriminant function analysis. As in Figure 11, the analysis maximizes the separation of groups, but in this case on the basis of beak dimensions, tarsus length, and wing length of males. Eight populations of *G. magnirostris* form one group (MAG), and populations of *G. conirostris* (CON) on Genovesa (G) and Española (E) form two other groups. Enclosed in the broken line are 16 male specimens from I. Darwin, not used in the analysis but projected onto the axes; half are closest to *G. magnirostris*, and half are closest to *G. conirostris*. The first axis represents size, increasing from bottom to top, and the second axis represents a shape factor, with the bill becoming blunter (deeper, but relatively shorter) from left to right. From Grant et al. (1985).

and found to have a deep bill like that of a typical *G. magnirostris*. The other was not captured but was seen to have a longer and more pointed bill like that of *G. conirostris*. The enigma remains.

These problems have been discussed to show that fourteen is a probable but not certain number of species living today. The number could be higher or lower by one or two. New techniques are needed to resolve the taxonomic status of the well-differentiated allopatric populations.

PLATE 28. Upper: Medium tree finch, *Camarhynchus pauper*. Floreana (*photo by M.P. Harris*). Lower: Large tree finch, *Camarhynchus psittacula*. Santa Cruz (*photo by R. Å. Norberg*).

Upper: Small tree finch, *Camarhynchus parvulus*. Floreana (*photo by M. P. Harris*). Lower: Medium tree finch, *Camarhynchus pauper*. Floreana (*photo by Tj. De Vries*).

SUBSPECIES

Subspecies have been described for several of the species (Swarth 1931, Lack 1945, 1947, 1969). Since the basis for recognizing subspecies has largely been variation in size, which is treated as a subject in its own right without regard to taxonomy in the next chapter, I have ignored subspecific names here. See Lack (1945, 1947, 1969) for the most recent treatment of subspecific variation; he recognizes 35 subspecies for the Galápagos (Lack 1969), and the Cocos Finch would bring the total to 36.

DISTRIBUTIONS

Distributions of finches were nearly completely known by the end of the year-long expedition of the California Academy of Sciences (1905–1906). Since 1973 we have visited all of the major islands at least once. Finch distributions among these islands are listed in Table 2, which differs only to a minor extent from the equivalent table in Lack (1947). We have also visited all the small islands from which finches had been reported previously, except for Eden and Caldwell, and finch distributions among these islands are listed in Table 3.

These recent visits have confirmed the breeding status of many species inferred previously from occurrence data only. Therefore, modern lists are more reliable than older ones, although breeding by the few birds of a rare species on some islands may still have been overlooked. *Camarhynchus parvulus* has been recorded on Pinta several times, but although it has always seemed rare it probably breeds there in small numbers (Table 2). More doubtfully, *G. scandens* may breed in very low numbers in parts of Fernandina not normally visited by scientists; an immature bird was seen by D. Schluter (pers. comm.) in the dry season of 1981. As an illustration of the difficulties of establishing the status of such small populations, we would not know that *G. magnirostris* bred on Santa Fe in the period 1966–1970 if Tjitte de Vries had not followed the fate of the single pair during his study of the Galápagos Hawk. It would certainly have been missed in most types of field study. I have included it in Table 2 because breeding was continuous from year to year. In contrast, I have not included a couple of species that are usually absent from two islands but present on rare occasions as breeders. These are *G. magnirostris* on Daphne Major, where it is known to have bred four times (Beebe 1924, Grant et al. 1975, and Chapter 8), and *G. conirostris* on Wolf, where it has bred once (Curio and Kramer 1965a). In each case only a single pair bred.

There is more uncertainty about the past status of some populations, especially on those islands, such as San Cristóbal, Floreana, and Isabela, that

TABLE 3. Species of Darwin's Finches breeding on the small islands of the Galápagos. All are populations of ground finches, and none are known to have gone extinct. *G. fulginosa* and *G. magnirostris* have bred occasionally on Daphne Major (Table 11). *Certhidea olivacea* breeds on Gardner (by Española), and may breed occasionally or regularly on a few of the others.

	G. fortis	G. fuliginosa	G. scandens	G. conirostris
Plazas	B	B	B	
Gardner (by Española)		B		B
Bartolomé		B		
Daphne Major	B		B	
Daphne Minor	B	B		
Tortuga		B		
Hermanos		B		
Eden		B		
Bainbridge		B		
Beagle		B		
Cowley		B		
Gardner (by Floreana)		B		
Enderby		B		
Champion	B		B	
Caldwell		B		

suffered habitat destruction in the last century. The question is whether rare specimens were immigrants or residents. In some cases one, or only a few, birds in adult plumage were collected in the last century or early part of this century, and have never been seen since; e.g. *G. difficilis* on Isabela, Floreana, and San Cristóbal. They were probably some of the last remnants of breeding populations. It is quite possible that the tree finches *Platyspiza crassirostris* and *Camarhynchus psittacula* were resident on San Cristóbal in the last century but became extinct as a result of forest removal. *P. crassirostris* has never been recorded from the island, and only one specimen of *C. psittacula* has been collected there. The California Academy expedition collected 6 specimens of *G. fuliginosa* on Wolf and 15 specimens of *G. fortis* on Española, all apparently in immature plumage. They may have been born on those islands (Bowman 1961, Harris 1973), but in this case I think, like Lack (1945, 1947, 1969), that it is more likely they were immigrants, since none were in the black plumage which adult males usually acquire, and most vagrants to islands are immature birds (Grant 1984b). A single immature *G. fortis* was seen on Española by Ian Abbott in 1973 (Abbott et al. 1975), but otherwise neither species has been seen on the respective islands in either the breeding or nonbreeding season since 1906.

No species has been known to newly colonize a large island in the last 150 years, with the possible exception of *G. scandens* on Pinzón (see below).

PATTERNS AMONG THE ISLANDS

There are three main features of the finch distributions. First, there is great variation shown by the species, ranging from the widespread *G. fuliginosa* which occurs on almost all islands, sometimes as the only breeding species (Table 3), to *Camarhynchus pauper* which is restricted to Floreana. Second, tree finches are restricted to the larger and higher islands. Third, in general the larger, higher, and more centrally located the island, the larger is the number of species it supports, the maximum being ten out of a possible thirteen. The last two features are largely explained by the different habitat affinities of the species. When breeding, tree finches occur most frequently at medium elevations in transitional forest, also in moist forest, but less often in dry forest at low elevations (Lack 1947). That they can persist nevertheless in dry forest is shown by the fact that two of them, *Camarhynchus psittacula* and *Platyspiza crassirostris*, occur on the large island of Marchena which has only arid zone vegetation. The ground finches, in contrast, are common in both arid and transitional zone habitats, and least common in moist forest, in the breeding season. They are not restricted to the large islands (Table 3). The warbler finch is less tied to upland forest for breeding purposes than are the tree finches, and is much less restricted geographically.

It has been pointed out repeatedly that the small outlying islands have few species but a large proportion of them are endemic subspecies (Rothschild and Hartert 1899, Swarth 1934, Lack 1945, 1947). This is illustrated in Figure 14. The significant factor is the isolation of those islands, not their size. Hamilton and Rubinoff (1963, 1964, 1967) have shown with multiple regression analysis that variation in the number of endemics among islands can be largely accounted for, statistically, by variation in their isolation as measured by the distance to the nearest island. Near islands have similar species compositions, while more distant islands have fewer species or sub-

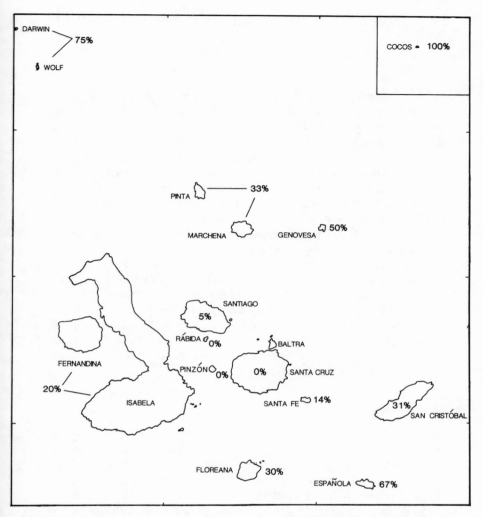

FIG. 14. Proportions (percent) of finch species on the major Galápagos islands that are endemic subspecies. Redrawn from Lack (1947) as modified slightly by Sulloway (1982a).

species in common (Rothschild and Hartert 1899, Power 1975, Abbott et al. 1977).

EXTINCTIONS

The few changes that have occurred in the distributions of species appear to be largely the consequence of direct and indirect human activity. While no species has become extinct, populations have (Table 2), although the exact causes of extinction have not been established in any one instance. As mentioned previously, the unusually large form of *G. magnirostris* on Floreana and San Cristóbal became extinct some time after Darwin's visit. Populations of *G. difficilis* became extinct on Floreana, and possibly San Cristóbal, some time after 1852 (Sulloway, 1982a) and on Santa Cruz after 1932 (Lack 1947). A population of *G. scandens* on Pinzón appears to have become extinct after 1906 (Harris 1973), perhaps associated with destruction of *Opuntia* cactus by goats. However, in this case the demise of the population, if that was its fate, was not permanent. In February of 1984 H. L. Gibbs (pers. comm.) saw at least five *G. scandens* individuals on the island, including two in black plumage, and captured and measured two of them. Either the population had remained low and undetected for nearly eighty years and then recovered, or else the island had been reinvaded. Whichever occurred, the removal of the goats in the 1970s, allowing a partial recovery of the cactus, may have combined with the extensive rains of 1982–1983 to make conditions favorable for the return of the species.

OTHER LAND BIRDS

Darwin's Finches have to contend with other species as potential predators, competitors and, conceivably, reservoirs of diseases. The full list of twenty-one species of other land birds in the Galápagos is given in Table 4.

Hawks and short-eared owls (Plate 38) are known to prey on finches (Curio and Kramer 1965b, de Vries 1975, 1976, Grant et al. 1975, Grant and Grant 1980a). One or other of these predatory species has been recorded on all the main islands except Wolf. Other occasional predators are mockingbirds, great egrets, yellow-crowned night herons, and lava herons (Boag and Grant 1984a). It is possible that cuckoos also prey on finches, but scarcely anything is known about this species on Galápagos. The barn owl is mainly a predator of mammals (Abs et al. 1965). Competitive relations between Darwin's Finches and some of the land birds are possible, but unknown. For example, the yellow warbler and warbler finch (Plate 31) sometimes exploit similar foods in places, such as on Genovesa, where they occur together (Grant and Grant 1980a). Little is known about the habits of individuals and the demography of land birds other than Darwin's Finches,

TABLE 4. Land birds on the Galápagos, broadly categorized but excluding Darwin's Finches (after Harris 1973, and Grant 1984b). An asterisk signifies an endemic species.

Scientific Name	English Name
Ardea herodias	Great Blue Heron
Casmerodius albus	Great Egret
Bubulcus ibis	Cattle Egret
*Butorides sundevalli	Lava Heron
Butorides striatus	Green-backed Heron
Nyctanassa violacea	Yellow-crowned Night Heron
*Buteo galapagoensis	Galápagos Hawk
*Laterallus spilonotus	Galápagos Rail
Neocrex erythrops	Paint-billed Crake
*Zenaida galapagoensis	Galápagos Dove
Coccyzus melacorhyphus	Dark-billed Cuckoo
Tyto alba	Barn Owl
Asio flammeus	Short-eared Owl
Pyrocephalus rubinus	Vermilion Flycatcher
*Myiarchus magnirostris	Large-billed Flycatcher
*Progne modesta	Galápagos Martin
*Nesomimus parvulus	Galápagos Mockingbird
*Nesomimus trifasciatus	Charles Mockingbird
*Nesomimus macdonaldi	Hood Mockingbird
*Nesomimus melanotis	Chatham Mockingbird
Dendroica petechia	Yellow Warbler

NOTE: The Smooth-billed Ani, *Crotophaga ani*, has become established recently on Santa Cruz, perhaps having been introduced (F. Köster, pers. comm.). It is listed by Harris (1974) as the Groove-billed Ani, *C. sulcirostris*.

mockingbirds, and hawks (Grant 1984c). And virtually nothing is known about parasites and disease beyond the discovery of parasitic worms in a cactus finch (Salvin 1876), the occasional observation of worms in the feces of ground finches (D. Schluter, pers. comm.), and the observation that many finches develop symptoms resembling those of avian pox, from which some birds recover.

On Cocos Island, the finch coexists with only three other species of land birds: yellow warbler, a cuckoo, and a flycatcher (Sherry 1985).

SUMMARY

The fourteen species of Darwin's Finches fall into four groups: ground finches, tree finches, a warbler finch, and the Cocos finch. The tree finches comprise three genera, and the remainder comprise one genus each. The groups (and genera) are recognized in part by their habits, as implied by their names, and in part by their appearance (plumage). Adult male ground finches and the Cocos finch are black, while females are brown and streaked. The warbler finch is a greenish-grey color and unstreaked in both sexes. The tree finches constitute a heterogeneous group that displays intermediacy in these characteristics.

Species are distinguished from each other by bill size and shape and by body size, and most can be recognized clearly, but some allopatric populations are difficult to classify. For example, two marked forms (subspecies) of *G. conirostris* could in fact belong to two other species. Some well-differentiated populations of *G. difficilis* could be different species. Conceivably, therefore, there are one or two more species, or fewer species, than the fourteen currently recognized.

Distributions of species in the Galápagos archipelago range from the highly restricted to the widespread: from *Camarhynchus pauper* on only one island (Floreana) to *Geospiza fuliginosa* which breeds on fourteen of the seventeen major islands as well as on thirteen smaller islands. Some of these islands support breeding populations of ten of the thirteen species. No species is known to have become extinct, but several populations have probably become extinct, largely as a direct or indirect result of human influence on habitats. *G. magnirostris* became extinct on Floreana and San Cristóbal after Darwin's visit (1835); *G. difficilis* became extinct on Floreana after 1852 and on Santa Cruz after 1932; and *G. scandens* became extinct on Pinzón after 1906, but has been rediscovered on the island recently (1984). *Platyspiza crassirostris* and *Camarhynchus psittacula* became extinct on Pinzón some time in this century.

PLATE 29. Small tree finch, *Camarhynchus parvulus*. Santa Cruz. Upper: Male. Middle: Female or young male. Lower: Immature (*upper and middle photos by U. M. Norberg, lower photo by R. I. Bowman*).

PLATE 30. Upper: Woodpecker finch, *Cactospiza pallida*. Santiago. Lower: Mangrove finch, *Cactospiza heliobates*, Bahía Elizabeth, Isabela (*photos by D. Day, H. Snell*).

PLATE 31. Convergent evolution. Upper: Warbler finch, *Certhidea olivacea*, perched on *Waltheria ovata* and feeding on spiders on *Chamaesyce amplexicaulis*. Genovesa. Lower: Yellow warbler, *Dendroica petechia*. It is not a Darwin's Finch. Santa Cruz (*photos by A. Simon, D. Day*).

PLATE 32. Sexual dimorphism: Upper: Large cactus finch, *Geospiza conirostris*, Española. The black male (right) is larger than the sooty-brown female (left). The two individuals also differ in bill shape, but this does not exemplifly a consistent difference between the sexes. Lower: Sharp-beaked ground finch, *Geospiza difficilis*. Pinta (*photo by D. Schluter*).

PLATE 33. Individual variation in beak size and shape: large cactus finch, *Geospiza conirostris*, Genovesa. The lowest bird has a longer beak with a more pronounced curvature to the culmen. Compare with Figure 62, which provides another illustration of individual differences, and with Plate 32 for an example of inter-island differences in beak size and shape.

PLATE 34. Individual variation in the beak size and shape of medium ground finches, *Geospiza fortis*. Upper: Daphne Major. Middle and Lower: Bahía Academía, Santa Cruz. Scales slightly differ; the middle and lower birds are actually larger than the upper ones. The middle bird has a beak size almost the same as the upper left one, and the lower bird has a correspondingly larger beak (*upper photo by P. T. Boag*).

PLATE 35. Similarity of allopatric species. Upper: Large cactus finch, *Geospiza conirostris*, Genovesa. Lower: Cactus finch, *Geospiza scandens*, Champion.

PLATE 36. Similarity of resident and immigrant species. Upper: Medium ground finch, *Geospiza fortis*, resident on Daphne Major. Lower: Small ground finch, *Geospiza fuliginosa*, resident on Santa Cruz, but a frequent and temporary immigrant to Daphne Major (*photos by D. Day, P. T. Boag*).

PLATE 37. An assemblage of ground finch species. Genovesa has three species, differing in size and beak shape: a small species (12 g), the sharp-beaked ground finch, *Geospiza difficilis* (upper left); a medium species (25 g), the large cactus finch, *Geospiza conirostris* (upper right); and a large species (36 g), the large ground finch, *Geospiza magnirostris* (lower) (*photos by R. L. Curry*).

PLATE 38. Finch predators. Upper: Galápagos hawk, *Buteo galapagoensis*. Daphne Major. Lower: Short-eared owl, *Asio galapagoensis*. Santa Cruz (*photos by P. T. Boag, R. Å. Norberg*).

CHAPTER FOUR

Patterns of Morphological Variation

INTRODUCTION

Species of Darwin's Finches are small or large, and they have beaks that are long and thin, short and narrow, or long and deep. This variation is quantitative; differences between species are often matters of degree, as are the differences between populations of the same species. The purpose of this chapter is to describe the variation that occurs among populations of Darwin's Finches in order to show the main morphological changes that took place in the adaptive radiation.

THE MAJOR SIMPLE PATTERNS

I shall start with a single trait, bill depth (I use the terms bill and beak interchangeably). Quantitative variation in this trait is represented by a frequency distribution of measurements taken from members of a population. The distribution is approximately normal, and is largely characterized by two statistical parameters: the mean, or average, and the variance. Frequency distributions of different populations can be compared on a common axis, and in fact many of the important features of morphological variation among Darwin's Finches as a whole are captured in a single figure (Fig. 15). The irregularities in two of the frequency distributions in this figure should be ignored as they are probably the result of sampling error. For example, in April 1979, T. D. Price (pers. comm.) measured twice as many *G. fuliginosa* on Los Hermanos as occur in museum collections, and found that the measurements conformed more closely to a normal frequency distribution (Chapter 12, fig. 95) than do the measurements of museum specimens from this population used in Figure 15.

The first important feature is the difference between species in average beak depth, as illustrated in the figure by the three species on Santa Cruz. The difference between *G. fuliginosa* and *G. fortis* is about the same as the difference between *G. fortis* and *G. magnirostris*; in other words, the species are approximately equally spaced apart along this axis. The second feature is the intraspecific analogue; populations of the same species on different islands differ in their means, as shown in the figure by *G. fortis* and by *G. fuliginosa*.

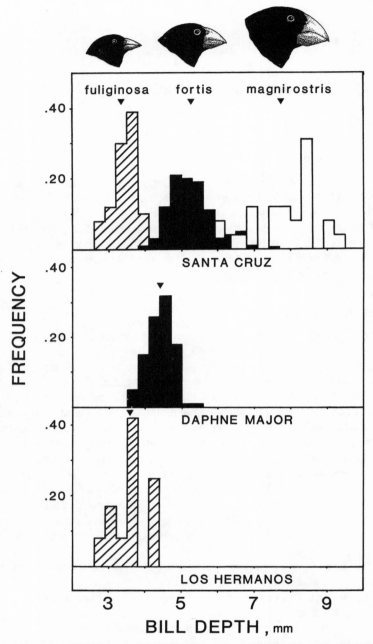

FIG. 15. Frequency distributions of beak depths (upper mandible) of adult males in populations of ground finches (*Geospiza*). Number of specimens: from Santa Cruz 134 *G. fuliginosa*, 156 *G. fortis*, and 26 *G. magnirostris*; from Daphne Major 89 *G. fortis*; and from Los Hermanos 12 *G. fuliginosa*. All specimens are in museum collections and they were measured by I. Abbott, except for 58 live birds measured on Daphne Major by P. T. Boag and the author and rescaled to the mean of the museum specimens from Daphne. Means are indicated by the solid triangles.

All populations exhibit variation around a mean value, but some populations vary more than others. *G. magnirostris* and *G. fortis* on Santa Cruz vary more than *G. fuliginosa*, and *G. fortis* vary more on Santa Cruz than on Daphne Major. Thus there are both interspecific and intraspecific differences in population variation, as well as interspecific and intraspecific differences in population averages.

All four sets of differences can be demonstrated statistically (Grant et al. 1985); that is, they cannot be attributed to sampling error, and therefore they represent real differences between the sampled populations. The biological significance of differences in the variances will be considered in Chapter 8. In the remainder of the present chapter I will discuss the differences and similarities in mean values.

THE MINOR SIMPLE PATTERNS

There are other forms of variation in beak depth among these (and other) species that are not captured by the figure. On the large islands of Santa Cruz and Isabela, *G. fortis* varies geographically in beak depth. As suggested by Lack (1945, 1947) and confirmed by subsequent statistical analysis (Grant et al. 1985), *G. fortis* have smaller beaks and other dimensions in the northern part of Isabela than in the south. A parallel trend is found on Santa Cruz (Abbott et al. 1977). An additional trend occurs along altitudinal gradients, with birds at high elevations tending to be larger than those at low elevations (Grant et al. 1985). These variations within the large islands are small, however, and do not contribute importantly to the larger variation around the mean that we find on comparing large islands with smaller islands such as Daphne Major (Fig. 15).

Another form of variation missing from the figure arises from differences between the sexes. Males are slightly larger on average than females by a few percent ($< 5\%$) in all dimensions (e.g. see Plate 32), not just in beak depth, although there is considerable overlap in the measurements of the two sexes. While the degree of sexual dimorphism is typical of small passerines in being not very pronounced, it is large enough to influence comparisons of populations where the sexes are unequally represented. For this reason Figure 15 shows variation in only one sex, and male measurements are used because there are more male than female specimens in museum collections.

CORRELATIONS BETWEEN TRAITS

That a single set of histograms can capture so much of the important variation within and between populations is due largely to strong positive correlations between traits. We have measured the length, depth, and width of upper and lower mandibles, and the length of wing and tarsus, on almost all museum specimens in existence (Grant et al. 1985) and on numerous live ground finches (Plate 39) captured and released on most of the islands during extensive field studies since 1973 (e.g. Abbott et al. 1977, Boag 1983, Grant 1983a, Boag and Grant 1984a, Schluter and Grant 1984a). All traits have been found to be positively correlated, often strongly so. Therefore, in comparisons between populations, the differences between means and variances in one trait, such as beak depth, are usually accompanied by differences in the same direction in another trait, such as beak length (Abbott et al. 1977, Boag 1984, Grant et al. 1985). The magnitude of the differences, however, varies according to the traits being compared.

In spite of the phenotypic correlations between traits, one trait does not serve as a perfect index for all of the rest. To take a simple example, when beak length is substituted for beak depth on the abscissa of Figure 15, and *G. scandens* is substituted for *G. magnirostris* on Santa Cruz, the four patterns currently shown in Figure 15 are broadly maintained, although variation of *G. scandens* values around the mean is less than the *G. magnirostris* variation (Fig. 16). But when the beak depth axis is preserved, with again *G. scandens* replacing *G. magnirostris*, a difference in the pattern is now seen (Fig. 16). The frequency distribution of *G. scandens* largely overlaps the distribution of *G. fortis*. On average *G. scandens* has a longer but slightly shallower beak than *G. fortis* (Plate 40). *G. scandens* also occurs on Daphne Major where its beak depth distribution is almost identical to that of *G. fortis* (Abbott et al. 1977). Thus *G. scandens* has a relatively long and shallow beak, and differs in beak proportions from both *G. fortis* and *G. magnirostris*.

In general, species differ from each other in some dimensions but not in others, or not to the same extent in others; they differ from each other in size, shape, or both. We need to consider size and shape separately.

SIZE

The size of a bird is its mass. Any dimension can represent overall size (Mosimann and James 1979), but the most direct measure is the fat-free weight. Birds from many populations of ground finches and a few populations of tree finches and the warbler finch have been weighed when they are likely to be relatively free of fat, in the dry season (Grant et al. 1985; see also Table 1, and Bowman 1961, 1983). Analysis of the ground finch weights shows that each species varies significantly among islands. The most conspicuous variation, that occurring among populations of *G. difficilis*, has been mentioned already (p. 52). Taking all populations into account, no two species of ground finches have the same average weight. Thus size varies within and between species.

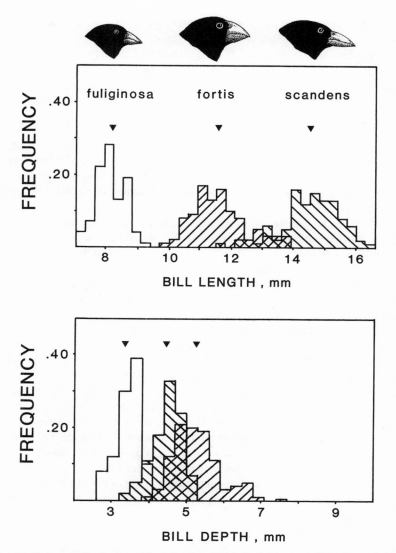

FIG. 16. Frequency distributions of beak length and beak depth (upper mandible) in populations of adult male ground finches (*Geospiza*) on Santa Cruz. Measurements of museum specimens were made by Ian Abbott: 134 *G. fuliginosa*, 156 *G. fortis*, and 123 *G. scandens*. Means are indicated by the solid triangles. Note that *G. fortis* has a shorter beak than *G. scandens*, on average, but a deeper beak.

In the absence of weight, wing length (Fig. 17) is a good index of size because mean wing length and mean weight of ground finch populations are tightly correlated ($r = 0.96$, df 38, $P < 0.001$). The tree finch values are embedded within and indistinguishable from the cloud of ground finch values in this figure.

Allometry

The size of a dimension such as beak length is expected to be larger in large species than in small ones. The question is whether the size of a dimension varies as a constant proportion of body size or of some other trait, or disproportionately; put another way, one can ask whether shape itself remains constant or changes.

When the proportions of two traits remain the same over a range of absolute size variation, the slope of the relationship between the traits on a double logarithmic scale is 1.0. This is referred to as isometry. When the slope differs from 1.0, because proportions change systematically, the relationship is referred to as allometric. Shape may change regularly, as in this case, or irregularly. Irregular change is recognized by departures or deviations from a single line of allometry.

Geometric mean regression analysis can be used to see if a pair of traits varies isometrically or allometrically among species. This is interspecific allometry. Populations of the same species may be compared in the same fashion (interpopulation allometry). The slope of the line that characterizes the relationship, the allometric coefficient, may then be compared with coefficients for other species to see if they differ or are the same. A third type of allometry, intrapopulation allometry, describes the relationship between traits among individuals (usually adults) within a single population. Intrapopulation allometry is sometimes referred to as static allometry, to contrast it with dynamic allometry which describes the way in which two traits of individuals change through time as a result of growth. Dynamic allometry will be discussed in the next chapter on growth. Here I will concentrate on the differences between species in shape as shown by their static allometric relations.

Shape

Internal structures are generally expected to vary regularly with size, from simple physiological and anatomical considerations (e.g. Calder 1984). Deviations from allometry are interesting because they suggest the possibility of adaptive modification of the part independent of other parts or of the whole.

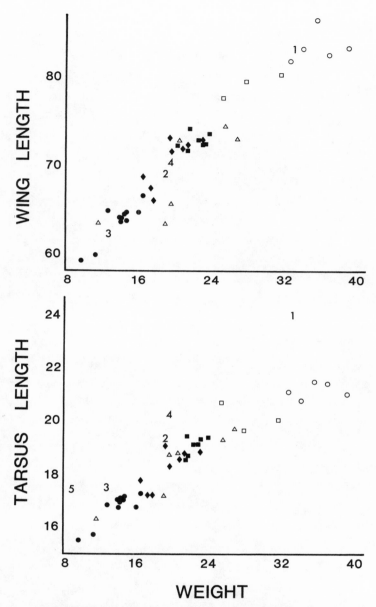

FIG. 17. Relationships between mean wing length (mm) and mean weight (g), and between mean tarsus length (mm) and mean weight, among populations of the six ground finch species, males only. Symbols: *G. magnirostris* (○), *G. conirostris* (□), *G. scandens* (■), *G. fortis* (♦), *G. difficilis* (▲), and *G. fuliginosa* (●). Numbers refer to average values for tree finch species, males and females combined: 1, *Platyspiza crassirostris*; 2, *Camarhynchus psittacula*; 3, *C. parvulus*; 4, *Cactospiza pallida*; 5, *Certhidea olivacea*. *C. olivacea*, not shown in the upper diagram, has a mean wing length of 53.2 mm. From Grant et al. (1985).

PLATE 39. Measurements of the large cactus finch, *Geospiza conirostris*, on Genovesa. Upper: Bill length. Lower: Bill width.

Upper: Bill depth. Lower: Tarsus length (the leg carries a numbered metal and a black plastic band).

Bowman (1961) measured and weighed many bones and organs of nine species of Darwin's Finches and found that, in general, the larger the species in overall size the larger were its component parts. I have analyzed some of the relationships with geometric mean regression of logarithmically ($\log_e$) transformed means to provide the best estimate of each slope, or reduced major axis (see e.g. Harvey and Mace 1982), and least squares (predictive) regression to calculate 95% confidence limits. The allometric (slope) coefficients show that gizzard weight (1.12 ± 0.40) and heart weight (0.95 ± 0.19) scale isometrically with body weight among these species; that skull weight (1.70 ± 0.50) varies allometrically with body weight, with the larger species (especially *G. magnirostris*) having disproportionately large heads; and that the weight of the entire skeleton has an allometric coefficient (1.28 ± 0.24) that is barely significantly different from 1.0 ($P \sim 0.05$), so that no clear decision can be made on the scaling relationship in this case. Inspection of the residual values around the regression lines reveals two striking deviations: *P. crassirostris*, the vegetarian finch, has a disproportionately large gizard (Fig. 18), but a disproportionately small heart. It also has a disproportionately long intestine. The sizes of gizzard and intestine are readily interpretable in terms of its unique diet of relatively indigestible leaves and buds. The small heart suggests a profound physiological difference in this species, perhaps affecting both metabolic and general activity.

External structures may also vary regularly with body size, but are less likely to do so than are internal structures because they are less subject to physiological constraints. Wing length is isometric with body size; the relationship depicted in Figure 17 has a slope close to 1.0 on a logarithmic scale. The length of the tarsus element of the leg, and the length of the hind toe plus claw (hallux), also vary approximately isometrically with body size among the ground finches, and among the tree finches, but there are some interesting deviations (Fig. 17). For a given body size tree finches have longer tarsi than do ground finches. Among the ground finches, those that climb and scratch the most have the proportionally longest hallux. These include populations of *G. scandens*, the species which spends most time climbing on cactus (Plates 4, 12, and 41), and the population of *G. conirostris* on Genovesa which, unlike the conspecifics on Española and nearby Gardner, is a similar cactus specialist (Grant and Grant 1981, 1982). They also include populations of *G. difficilis* (Plate 42), which forages mainly on the ground by scratching in the litter with its feet (Schluter 1982a). Interestingly, the population of *G. difficilis* that does this to the least extent (Genovesa) has the shortest hallux, both absolutely and relatively (Grant et al. 1985).

The most conspicuous shape variation is in the beak. For example, bill

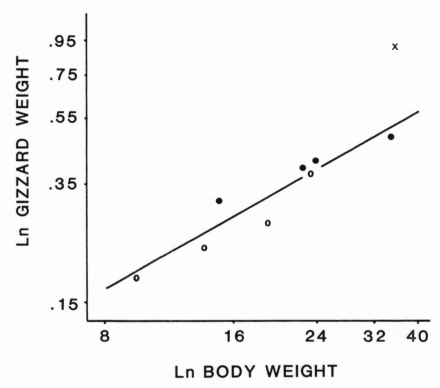

FIG. 18. Gizzard weight (g) as a function of body weight (g) among nine species. Ground finch species are indicated by solid circles, tree finches are indicated by open circles, except for the vegetarian finch (cross). The vegetarian finch, *Platyspiza crassirostris*, was not included in the geometric mean regression analysis of the relationship.

length does not scale in a simple manner to body size, either among the ground finches (Fig. 19) or tree finches; mean bill lengths of the six ground finch species do not lie on a single straight line, and bill length increases with body size among populations faster in some species such as *G. scandens*, than in others such as *G. fortis*. Therefore beak size varies allometrically, both among species and among some populations of certain species. Also, species differ most strongly from each other in bill shape, as represented by the relationship between bill length and bill width, for example (Fig. 19). Again, all points for the six ground finch species do not lie on a single line, and bill shape changes with increasing size differently among populations of the different species. Notice that the points for the *G. difficilis* populations have the greatest scatter, indicating a large variation among the populations in bill shape.

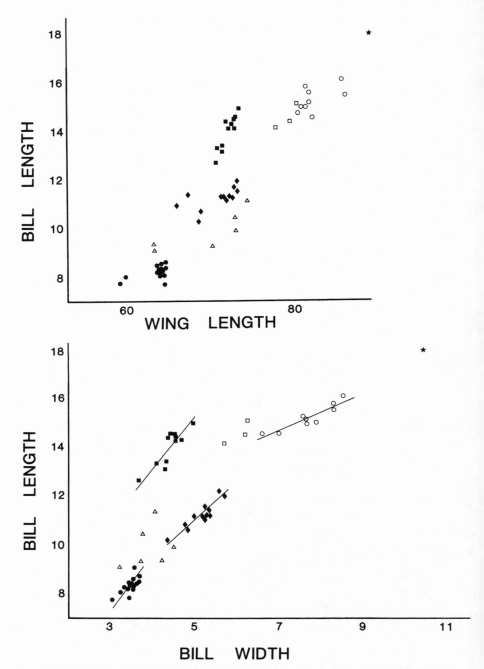

Fig. 19. Relationships between mean bill length and mean wing length, and between mean bill length and mean bill width (all in mm), among populations of the six ground finch species, males only. Symbols as in Figure 17. From Grant et al. (1985).

The contrast between changes in proportions of bill features but constancy of shape in other traits is also manifested within individual populations. Boag (1984) analyzed morphometric relations in populations of *G. fuliginosa*, *G. fortis*, and *G. magnirostris* on Santa Cruz, and in populations of *G. fortis* and *G. scandens* on Daphne. Measurement error in the six primary traits was low (Boag 1983, 1984). He found that weight, wing length, and tarsus length varied isometrically in pairwise combinations in each population, and that the three bill dimensions were another isometric set, but bill dimensions varied allometrically with any member of the other set, especially in the two *G. fortis* populations (Fig. 20). These two populations, of very different mean sizes (e.g. Fig. 15), have very similar allometric relations, as indicated by almost identical regression coefficients (intercept and slope) for a given pair of traits. The high slopes of bill dimensions regressed on body size reflect the greater relative (scale-corrected) variation in the bill dimensions. It can also be seen in the figure that *G. scandens* has a very different bill shape from the rest, but the rest are not identical except for their size. The lines of allometry are almost parallel but do not run into each other; instead they are slightly offset. Thus, for example, *G. magnirostris* has a wider beak in relation to beak length than the other species.

MULTIVARIATE SHAPE VARIATION

Since much of the variation in proportions among the species accompanies variation in size, the question arises as to how much shape variation is independent of size. Principal Components Analysis provides an answer by finding trends among several dimensions considered simultaneously. Variation among the dimensions is expressed on a small number of synthetic axes. When the first component adequately characterizes size variation, additional components represent residual shape variation.

Variation among the fourteen species along the first two component axes is shown in Figure 21. The first component, representing body size, statistically accounts for a large amount of the variance (90.4%) among populations and species. It is interpreted as a body size axis for two reasons. First, almost all factor loadings of the original dimensions on this component are positive, i.e. the dimensions are all positively correlated with component scores; the loadings are similar among species. Second, average body weights of 11 of the species are positively correlated with their mean PC I scores ($r = 0.91$, $P < 0.01$).

An additional 6.1% of the variance is accounted for by the second component. Factor loadings of wing and tarsus are almost all positive, whereas those for bill dimensions are of mixed sign. Therefore the second component represents a complex shape factor of bill shape in relation to overall size.

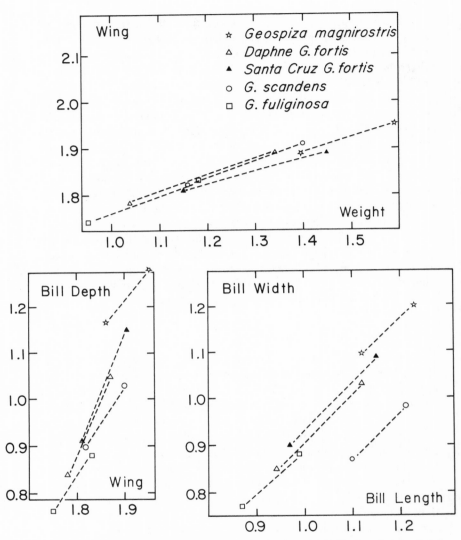

FIG. 20. Log$_{10}$–log$_{10}$ plots of static allometry for selected dimensions in five populations of adult ground finches. Symbols designate the minimum and maximum values for each functional regression line. Note the similarity of the lines for wing length in relation to weight, in contrast to the differences among bill widths in relation to bill length. From Boag (1984).

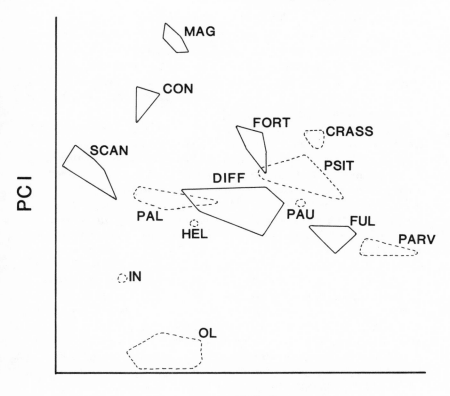

FIG. 21. Morphological relationships among all Darwin's Finch species on two principal components axes, the first representing overall size increasing away from the origin, the second representing a complex shape factor dominated by bill length decreasing away from the origin. Species names (see Table 1) are abbreviated. Polygons enclose populations of ground finch species (solid lines) and tree finch species (broken lines). Note that within each group of species no two polygons overlap. From Grant et al. (1985).

Thus species differ in bill shape relations after much of the size variation has been removed. They are broadly separated along the second component axis, and factor loadings on this component are heterogeneous among species (Grant et al. 1985).

The figure shows that all species are well separated in two-dimensional space, with overlap occurring between one ground finch species and one tree finch species in two cases, but never between species in the same group. Thus species are seen to differ in size, shape, or both, and no two species are the same. Separation between the ground finch and tree finch groups occurs along the beak shape axis (PC II), so for a given body size tree finches and ground finches have different beak proportions. But the two groups do not cluster in different parts of the two-dimensional plot; we do not find, for example, that all tree finches are in the bottom left corner and all ground finches are in the upper right corner. Instead they are broadly intermixed, which is why they were characterized in the previous chapter on the basis of plumage alone rather than by plumage and size.

Separate treatment of the two groups in Figure 22 provides a way of illuminating another feature in Figure 21: most ground finch species have counterparts in the tree finch group, as represented in these two-dimensional characterizations. There is a good correspondence between *G. magnirostris* and *P. crassirostris*; between *C. pallida* and *C. heliobates* combined and *G. scandens*; between *G. fuliginosa* and *C. parvulus*; and between *G. fortis* and *G. conirostris* combined and *C. psittacula*. Only *G. difficilis* lacks a tree finch counterpart. These comparisons are legitimate, if only approximate, because the synthetic axes being compared in the figure are constituted in a similar way; the factor loadings are similar (Grant et al. 1985). Thus there is some degree of morphological complementarity between the species of the two groups, although one group is not a simple replica of the other.

The area enclosed within each polyhedron in these figures gives an approximate measure of the degree to which populations of a species have become morphometrically differentiated on different islands. Among the ground finches, *G difficilis* (6 populations) is more differentiated than all others, as was previously indicated on the bivariate plots (e.g. Fig. 19) and by the multiple discriminant function analysis represented in Figure 11. Among the tree finches, *C. psittacula* (5 populations) is clearly more differentiated than its congener *C. parvulus*, which is represented here by the same number of populations. *P. crassirostris* (7 populations) has scarcely differentiated at all. *Certhidea olivacea* (14 populations) has undergone a large amount of morphometric differentiation in parallel with differentiation in color from grey to green (Lack 1945, 1947).

These various patterns of variation, both among populations of a species and between species, will be important in a later discussion of how the correlated dimensions of a phenotype change in the transformation of one species into another (Chapter 15).

GEOGRAPHICAL VARIATION IN SIZE

In large areas of continental regions the average size of birds often varies regularly with latitude, less often with longitude. In general, size increases with increasing latitude, hence with decreasing temperatures. Regular trends like this help us to identify environmental factors that might be responsible for variation in size.

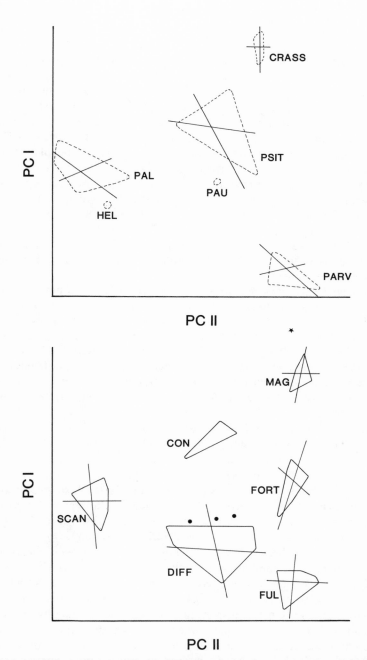

FIG. 22. Morphological relationships among the tree finches (above) and the ground finches (below) on two principal components axes. Lines are projections of the separate interpopulation (intraspecific) principal components I and II onto the principal components axes for all species combined in each group. The solid circles identify three specimens of the extinct *G. difficilis* from Floreana, and the star identifies a sample of 4 male specimens of the extinct form of *G. magnirostris* on Floreana and San Cristóbal. Species names are abbreviated. From Grant et al. (1985).

The pattern of size variation among populations of finch species on the Galápagos is more a mosaic one, with weak trends along a north-south axis. A good example is provided by *G. fortis*. Variation in the size of this species across the islands of Isabela and Santa Cruz has been mentioned already; birds are larger on the south of each island than on the north. Largest overall size occurs on Isabela and on the southern islands of San Cristóbal and Floreana, and mean size is much smaller in the most northerly populations on Pinta and Marchena. Thus the between-island trend is similar to the within-island trend. The regular pattern is disrupted, however, by the presence of the smallest birds on Daphne Major in the center of the archipelago.

Size differences among populations of *G. scandens* are the epitome of mosaic, geographical, variation (Fig. 23). For example, cactus finches on Daphne Major are large (unlike *G. fortis*), and while the smallest birds occur on Santiago a short distance to the north, the largest occur on Pinta and Marchena yet further north.

Do any species show regular patterns of size variation? Simple correlation analysis has been used to see if body size varies regularly with latitude and longitude, and also with island area, elevation, the number of plant species on an island as an index of its habitat diversity, and the average distance of an island from all others as an index of its isolation (Grant et al. 1985). The first principal component mean score was used as an index of body size for each of the ground finch species except *G. conirostris* (since it has only three populations), the warbler finch, and three tree finches (*P. crassirostris*, *C. psittacula*, and *C. parvulus*).

From six analyses performed for each of nine species (i.e. 54), two are expected to be significant at the 5% level by chance alone, and two were in fact found; body size of *G. parvulus* increases with degree of isolation, and body size of *G. fortis* increases with number of plant species on an island. But two other correlations were significant at the 1% level, so greater confidence can be attached to these; body size increases with degree of isolation in *Certhidea olivacea* and decreases with latitude northwards in *G. fuliginosa*. This last result parallels the trend described above for *G. fortis* but not demonstrated statistically. The result with *Certhidea olivacea* is interesting because it parallels the variation in plumage color (Swarth 1934, Lack 1945); widely separated islands have similar forms of this species, in both size and color, whereas intervening islands are occupied by different forms.

Thus Darwin's Finches display little geographical regularity in their variation in body size. The same is true for their variation in shape (PC II). Given the relatively small area of the archipelago, the distribution of islands north and south of the equator, and a relatively uniform climate among them, this result is not very surprising. It does not help us to identify the possible reasons for size variation. What is needed instead is more detailed ecological information from each island (Chapters 6–8).

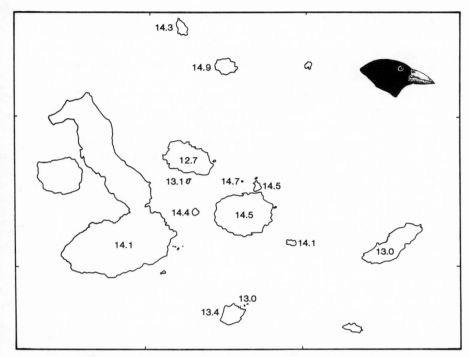

FIG. 23. Geographical variation in size, illustrated by mean beak lengths of *Geospiza scandens*. Redrawn from Lack (1947).

SUMMARY

All fourteen species of Darwin's Finches differ from each other in body size and/or bill size and shape.

Several populations of the same or different species differ in their mean size of traits and in the variation around the mean. Wing length, tarsus length, and weight covary isometrically among species, although for a given weight the tree finches have longer tarsi than do the ground finches. The greatest differentiation among populations of the same or different species occurs in bill dimensions, which vary allometrically among themselves as well as in relation to body size. The tree finch and ground finch groups differ more in bill shape than in body size, but are sufficiently similar that morphological counterparts of species in one group can be identified in the other group.

Deviations of population values from a regression relationship between a morphological structure and body size suggest the possibility of adaptive modification of the structure independent of body size. *P. crassirostris*, the vegetarian finch, has an exceptionally large gizzard but unusually small heart. *G. magnirostris*, the largest of the ground finches, has an exceptionally large skull. Those ground finches which spend a particularly large amount of time climbing or scratching have exceptionally long hind toes and claws.

There are some weak geographical trends in size variation. *G. fortis* tends to be smaller in the northern parts of two large islands than in the southern parts, and smaller on the northern islands than on the southern ones, but the only body size relationship with latitude that can be demonstrated statistically is shown by another species, *G. fuliginosa*. In general, the pattern of variation in size among islands is mosaic, and is of little help in identifying the factors that might be responsible for size variation.

PLATE 40. Adult males of the two resident species on Daphne Major. Upper: Cactus finch, *Geospiza scandens*. Single white feathers are formed occasionally on the heads of old birds. Lower: Medium ground finch, *Geospiza fortis*. This was the only individual born in 1976 that was known to survive the drought of 1977; it has an exceptionally deep beak.

PLATE 41. The cactus finch, *Geospiza scandens*. Upper: Opening a flower bud on *Opuntia echios*. Daphne Major. Lower: Hanging from cactus flower and feeding on pollen. Daphne Major (*photos by P. T. Boag, Y. Lubin*).

PLATE 42. Differences in legs and feet associated with differences in foraging. Upper: Legs, feet and claws of the small ground finch, *Geospiza fuliginosa* (left), and the sharp-beaked ground finch, *Geospiza difficilis* (right). Sharp-beaked ground finches spend much more time scratching in the leaf litter than do small ground finches. Pinta. Lower: Sharp-beaked ground finch, *Geospiza difficilis*, feeding on nectar from the flowers of *Waltheria ovata*. Genovesa. The species spends much less time scratching in the leaf litter than it does on Pinta, and it has the legs and feet more typical of small ground finches on other islands (*photos by D. Schluter, A. Simon*).

CHAPTER FIVE

Growth and Development

INTRODUCTION

How do the differences between species in the size and proportions of adults arise during development? There are several possibilities. At one extreme all differences between species might arise in the egg, with the result that at hatching one species is as different from another as are the adults, when allowance is made for the difference in scale. At the other extreme all species might be identical until the last stages of fledgling development, when they each branch off along unique pathways to reach final size and form. And of course there are intermediate possibilities. A study of nestling and fledgling development enables us to distinguish among the various possibilities, and to identify those characteristics of growth that have been evolutionarily labile.

I shall use the formalism of Alberch et al. (1979) as a framework for the discussion, beginning with a standard growth curve for an individual or population (e.g. Fig. 24). The temporal pattern of growth has four main features (Alberch et al. 1979); the timing of the start (onset) and the end (offset) of growth, the duration, and the rate of growth which may or may not be constant over the total period of growth. Species may differ in any of these four features. Figure 24 illustrates just three of the possible patterns. Onset, offset, and duration are the same in this figure, but embryonic growth rates differ among the three cases, and there is a further difference in nestling growth rates between *a* and *b* on the one hand and *c* on the other. As a consequence of growth rate differences alone, adult sizes are very different. Additional differences in onset, offset, and duration could complicate the patterns considerably.

The first part of this chapter describes the patterns of absolute growth. The second part deals with patterns of relative growth, that is, the growth in one trait in relation to growth in another trait. I shall show how the differences between species in adult shape discussed in the previous chapter, for example bill shape, are brought about in development by different patterns of relative growth.

VARIATION IN EGG SIZE

The logical starting point for an investigation of growth is the egg. It has often been found in studies of birds that small species lay small eggs and

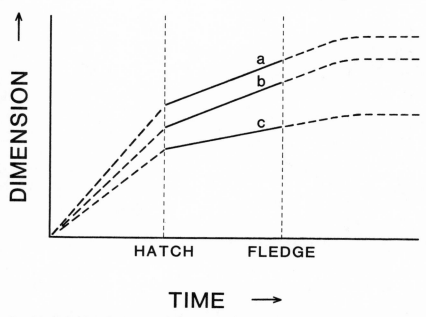

FIG. 24. Alternative growth patterns in the dimension of a trait. Embryonic growth differs among all three, whereas *a* and *b* have identical nestling growth yet differ from *c*.

large species lay large eggs (Lack 1968). Darwin's Finches are no exception. The relationship between mean egg size and mean body size for the ten species that have been studied is linear (Fig. 25), and the correlation is strong ($r = 0.97$, $P < 0.005$). Note in the figure that populations of a species differ from each other in the same way that species differ from each other. Thus egg size is a simple function of body size (Grant 1982, 1983b).

Nothing is known directly about the first stage of growth from the point of incubation to hatching. The length of this period seems to be roughly the same for all species. The warbler finch and the ground finch species have been studied in detail, and all have a modal incubation period of 12 days. Interspecific differences in embryonic growth rates can be inferred nevertheless from post-hatching characteristics, as will be discussed later.

A feature of Darwin's Finches worth mentioning here is the particular form of the relationship between egg size and body size. The slope of the relationship (0.55 ± 0.06) is low in comparison with the slope (0.81 ± 0.17) for six species of finches from the South American mainland (Fig. 25). Among Darwin's Finches it is the small species which are unusual in laying large eggs in relation to their body size. The systematic difference between continental finches and Darwin's Finches suggests that in the early

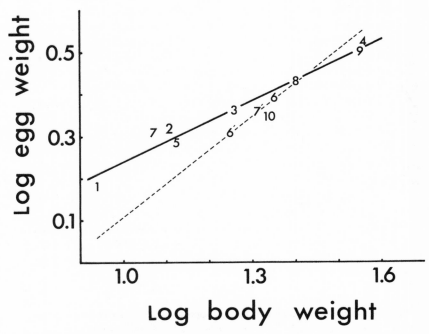

FIG. 25. Egg weight as a $\log_{10}$ function of adult body weight among ten species of Darwin's Finches (solid line). Identity of each species, indicated here by a number, is given in Grant (1982). 6 and 6' refer to two populations of *G. fortis*, and 7 and 7' refer to two populations of *G. difficilis*. The broken line is the line of best fit for six species of emberizine finches in continental South America.

stages of the evolution of Darwin's Finches a change took place in the influence of body size on egg size.

ABSOLUTE GROWTH

Nestling growth has been studied in detail in the six *Geospiza* species and in *Certhidea olivacea*. Weight, wing feather length, tarsus length, and bill length, depth, and width were measured on all nestlings in samples of ten or more nests of *G. fortis* and *G. scandens* on Daphne Major (Boag 1984), and *G. magnirostris*, *G. conirostris*, *G. difficilis*, and *Certhidea olivacea* on Genovesa (Grant 1981a). Nestlings in six nests of *G. fuliginosa* on Marchena and in eight nests of *G. difficilis* on Wolf were also measured. Nestlings are prone to desert the nest if handled after their 10th day; they fledge naturally between days 13 and 16. Starting with the day of hatching as day 0, nestlings were measured every other day up to day 7 or 8, when they were banded for subsequent identification. Only a small number of measurements were made during the last few days of the nestling period. Almost all measurements were made in 1978, a year of plentiful food, extensive breeding, and generally good conditions for growth. This is important for making comparisons between species, because Price (1985) has shown in other years that growth rates of *G. fortis* on Daphne Major vary in relation to feeding conditions. In 1978, 1979, and 1981 they were very similar, but in the poor year of 1980 they were distinctly slower.

Figure 26 provides an illustration of the change in appearance of the head, and in particular change in the size of the bill, of three *Geospiza* species. Feather development is also indicated in the figure; it is typical of the pattern shown by passerine birds elsewhere (Snodgrass 1903, Orr 1945).

It is to be expected that large species hatch at a larger size than do smaller species because the large egg provides the embryo with a large quantity of food for growth as well as a large space. This expectation is realized, but the large species hatch at a small size in relation to adult size. The negative correlation between adult weight and percent of adult weight at hatching is significant for the six species of *Geospiza* ($r = -0.92, P < 0.01$).

After hatching, the growth of nestlings is approximately linear in all dimensions for the first 10 days (e.g. Fig. 27). The slope of the regression of a variate on age provides an estimate of the rate of growth. It is best to characterize a population by first estimating growth rates of individuals, with so called "longitudinal" data (Cock 1963), then integrating them (Gould 1966), but four to six measurements on each bird do not yield a sufficiently reliable estimate of its growth rate to do this. Instead, data have been simply combined from all individuals measured repeatedly.

In all species, tarsus length, wing length, and weight increase most rapidly, and bill dimensions increase slowly and take the longest to mature (e.g. see Fig. 27). The growth rates of these dimensions and weight are typical of passerines. Boag (1984) has calculated that the rate of gain in weight shown by *G. fortis* and *G. scandens* is similar to, but slightly slower than, weight gain in the nestlings of five species of continental tropical finches studied by Ricklefs (1976), and markedly slower than temperate zone finches.

Darwin's Finch species differ in their rate of nestling growth. In terms of absolute rates, the largest species grow fastest. For example, the rank order of nestling weight growth rates of the six *Geospiza* species exactly matches the rank order of their adult body sizes, a correspondence that cannot be attributed to chance ($r_s = 1.0, P < 0.01$). However the higher rates of growth apparently do not fully compensate for relatively low hatching sizes of the large species. In other words, in their progress towards adult size the large species do not catch up with the small species during the nestling period. For example, by day 10 *G. magnirostris* nestlings have reached 64.7% of

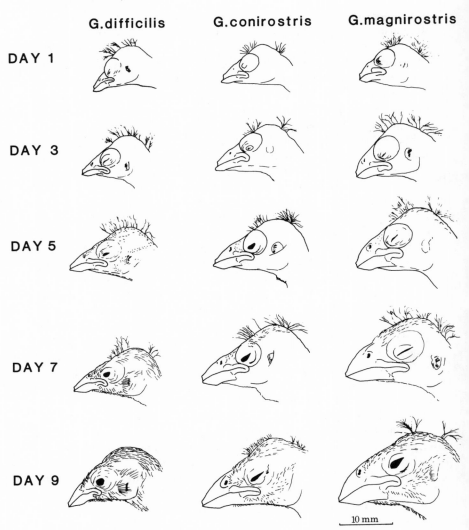

FIG. 26. Changes in the appearance of the head during nestling growth of three species of ground finches on Genovesa. Differences among the species in beak size and shape can be detected at hatching, even though they are small. From Grant (1981a).

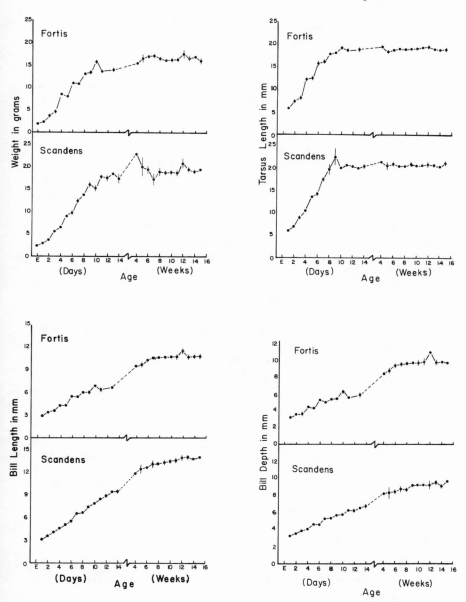

FIG. 27. Growth curves for the medium ground finch (*G. fortis*) and the cactus finch (*G. scandens*) on Daphne Major. Means (solid circles) and one standard error (vertical bar) are shown. Broken lines signify a period when birds were not measured. From Boag (1984).

adult weight on average, but *Certhidea olivacea* nestlings have reached 93.7%. Nor is catching up accomplished by a longer nestling period in the larger species. *G. magnirostris* do have a longer nestling period, 15 days on average, but it differs only slightly from the others (13–14 days).

For most species size at fledging is not known because birds are not measured at that time, and they cannot easily be recaptured for measuring until two to three weeks later. There are enough measurements upon recapture to permit a complete characterization of growth of only three species. These are *G. fortis* and *G. scandens* on Daphne Major and *G. conirostris* on Genovesa. The measurements show that the linear nestling growth trends continue unaltered in the fledgling stage until an asymptote is approached (Fig. 27). Size asymptotes are reached by *G. conirostris* at ages ranging from four to eleven weeks. Tarsus is the first dimension to reach adult size, being virtually at that size by fledging, and bill dimensions are the last (10–11 weeks). *G. scandens* grows in a similar manner, whereas the smaller *G. fortis* has reached adult bill size by about day 60 (8–9 weeks). This is another illustration of the fact that the larger species take longer to grow than do the smaller species. A very small amount of growth occurs later in life, perhaps in the second year (Price and Grant 1984). For example, the asymptotic sizes of weight and bill length of *G. conirostris* are below the mean sizes of their parents (Grant 1983a), and the difference disappears in the next year.

In summary, and with regard to Figure 26, onset of growth is unknown but is assumed to be the same for all species, offset varies to a small extent among species and so does the duration of growth, but the greatest variation among species (and dimensions) occurs in growth rates. Large species lay large eggs, hatch at large sizes, and grow fast, but need to grow for longer than small species to reach final size. Thus differences between species in the growth of a trait are characterized by the pair of lines *b* and *c* in Figure 26. Neither of the two extreme possibilities raised in the introduction, namely, all differences between species originating in the egg on the one hand, and none being manifested until late in the course of fledgling development on the other, is correct.

RELATIVE GROWTH

While studies of absolute growth help us to understand differences between species in size, relative growth needs to be known to understand their differences in proportions, or shape.

If one trait grows at the same rate as another, i.e. each doubling in size in the same interval of time, shape is constant and the growth is described as isometric. If one trait grows faster than the other, shape progressively changes during development and growth is described as allometric. The degree of allometry is measured by the exponent, α, in the power function $Y = b\,X^{\alpha}$ which relates growth in trait Y to growth in trait X, with b being a constant. For purposes of analysis it is customary to use the logarithmic transformation of the power function, as I did in the previous chapter when discussing static allometry. It is $\log Y = \alpha \log X + \log b$, where α is the slope of the regression line.

Some basic features of dynamic, or growth, allometry are illustrated in Figure 28. Three diagrams show the ways in which differences in adult proportions can be brought about during growth. In diagram 1 two individuals in a population grow along diverging allometric trajectories; the trajectories are allometric because shape changes progressively along each one. The individuals stop growing at different points, and have different proportions as adults as well as during growth. The line connecting their adult sizes describes the static allometry of adults in the population. It is steeper than the lines of dynamic allometry of individuals. In diagram 2 the individuals have the same dynamic allometries, but one is displaced, or transposed, relative to the other. The outcome is the same as in 1; static allometries are the same in the two diagrams, and different from the dynamic allometries. In diagram 3 growth trajectories of different individuals are the same, but their starting and stopping points are different. Since the growth trajectory is allometric, adults differ in their proportions. Nevertheless, in this case, static allometry among adults, the A_2–B_2 segment of the line, is the same as the dynamic allometry.

These comparisons apply to different species, and to different populations of the same species, as well as to different individuals within a population as described here. It will be shown in Figure 29 that there are examples of all three patterns of growth in Darwin's Finches.

Growth in relation to size. First I will consider the growth of dimensions in relation to weight as a measure of body size. Species are compared by analyses of covariance, using log-transformed data. The clear result from these analyses is that relative growth of wing and tarsus varies trivially among the species, conforming to the pattern illustrated in Figure 28 (2), but relative growth of bill dimensions varies substantially, as in Figure 28 (1). The common thread running through this variation is an association between growth rates and adult proportions; rates of relative growth are fastest in those dimensions that are of greatest relative size in adults. Bill length grows relatively quickly in the ground finch species with long and pointed beaks: the large cactus finch (*G. conirostris*), the cactus finch (*G. scandens*), and the sharp-beaked ground finch (*G. difficilis*). Growth of bill depth and bill width is relatively rapid in the large ground finch (*G. magnirostris*). As a result of this variation, species differ among each other at all stages of nestling

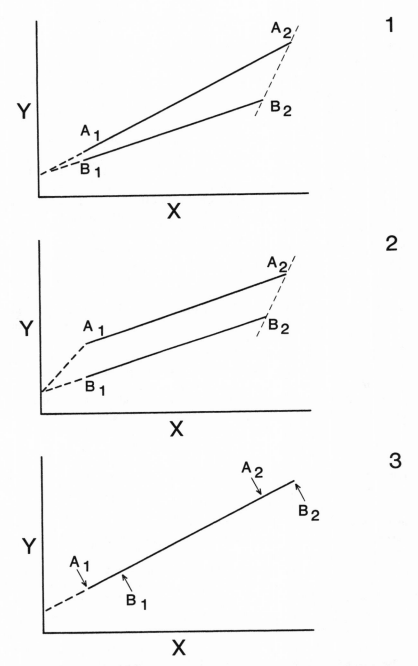

FIG. 28. Three hypothetical patterns of dynamic allometry on a logarithmic scale shown by a pair of species A and B. The resulting relationship between adults of the two species, the A_2–B_2 line of interspecific allometry, depends on the rate of nestling growth (solid line) and the sizes at hatching (A_1, B_1).

growth much more in bill dimensions than in other dimensions (Grant 1981a). Some differences in bill features can be seen in Figure 26.

Proportions at hatching. Adult proportions are determined not only by nestling growth rates but by the relative amounts of growth in different dimensions achieved by the time of hatching. Since species differ in their size at hatching they may also differ in their proportions, with the smaller species reaching the proportions of larger species one, two, or more days after hatching. This is the case for bill length in relation to weight (Fig. 29). The four species on Genovesa have approximately the same dynamic allometry, and the bill length growth point in relation to mass achieved by *G. magnirostris* at hatching is reached by the three other species some time later.

The remaining parts of the figure show that other dimensions grow very differently in relation to body mass. On the one hand, all species have approximately the same tarsus length when they hatch, in marked contrast to their differences in weight. On the other hand, species have very different bill depths and widths at the time of hatching, even in relation to weight. An important implication of these differences in patterns of relative growth is that embryonic growth leading up to sizes at hatching must differ among the species; large species are not simply scaled-up versions of small species, but have altered relative growth. For example, at hatching the bill lengths of *G. magnirostris* and *Certhidea olivacea* as proportions of adult sizes are similar, 20.1% and 24.9% respectively, which is remarkable in view of the extreme size differences between the adults of these two species; but proportional bill depths differ greatly at hatching, being 22.3% and 54.8% respectively.

Relative growth of nestlings. From the point of hatching onwards the different species have different growth trajectories. Since bill variation among species is much greater than variation in other traits, I will consider how one bill dimension changes in relation to another during growth. It can be inferred from Figure 29 that bill proportions change during growth, and that species diverge in bill form. The same inference can be made from the difference in proportional bill sizes of *G. magnirostris* and *Certhidea olivacea* at hatching, referred to above. As a further example, *G. scandens* differ slightly from *G. fortis* in bill dimensions at hatching, but these small differences become amplified during growth through higher rates of increase in bill length relative to bill depth or width in *G. scandens* than in *G. fortis* (Boag 1984, Price and Grant 1985).

G. magnirostris, *G. fortis*, and *G. fuliginosa* differ as adults largely in size, and to a lesser extent in bill proportions. Nevertheless, they cannot always be distinguished by their appearance, either as adults (Chapters 1 and

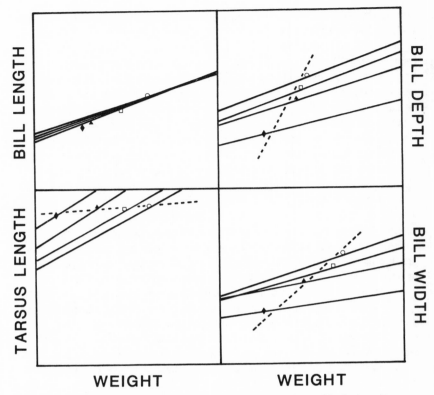

FIG. 29. Dynamic allometry of three species of ground finch and the warbler finch on Genovesa: *G. magnirostris* (○), *G. conirostris* (□), *G. difficilis* (▲), and *Certhidea olivacea* (♦). Symbols are placed on regression lines at points coinciding with the size of the traits at hatching. A broken line fitted by eye connects these points to emphasize the relationship among species of relative size at hatching. This is the hatchling isoline. Solid lines below or to the left of the hatchling isolines are extensions of the nestling regression lines; they all depart from the true but unknown embryonic growth paths which must start at the origin or just to the right of it. Weight is in g, other dimensions are in mm. From Grant (1981a).

4) or as juveniles: ". . . young birds of *G. fortis* have bills almost exactly duplicating those of adults of *G. fuliginosa*. We have examined young specimens which in fact could not be definitely assigned to either species" (Snodgrass and Heller 1904, p. 319; see also Snodgrass 1903). Here would seem to be a good example of species growing along the same allometric growth trajectory, but starting and stopping at different points (cf. Fig. 28, 3). Reasonable as this is, it is not correct. Bill length grows faster in relation to bill depth in *G. fuliginosa* than in the other two species (cf. Fig. 28, 1),

and whereas *G. magnirostris* and *G. fortis* have similar rates of relative growth, the trajectory of *G. fortis* is displaced in the direction of greater bill length for a given bill depth (cf. Fig. 28, 2). The differences between species, although statistically significant, are small. Variation around mean growth trajectories overlaps, so that some young individuals of similar species cannot be identified with certainty. For instance, I have misidentified, in the hand, one fledgling *G. magnirostris* and one fledgling *G. conirostris* of approximately comparable age. They were banded as nestlings, and I discovered the mistake only later when checking the records to identify their nests and parents.

In two cases it is known that different populations have the same or similar growth trajectories. In the first instance the different populations belong to the same species, *G. difficilis*. *G. difficilis* on Wolf have larger, and more elongated, bills than do *G. difficilis* on Genovesa. They also differ in size: adults are about 20 g on average on Wolf, but only about 12 g on Genovesa. They hatch at different points on a common relative beak growth trajectory, and stop at different points. Since relative growth is allometric, the adults of the two populations differ in bill proportions as well as in overall size.

In the second instance the populations belong to separate species, *G. difficilis* on Genovesa and *G. fuliginosa* on Marchena. There are subtle morphological features that distinguish the two species (Bowman 1961), but in the morphological features I have been considering the two species are very similar and convergent in adult morphology (Lack 1947, Schluter and Grant 1982, 1984a). They might be expected to grow along converging trajectories from different starting points, but in fact they have the same trajectory of bill length growth in relation to bill depth, and, unlike the two *G. difficilis* populations described above, they both start and stop at the same position. They have similar growth trajectories of bill or tarsus in relation to weight too.

Relative growth of fledglings. After fledging, growth occurs mainly in bill dimensions along trajectories established during the nestling stage (Fig. 27). There is one exception. *G. magnirostris* is the only species to change twice in bill proportions after fledging. Initially bill depth increases at a relatively faster rate, and later it is bill length which increases relatively faster (Grant 1981a).

SUMMARY

Morphological differences among Darwin's Finch species arise early in development, for they are already manifest at the time of hatching. At this time the species differ in both size and shape. The differences are magnified through further growth during the two weeks in the nest and in the following nine weeks. At all times during development, as well as when adult, the species differ more in bill dimensions than in weight, wing length, or tarsus length.

Large species lay large eggs. The chicks of large species hatch at large absolute size, but they are small in relation to adult size compared with the chicks of small species. They grow faster than do chicks of the smaller species, but do not fully compensate during the nestling phase for their relatively small size at hatching. Instead they grow for longer, as fledglings. In general, though, species differ less prominently in their durations of growth than in their rates of post-embryonic growth, the growth being approximately linear and continuing unaltered from nestling to fledgling phase.

Allometric growth of the bill dimensions, either in relation to each other or in relation to weight, varies substantially among the species and contributes importantly to their differences in adult bill form. Growth of depth and width of the bill varies much more in relation to body size among species than does growth of bill length. Considering just the bill dimensions, relative growth is fastest in those dimensions most pronounced in adults. Three patterns are shown by ground finch species. First, bill length grows faster in relation to bill depth in *G. fuliginosa* than in *G. fortis* and *G. magnirostris*. Second, *G. fortis* and *G. magnirostris* have the same relative growth rates of these two dimensions, but for a given bill depth *G. fortis* has a longer bill; the growth curve is transposed. Third, *G. fuliginosa* on Marchena and *G. difficilis* on Genovesa have identical relative growth trajectories and stop at identical points; hence adults of the two species have the same bill proportions. Another population of *G. difficilis*, on Wolf, grows along the same relative growth trajectory but between different starting and stopping points, with the result that adults in the two populations have different bill proportions.

The growth characteristics of the tree finches are not known, and no direct study of embryonic development has been made for any of the species. Among the ground finches, *G. magnirostris* appears to be unique in growing along an altered trajectory during the fledgling phase: initially bill depth increases faster than bill length as in the other species; then the order of these rates is reversed.

CHAPTER SIX

Beak Sizes, Beak Shapes, and Diets

INTRODUCTION

The previous two chapters described variation in morphology among adults of different species, and showed how it arises in development. The principal variation among species resides in beak dimensions at all stages of development. Birds use their beaks to feed; therefore there is likely to be a close relationship between beak size and shape on the one hand, and foraging activities and diet on the other. The association can be expected to occur at three levels of variation: at the level of species, populations, and individuals. Different species may have different diets, different populations of the same species may have different diets, and different individuals in a single population may feed in different ways and on different foods. I shall discuss them in that order of decreasing scale, beginning with some elementary remarks about feeding mechanisms.

FEEDING MECHANICS

The beak of a bird is an instrument for picking up food items and dealing with them in such a way, by crushing or partitioning them, that they can be partly or wholly swallowed. Different beak sizes and shapes can be thought of as different designs of the same instrument, with different mechanical properties and hence different efficiencies at performing the same task, or, alternatively, as best suited for performing different tasks. Bowman (1963) nicely illustrated this concept in an analogy with pliers of different shapes and sizes (Fig. 30).

The analogy is limited, however, because the same human hand may use all of the pliers. Very different forces are required to squeeze the handles of the different types of pliers, and similarly, different bill forms require different forces arising from different muscle masses to operate them. Not surprisingly, therefore, the muscles of adduction vary considerably among the species in a parallel manner to the variation in beak size and shape. These differences are illustrated in Figure 31 with three species of the genus *Geospiza*. The figure shows that the largest species has the greatest development of the adductor muscles; in particular, notice the massive *M. adductor mandibulae externus superficialis*, and the extensive development of *M. ptery-*

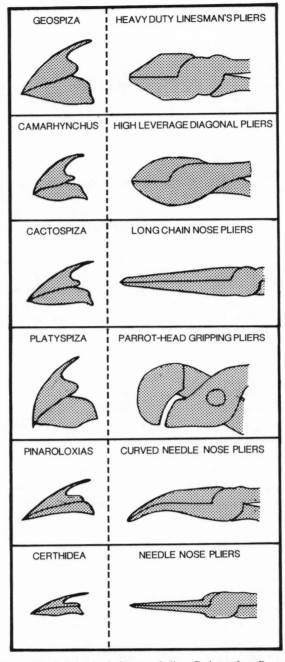

Fig. 30. An analogy between beak shapes and pliers. Redrawn from Bowman (1963).

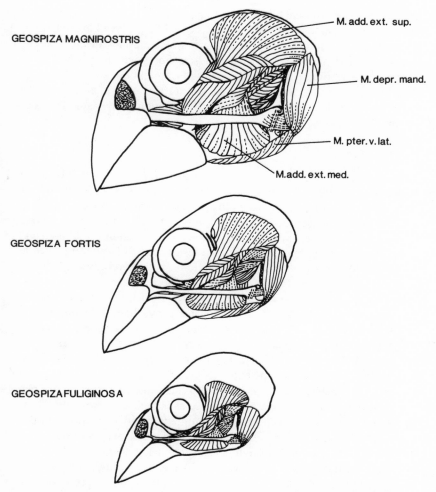

FIG. 31. Superficial jaw muscles of three species of ground finches. The largest species, *G. magnirostris*, has the largest muscle mass, both relatively and absolutely. Muscles identified by abbreviations are *M. adductor mandibulae externus (superficialis* and *medialis), M. depressor mandibulae*, and *M. pterygoideus ventralis lateralis*. Redrawn from Bowman (1961).

goideus ventralis lateralis. Associated with differences in muscle mass among the species are differences in the sites of origin of the muscles, and differences in skull architecture. The mechanical potentialities of these different muscle-bone systems have been studied in great detail by Bowman (1961), and integrated with additional information on the pneumatization of the skull and the architecture of the horny palate to determine the mechanical properties of different bill sizes and shapes. Bock (1963) has pointed out that further refinements of these analyses could be achieved by including, for example, measurements of the angles at which muscles attach to bones, and the internal arrangement of the muscle fibers within the muscle. It would be interesting to test the properties of different bill sizes and shapes directly with a strain gauge such as Lederer (1975) used in a study of fly-catchers. An indirect alternative is to use information on the natural diets of the birds (see below).

FEEDING TYPES

The great variation among species in the feeding functions of their bills can be reduced to three classes: probing, tip-biting, and base-crushing (Bowman 1961). The long and pointed bills of *C. olivacea*, *P. inornata*, *C. pallida*, and *G. scandens* are suitable for probing flowers, foliage, or woody tissues. Bills with a convex curvature of both upper mandible (culmen) and lower mandible (gonys) are appropriate instruments for applying force at the tip; the *Camarhynchus* species have tip-biting bills of this shape (Figs. 2 and 30). Bills of the *Geospiza* species are, to a greater or lesser extent, relatively deep at the base and hence suitable for crushing hard foods there. They have a less pronounced curvature to the gonys than do the bills of tree finches. *P. crassirostris* does not fit neatly into this scheme because its short, stubby bill combines the tip-biting characteristics of *Camarhynchus* species with the base-biting characteristics of *Geospiza* species. It is unique among the finches in using the entire length of its bill, from base to tip, for crushing (Bowman 1961). *G. difficilis* also shares the characteristics of two feeding groups. Its bill is deep at the base as is typical of members of the genus, but the culmen has a straighter profile than any congener, and this may be related to a probing function for nectar in flowers (Plate 42) or arthropods in leaf litter and epiphytic moss (Schluter and Grant 1984a).

These functional considerations warn us that simple measures of beak size based on length, depth, and width, can only capture a portion of the beak's architecture, and may miss something important, such as curvature (see also Bock 1963). In Bowman's (1961, pp. 155–156) estimation: ". . . differences in curvature of the culmen and gonys, in depth of upper and lower mandibles, and in total length of bill, no matter how small and insignificant they may appear to be, are, in fact, extremely important in determining the mechanical potentialities of the bill." Furthermore, *C. pallida* and *G. scandens* have somewhat similar bill shapes but feed differently, the former tip-biting and the latter base-crushing, as a result in part of their differences in adductor musculature (Bowman 1961). Comparisons of simple indices of bill shapes among species are therefore most meaningful when restricted to members of a closely related group such as the ground finches.

ECOLOGICAL SIGNIFICANCE OF BEAK DIFFERENCES BETWEEN SPECIES

In a general way, the significance of the differences is obvious: birds of different beak sizes and shapes feed on different things. The warbler finch (*C. olivacea*) feeds on small arthropods (Bowman 1961) and nectar (Grant and Grant 1980a), the vegetarian finch (*P. crassirostris*) feeds largely on fruits, leaves, and buds, other tree finches excavate arthropods from dead woody tissue, and ground finches crack and consume seeds. Their beak sizes and shapes are appropriate for these feeding tasks. Differences among the genera are very clear and interpretable.

It is less obvious that morphologically similar species in the same genus feed on different foods, and past views have been sharply conflicting on the significance of differences between species, be they small or large. Thus Beebe (1924) wrote, "Birds utterly dissimilar in relative proportions of mandibles were feeding upon identical food, and food which usually showed no signs of being crushed." He was referring to the three species in Figure 31, whose bills, skull architecture, and muscles differ so enormously. David Lack confirmed this result by observation, and concluded that "the slight differences between the diet of these three species cannot be the cause of their marked bill differences" (Lack 1945, p. 43). This became expanded to "there is no evidence that in closely related species the bill differences are related in any way to differences in food, feeding habits, or other differences in the ways of life of the species. In particular, *Geospiza magnirostris* and *G. fortis* occupy identical habitats and seem to have identical habits and food. This also seems true of the less-studied *Camarhynchus psittacula* and *C. parvulus*. *Geospiza fortis* and *G. fuliginosa* also show no significant differences" (Lack 1945, p. 119: he later changed his views; see Chapter 11).

But Lack was wrong to declare no evidence. Much earlier, Snodgrass (1902) had examined the stomach contents of 209 specimens of 8 species collected on 10 islands. Although he was more impressed by dietary similarities than by differences among species, he concluded from his analysis that species with small bills eat only small seeds whereas species with large bills eat both small and large seeds. The species he examined included the three *Geospiza* species mentioned by Lack.

These earlier findings have been confirmed and amplified by subsequent quantitative studies. As Allen (1925, pp. 81–82) pointed out, "It would be thought that the heavy-beaked birds would feed on larger or harder seeds that required strength for crushing, and that the smaller-beaked species would eat thinner-shelled seeds," thereby drawing attention to the hardness as well as the size of the food item. Bowman (1961) analyzed stomach contents of several species from Santa Cruz, and concluded that soft parts of the largest fruits are eaten by both small-billed and large-billed finches, but the hard parts of the largest fruits (e.g. *Cordia*, *Bursera*, and *Tournefortia*) are avoided by all but the largest-billed species. The *Camarhynchus* and *Cactospiza* tree finches do not crush large seeds, nevertheless the bill difference between *C. psittacula* and *C. parvulus* is accompanied by a difference in arthropod prey. The two species feed on caterpillars of different mean sizes but also largely in different taxa and hence presumably in different microhabitats (Bowman 1961). The significance of the bill size difference here is not so much in the association with different prey sizes as in the association with different woody microhabitats which require different bill-related skills to penetrate. *C. psittacula*, with the larger bill, can excavate more deeply and reach the larger prey.

DIETARY DIFFERENCES BETWEEN SPECIES

Quantitative analysis of diets from stomach contents is limited by two short-comings of the data: identification problems and small samples. Seeds and arthropods are swallowed, if small, or crushed; if crushed, the fragments can be difficult or impossible to identify, and hence valuable data are missed. Other food types unlikely to be recorded are nectar and pollen. Since only one set of data are provided by one stomach (bird), sample sizes are small. Much more information can be obtained by observing birds feeding. Identification of food items is a problem here also, but not as serious as it might be because most birds are fairly tame and can often be observed at distances as short as 2–5 meters (Plate 43).

Feeding observations. In 1973 we made quantitative feeding observations on 21 populations of the 6 *Geospiza* species at 8 sites on 7 islands in the period late February to late June; this was the early part of the dry season, since the rains had ceased unusually early, at the end of January, and breeding had almost finished when the study began (Abbott et al. 1977). At each site measurements of all seed types were also made.

Characteristics of the observed food items can be related to bill sizes, providing that appropriate scales are chosen for the comparison. From the mechanical analyses of Bowman (1961), it is clear that bill depth (or bill width) is the important dimension which determines cracking efficiency with seeds that vary in size, shape, and hardness. For the food axis, some composite index of size and hardness of seeds is required, an index which takes into account the fact that seeds (or fruits) may be large yet soft, or small yet hard. Abbott et al. (1977) assumed that these two types of seeds would present functionally equivalent problems to birds, and so they multiplied the hardness of a seed by its second longest dimension (depth), then took the square root to give a geometric mean index ($\sqrt{DH}$). The assumption of functional equivalence could be wrong in certain instances, but the two physical properties of the seeds probably give more meaningful information than does one alone, and of the various ways of combining them the above index employs one of the simplest (in Chapter 12 I will discuss an alternative). A pliers device with an associated scale in kilograms force (kgf) was invented for the purpose of making the hardness measurements (Plate 44).

Snodgrass's main result was confirmed by this study: all species ate small seeds but only the large species ate the large ones. The maximum size and hardness of a seed type dealt with by a species was a function of its beak size (Abbott et al. 1977). For example, the maximum $\sqrt{DH}$ values for seeds consumed by the three species depicted in Figure 31 are ~4 for *G. fuliginosa*, ~9 for *G. fortis*, and ~14 for *G. magnirostris*. Subsequent studies on Pinta (Schluter 1982a, 1982b) and Daphne Major (Boag and Grant 1984a) have yielded similar results.

The difference in diets between any two sympatric species was found to be a simple function of the difference in their beak depths (Fig. 32; $r = 0.62$, df 16, $P < 0.01$). This follows as a consequence of the relationship

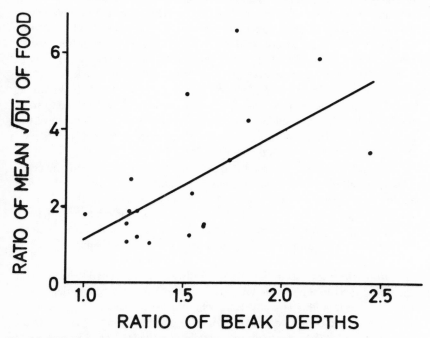

FIG. 32. Differences in diet between sympatric populations of species of ground finches as a function of the differences in their beak sizes. $\sqrt{DH}$ is the square root of the product of depth and hardness of seeds in the diet. From Abbott et al. (1977).

between the beak size of a species and the maximum size and hardness of the seeds it cracks. The corollary is that similarity in diets is inversely related to beak differences. Using a different measure of similarity, Abbott et al. (1977) found this to be true for differences in beak depth ($r = -0.66$, $P < 0.01$) and beak width ($r = 0.65$, $P < 0.01$) but not for differences in beak length ($r = -0.15$, $P < 0.1$) which, for mechanical reasons, is to be expected. There is a technical problem here in assessing the significance of the correlations; on some islands a species (i) enters the analysis twice, in comparisons with species j and with species k. Correcting for lack of independence, by reducing the degrees of freedom, lowers the level of significance from 0.01 to 0.05 but does not alter the conclusions.

Diet diversity is also a function of beak size. Seeds were classified into fourteen $\sqrt{DH}$ categories, and the number of such categories represented by seeds in the diet of a species is a measure of diet diversity. The correlation between beak depth and diet diversity was found by Abbott et al. (1977) to be positive but not strong ($r = 0.43$, df 19, $P \sim 0.05$). A better test of the relationship can be made by using the range of size-hardness categories represented in the diet, rather than the number, since some categories within the range may be missing because the appropriate seeds are not available. When diet range is used, the rank correlation with beak depth is strong among the 21 populations ($r_s = 0.85$, $P < 0.01$), and perfect for the 6 species ($r_s = 1.0$, $P < 0.05$). The largest species, G. magnirostris and G. conirostris, have the most diverse diets, as only these can crack the seeds (stones) with the largest $\sqrt{DH}$ values, those of Cordia lutea. It should be added that in terms of taxonomic diversity of seed types, as opposed to size-hardness diversity, the smaller species G. fuliginosa and G. fortis have the broadest diets. This is because there are many more species of plants producing small and soft seeds than large and hard ones.

These relationships are specific to seed-eating, and ignore both the complexity of bill form and some elements of the diet. G. fuliginosa and G. difficilis, for example, differ to a small extent only in beak depth and in the ranges of seed sizes consumed. But they differ more conspicuously in beak shape (Figs. 11 and 12) , and their diets differ as a result of G. difficilis feeding extensively on arthropods (Schluter and Grant 1982, 1984a).

Handling efficiency. Seed "handling" skills which are related to bill size set the upper limit to the size and hardness of seeds that a bird can deal with. One can see how this is so by comparing the performance of two species with a large and hard seed or fruit. For example, the woody mericarp (fruit part) of Tribulus cistoides (caltrop) contains 4–6 seeds (Plate 45), and being about as long as the bill of G. fortis, presents it with a challenge (Fig. 33). Timed observations with a stopwatch on Daphne Major have shown how the

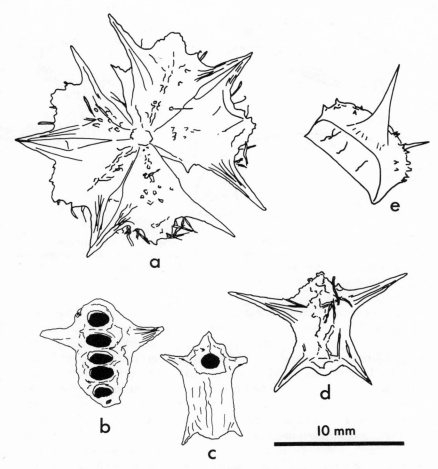

Fig. 33. A fruit (*a*) and individual mericarps (*b–e*) of caltrop (*Tribulus cistoides*). Each mericarp has 3–6 seeds, one per cell. Seeds have been removed by finches from *b* and *c*. See also Plate 45. From Grant (1981b).

larger *G. magnirostris* is much more efficient at meeting this challenge than is *G. fortis* (Grant 1981b). To crack a mericarp transversely, as *G. magnirostris* usually does, requires an average force of 25.8 ± 1.4 (SE) kgf (*N* = 40). *G. fortis* bites and twists the lower surface of the mericarp (Fig. 34), and the force necessary to crack the woody exterior in this manner is much less—only 5.8 ± 0.8 kgf (*N* = 20). But even though *G. fortis* exerts much less force to gain its reward, it is at a disadvantage compared with its congener. *G. magnirostris* takes, on average, only 2 seconds to crack a meri-

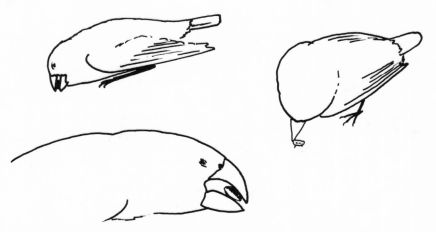

Fɪɢ. 34. Feeding of finches on the mericarps of *Tribulus cistodes*. *G. magnirostris* (lower) cracks the whole mericarp. *G. fortis* (upper) occasionally attempts this by bracing the upper mandible against a rock and squeezing with the lower manidible, but they rarely succeed; instead they usually pick, twist, and crush parts of the mericarp (right). From Grant (1981b).

carp, and a further 7 to extract a seed, and it often continues until it has eaten all seeds, then spends another 6 seconds before picking up another mericarp. *G. fortis*, in contrast, takes an average of 7 seconds to crack a mericarp, a further 15 to extract a seed, usually eats only one or two seeds and spends 24 seconds between cracking successive mericarps. This last interval is long because birds discard many mericarps without cracking them: they reject 45–95% of those they pick up, whereas *G. magnirostris* reject no more than 20%. Overall, even allowing for its greater metabolic requirements, *G. magnirostris* has a large energetic advantage over *G. fortis* when feeding on the seeds of *Tribulus*; it gains 2.6 times as much seed energy per unit time as does *G. fortis*, which more than compensates for its 60% greater energy needs (Grant 1981b). Providing *G. magnirostris* does not take much longer than *G. fortis* to digest the seeds, and I have no reason to believe that it does or that it assimilates less energy from them, it gains significantly greater energy rewards from this seed type.

This is a specific example in detail of a general trend established by Abbott et al. (1977) in their study of ground finches at several sites; all species take about the same time to crack the small and/or soft seeds, but large species take a shorter time than smaller species with larger and/or harder seeds. Interspecific differences have been confirmed by further observations of some of the same species (Schluter 1982a, Boag and Grant 1984b), as well as with *Platyspiza crassirostris* in comparison with *G. fortis* and *G. fuliginosa* (Downhower and Racine 1976) when feeding on the fruits of *Croton scou-*

leri (Plate 46). With regard to diets, since the small seeds have small kernels with relatively low energy rewards to the consumers, large species such as *G. magnirostris* tend to avoid them. Even the smallest species, *G. fuligi-nosa*, concentrates on the largest seeds that it can crack because the energy reward per unit handling time, or profitability, of those seeds exceeds the rewards from any of the smaller seeds (Schluter 1982a). Hence the propor-tions of various seeds in the diets of *G. magnirostris*, *G. fortis*, and *G. fu-liginosa* reflect the profitabilities of the seed types (Fig. 35). To some degree the relative availabilities of the seed types exercise a modifying influence. Rare seed types, however much preferred because they are individually prof-itable, cannot be common in the diet. Conversely, common and profitable seeds can dominate the diet. This happens when individually profitable seeds occur in patches, for then the time spent searching for seeds is mini-mal. For these reasons Figure 35 shows the profitability of naturally occur-ring patches or clusters of seeds of different types, for the three species of finches.

In summary, the bill size and shape of a bird sets the potential range of food types in the diet. The abundance of the food types and their profitabil-ities, as determined by the handling efficiency of the bird and by the food reward, are largely responsible for which particular foods within the poten-tial range actually contribute to the diet at a given time, and in which pro-portions.

Seasonal changes in diet. As the dry season progresses, the number of fruit and seed types (species) available for finches to feed on falls in parallel with a decline in overall food abundance. This is known from surveys conducted on Genovesa, Daphne Major, and Santa Cruz (two sites) in the early (Ab-bott et al. 1977) and late (Smith et al. 1978) dry season of 1973, on Pinta and Marchena in 1978 and 1979 (Schluter 1982a, 1982b), on Daphne Major in the years 1976–1978 (Boag and Grant 1984a) and subsequently, and on Genovesa in 1978 (Grant and Grant 1980a) and in subsequent years. These seasonal changes affect diets and the relationship between bill sizes and diets.

A finding common to all these studies is a divergence of diets through the dry season. For example, Smith et al. (1978) used a similarity index to cal-culate the degree to which the diet of each species is different from its sym-patric congeners at early and late dry season times. The index $\Sigma|p_{1i} - p_{2i}|$, simply sums the difference in proportions (p) of the ith food item in the diets of a species (1) and in the diets of all other species combined (2). Diets of all species at the four sites studied were more distinctive in the late dry sea-son than in the early dry season (Wilcoxon's test, $N = 10, P < 0.005$). All species shifted from a common diet of soft, easy-to-handle, seeds and fruits

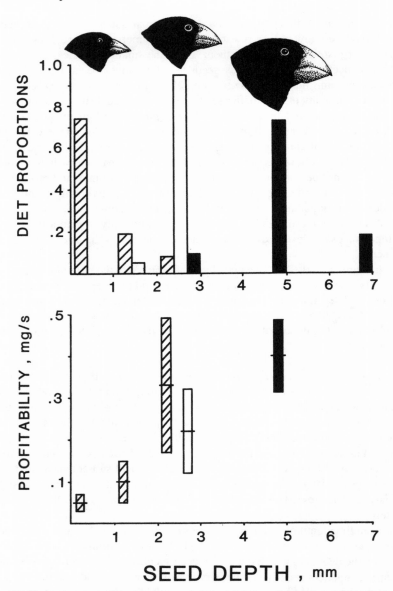

FIG. 35. Proportions in the diet of various seed sizes, and their profitabilities (see text). The upper figure shows the diets of the large ground finch, *G. magnirostris* (solid bars), the medium ground finch, *G. fortis* (open bars), and the small ground finch, *G. fuliginosa* (hatched bars). The lower figure shows the profitability of each seed class for each species, with the same symbols; the vertical bars span the range of the 95% confidence intervals for the estimates of mean profitabilities. Redrawn from Schluter (1982a).

(and caterpillars) to different diets reflecting the bill characteristics of each species (Smith et al. 1978).

An example is given in Figure 36 of the differences in diets among three common species of ground finches at one of the sites studied by Abbott et al. (1977). Late in the dry season of 1973 all species fed extensively on seeds on the ground (Fig. 37); the foods they obtained from the ground differed, but to an unknown extent owing to problems of identifying many of the seeds. The principal known differences among the species were as follows: the small ground finch (*G. fuliginosa*) fed on the small and soft fruits of *Commicarpus*, and extensively on nectar from the vine *Sarcostemma* (Plate 10); the medium ground finch (*G. fortis*) fed on the medium-sized fruits and seeds of *Scutia*, *Rhynchosia*, and *Passiflora*; and the cactus finch (*G. scandens*) used its longer bill to feed almost exclusively on *Opuntia* nectar and pollen (Smith et al. 1978; see also Plate 41). In no case did the diet of a species simply match the relative availability of different food types (Fig. 36).

On Daphne Major the similarity in the diets of the two resident species, *G. fortis* and *G. scandens*, diminishes as a result of different responses to a changing food supply. When the diversity of available food types declines, *G. fortis* broadens its diet whereas *G. scandens* narrows its diet and specializes on different parts of *Opuntia* bushes (Boag and Grant 1984a, 1984b). The process of diet divergence is reversed at the end of the dry season when *Opuntia* flowers become abundant; the two species then converge in their feeding on the nectar and pollen of these flowers (Grant and Grant 1980b, 1981).

The maximum convergence in diets occurs at all sites in the wet season, the time of greatest abundance and diversity of food, and the time when finches are breeding (Grant and Grant 1980a, 1980b, Schluter 1982a, Boag and Grant 1984a). This can be appreciated by comparing Figures 36 and 37; the diets of the three species depicted match each other more closely in the wet season (Fig. 37) than in the dry season (Fig. 36).

Early views on the similarity of diets of species with different bill sizes were conditioned by observations made only or largely in the wet season. Beebe's observations on the identity of the diets of *G. fuliginosa*, *G. fortis*, and *G. magnirostris* were made on Daphne Major in the wet season. Likewise Lack's visit to the Galápagos in 1938–1939 coincided with the wet season (of an El Niño year!). Specimens analyzed by Snodgrass (1902) were collected between December 1898 and June 1899; some were obtained at the end of the dry season but most, apparently, were taken immediately after or during the breeding season, which began in January (Snodgrass and Heller 1904). The past conflict of views regarding the bill-diet association is resolved when allowance is made for seasonal variations in feeding.

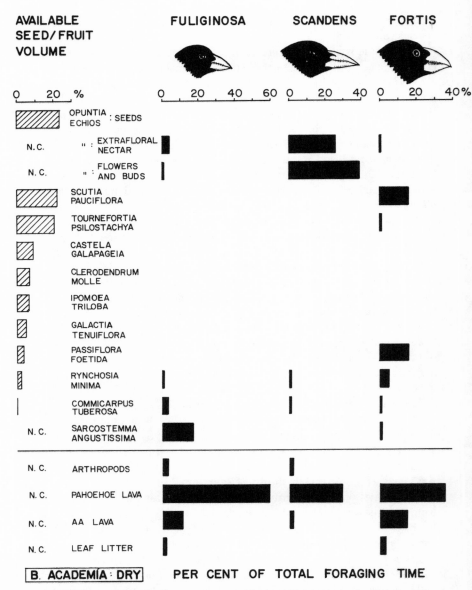

FIG. 36. Diets of three common species of ground finches at Bahía Academía, south Santa Cruz, at the end of the dry season. Items below the horizontal line were identified to general food type or foraging substrate only. Relative availability of each type is shown on the left. N. C. means food availability was not censused. From Smith et al. (1978).

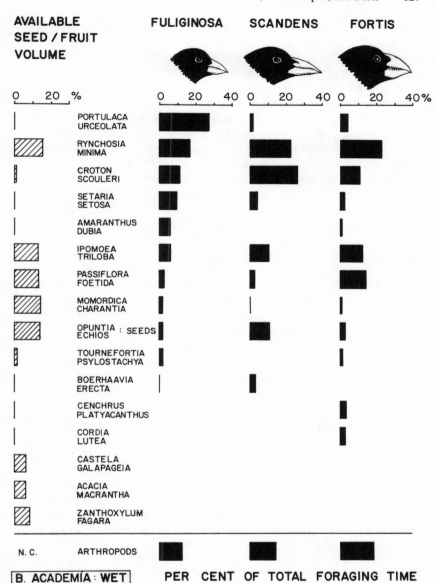

FIG. 37. Diets of three common species of ground finches at Bahía Academía, Santa Cruz, in the wet season. Note that diets do not simply match relative availabilities of the different food types, and that the diets of the three species are more similar to each other at this time than in the dry season (Fig. 36). N. C. means not censused. From Smith et al. (1978).

The development of diets. Nestlings of different species receive a basically similar diet of caterpillars, small seeds, and pollen from their parents (Downhower 1976, Grant and Grant 1980a). Nevertheless, there are subtle but important differences between the species. On Genovesa in 1978 the seeds of *Bursera graveolens*, which are difficult to crack ($\sqrt{DH} \sim 4.0$), were absent from the nestling diets of the smallest ground finch there, *G. difficilis*, and were most frequent in the nestling diets of the largest, *G. magnirostris* (Grant and Grant 1980a). Only *G. conirostris* fed nestlings on the pollen of *Opuntia* flowers. These three species also fed size-selectively on spiders in proportion to their beak sizes, as shown by measurements of spiders taken from the crops of nestlings (Fig. 38).

For the first two to four weeks after they leave the nest fledglings are dependent upon their parents for food. At this age their beaks are too soft to do little more than pick up small objects. They peck, apparently haphazardly, at leaves, bark, flowers, even the ground (Grant and Grant 1980a). Distinctive adult, species-specific, feeding behaviors appear early but develop slowly (Plate 47). On Genovesa these include feeding from *Waltheria* flowers by *G. difficilis*, and bark-stripping to reach hidden arthropods by *G. conirostris*. Observations on birds that had been banded in the nest at precisely known ages established that these behaviors were first manifested in the period 12–25 days after fledging, i.e. at post-hatching ages of about 25–40 days when beaks would be 90–95% of full size (Chapter 5). This is a period when fledglings of different species begin to associate with each other, and parental dependence is waning. They appear to copy each other's feeding behaviors, because they attempt to feed in ways that adults of the same species do not show. But after achieving independence they drop these heterospecific behaviors from their repertoires, probably as a result of inefficiency and poor rewards. The acquisition of seed-handling skills at an adult level takes a long time, as it is dependent on both learning and the maturation of bill, skull, and associated muscles. For example *G. scandens* individuals take more than a year to reach the level of efficiency at dealing with *Opuntia* seeds shown by birds two years and older (Millington and Grant 1984), and *G. magnirostris* take a similarly long time to achieve maximum efficiency at cracking the large and hard stones of *Cordia lutea*.

DIETARY DIFFERENCES BETWEEN POPULATIONS OF THE SAME SPECIES

Populations of the same species on different islands are more similar to each other, morphologically, than are populations of different species on the same island. Correspondingly they should be more similar in diets. From studies conducted on different islands at the same time in the dry season, this appears to be true. Nevertheless there are a few clear examples of

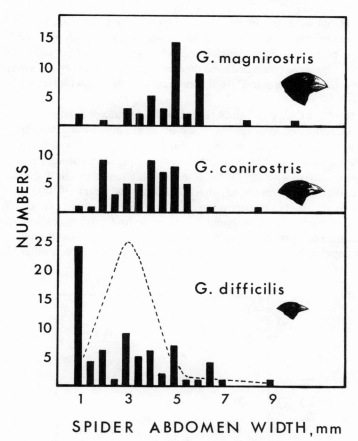

FIG. 38. Numbers of spiders of different sizes fed to the nestlings of three species of ground finches on Genovesa. The broken line represents the frequency of spiders of different sizes in the vegetation where finches were foraging. From Grant and Grant (1980a).

marked differences in diets between populations of the same species. The outstanding one is provided by *G. difficilis* which drink the blood of seabirds (Plate 5) on the islands of Darwin and Wolf only (Gifford 1919, Bowman and Billeb 1965, Schluter and Grant 1984a). The fact that these populations have the longest and most pointed beaks suggests a functional association between beak form, habit, and diet.

The most extensive inter-island comparisons of populations of the same species have been made with *G. fuliginosa* and *G. difficilis* (Schluter and Grant 1984a). Six populations of each species, and for additional comparison, two populations of *G. scandens*, were studied in the dry season. The observations were cast into nine diet categories and subjected to principal components analysis because, unlike the seed data, they are too heterogeneous to be expressed on a single axis such as seed depth. Populations of all three species were combined, and then principal components were derived from covariances computed from their diet proportions (Schluter and Grant 1984a).

As shown in Chapters 3 and 4 (Fig. 12), *G. difficilis* populations are much more differentiated in morphology, including bill dimensions, than are populations of *G. fuliginosa*. Figure 39 shows that populations of *G. difficilis* are well differentiated also in diet. They are spread out along the second component axis, whereas *G. fuliginosa* populations are tightly clustered, with a single exception (on Española) (Schluter and Grant 1984a).

The figure shows stronger interspecific differences in diet. *G. scandens* is separated from all other populations on the first axis, which represents variation among populations in the use of *Opuntia*. *G. fuliginosa* and *G. difficilis* populations are almost completely separated from each other along

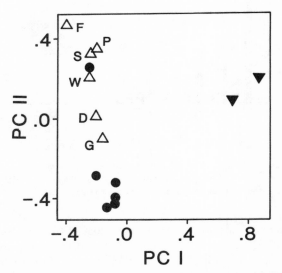

FIG. 39. Dietary relationships among three species of ground finches in the dry season, represented on two principal components axes that summarize variation among many dietary categories. The species are *G. scandens* (▼), *G. fuliginosa* (●), and *G. difficilis* (△) whose populations are indicated by F (Fernandina), P (Pinta), S (Santiago), W (Wolf), D (Darwin), and G (Genovesa). *G. scandens* is distinctive in its exploitation of *Opuntia* foods (PC I), while *G. difficilis* and *G. fuliginosa* differ from each other (PC II) largely in their proportional exploitation of exposed seeds (*G. fuliginosa*) and concealed arthropods (*G. difficilis*). From Schluter and Grant (1984a).

the second axis, which represents variation in the exploitation of concealed invertebrates and seeds by scratching in the soil or litter (especially by *G. difficilis*), and exploitation of exposed seeds by picking them up from the soil surface or directly from plants (especially by *G. fuliginosa*).

The population of *G. difficilis* on Genovesa is the most similar in body size and beak morphology to populations of *G. fuliginosa*, and this corresponds well with their dietary similarity (Fig. 39). The populations of *G. difficilis* on Darwin and Wolf, on the other hand, are sufficiently similar in beak shape to *G. scandens*, including a slightly curved culmen profile, to have been once classified as that species, but in this instance there is no close correspondence with dietary similarity. Lack (1945, 1947, 1969) suggested the morphological similarity reflected a similarity in specialized use of *Opuntia* believing, as did Bowman (1961), that other populations of *G. difficilis* did not exploit *Opuntia* cactus for food. This is now known to be only partly correct: *G. difficilis* on Darwin and Wolf certainly exploit *Opuntia* flowers for food (Grant and Grant 1981, Schluter and Grant 1984a), but so do shorter-billed birds of this species on Genovesa (Grant and Grant 1980a, 1981). Furthermore, the analysis represented in Figure 39 does not show a close similarity on the *Opuntia* feeding (first) axis between *G. scandens* and any of the *G. difficilis* populations. Instead, the significance of the long bills of the Darwin and Wolf populations may lie in the unique habits of causing bleeding of seabirds and breaking open their eggs (Schluter and Grant 1984a), habits which have not been witnessed on Genovesa despite intensive studies of *G. difficilis* near seabirds (Grant and Grant 1980a).

The lack of a close correspondence between bill form and diets when bills are rather similar suggests that bills of different sizes and shapes (e.g. those of *G. scandens* and of *G. difficilis* on Genovesa) adapted to different tasks can nevertheless be used to perform some similar tasks, such as *Opuntia* flower probing. The corollary is that the same bill can be used moderately efficiently for different tasks, for example nectar-feeding and seed-crushing. So the differences in diets between populations of the same species at one time may not correspond well with differences in morphology because they reflect more the different food availabilities on the islands at the time of sampling (Snodgrass 1902). As an example, on Española *G. fuliginosa* were feeding largely on concealed seeds in the soil during Dolph Schluter's visit in November 1981 (see Fig. 39), but on exposed seeds during my visit in August 1980, the difference being apparently due to a difference in the availability of exposed seeds. With seed-crushing, however, there is less ambiguity in the comparison of populations on different islands. For example, *G. fortis* have larger beaks on Pinta than on Daphne Major (Grant et al. 1985). On both islands they crack the stones of *Bursera* (Plate 10) which are almost the same size and hardness on the two islands. Yet birds on

Daphne have greater difficulty performing this task than do their Pinta counterparts, and take as much as six minutes to crack open a single stone. Only 34 out of a total of 81 birds (42%) that were observed trying to crack the stones on Daphne were successful (Grant et al. 1976), whereas on Pinta the proportion successful was much higher, 24 out of 31, or 77% (Schluter 1982a).

DIETARY DIFFERENCES AMONG INDIVIDUALS IN A VARIABLE POPULATION

The bill size variation that has the least obvious dietary significance is the variation among adult individuals in a single population. It requires detailed observations of measured and banded individuals of a variable population to show that all phenotypes are not ecologically identical; they differ in subtle ways in how they exploit the food resources. These differences concern where they feed, on what, and with what efficiency.

At Bahía Borrero on the north shore of Santa Cruz, G. fortis individuals that were observed to forage in woodland habitat (Plate 20) were found to have slightly but significantly deeper and blunter bills than those that foraged in nearby parkland habitat (Grant et al. 1976). Parkland habitat (Plate 48) is occupied predominantly by the smaller G. fuliginosa, so the G. fortis were apparently exhibiting an adaptive habitat selection; in the parkland, a fuliginosa habitat characterized by small seeds, the most fuliginosa-like members of the fortis population occurred at a disproportionately high frequency. The larger fortis occurred more frequently in woodland habitat which was relatively richer in larger seeds.

Different individuals in a morphologically variable population have somewhat different diets. In 1973, by uniquely color-banding and measuring 100 G. fortis on Daphne and then observing the behavior of 50 of them, we were able to detect differences in diet and feeding skills that were correlated with differences in bill size. Those individuals subsequently seen to crack the moderately large and hard seeds ("stones") of Bursera malacophylla (Plate 10) had significantly larger bills than those seen to attempt but fail to crack them. Bowman (1961) had previously found similar evidence of selectivity: out of a total of 75 specimens of G. fortis collected at Bahía Academía on the south shore of Santa Cruz only 11 had the remains of Bursera seeds in the stomachs, and they were all above average in bill size.

Five of the birds on Daphne in 1973 were also seen to crack the smaller but moderately hard seeds of Opuntia echios; they too were significantly larger in all bill dimensions than those not observed to crack the seeds. In addition, the average time taken to crack a seed varied inversely among these five birds with the size of their bill: birds with the largest bills were the quickest, and hence most efficient.

These observations on habitat selection, food selection, and feeding (seed cracking) efficiency show that all individuals in a *G. fortis* population are not ecologically equivalent. How they exploit the environment is dependent on their bill size just as, in an analogous manner, species with different bill sizes feed on different seed types and crack certain ones with different efficiencies. But it is not always the case that efficiencies are highest when bill size is largest. A counterexample is provided by the work of Willson (1971). In a laboratory study of eight North American finch species she found that the species with the largest beak (Cardinal, *Richmondena cardinalis*) took the longest time to crack and consume the relatively small seeds of millet. This result (Abbott et al. 1975), our own, and the absence of statistical correlations between bill size of *G. fortis* individuals on Daphne and cracking times with small seeds, such as those of *Chamaesyce amplexicaulis* and *Heliotropium angiospermum* (Plates 49 and 50), led us to construct two alternative models to characterize the relationship between feeding efficiencies and seed characteristics for small, medium, and large phenotypes (Fig. 40; Grant et al. 1976). I am referring explicitly to different phenotypes

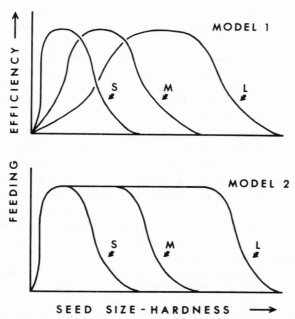

FIG. 40. Two models of the efficiency of small (S), medium (M) and large (L) finches dealing with seeds of different sizes and hardnesses. Efficiency represents the ability to crack and open the seeds, and thus secure the kernels, in a unit of time (e.g. number of seeds per minute). From Grant et al. (1976).

within a population, although the models also represent feeding relationships of different species with different bill sizes.

Feeding efficiency models. The basic feature for all phenotypes in both models is that feeding (cracking) efficiency rapidly rises to a maximimium, remains constant, and then falls as the size-hardness value of the seeds increases. Three other features are important. First, maxima differ among phenotypes in a regular way: large birds crack large seeds with greater efficiency than do small birds. Second, some large-hard seeds can only be cracked by the largest phenotype. Third, the range of seed types cracked by the largest phenotype is broader than the range of seeds cracked by the smallest phenotype. These three features were demonstrated in the preceding field study on Daphne. The first two have been mentioned already. The third, a scaling effect of size, is indicated by the fact that all members of the *G. fortis* population were observed to feed on small seeds, but only the individuals with the largest bills were seen to feed on the largest and hardest seed types.

The two models can be distinguished by observing the behavior of all phenotypes when dealing with the smallest and softest seeds (see Fig. 40). In practice it is very difficult to get the required data. The smallest seeds take less than a second to crack, and factors other than seed size, such as position of the seed (on the ground or plant), presence or absence of a wind, and social interactions, can strongly influence the performance of foraging birds in natural situations. Laboratory tests that control such factors are needed (e.g. Willson 1971, Pulliam 1980). From natural observation we do not know which model is the best.

By analogy with interspecific variation in bill size and feeding efficiency, model 1 may be preferable. The longer time taken by Cardinals than by smaller species to crack millet seeds in Willson's (1971) study is explicable in terms of model 1, but not by model 2, because the seeds are small and soft (Abbott et al. 1975). On the other hand Schluter (1982a) found that *G. fuliginosa*, *G. fortis*, and *G. magnirostris* on Pinta took about the same time to handle four types of small-medium seeds, and differed among themselves in efficiency only on the largest seed type in the diets of *G. fuliginosa* and *G. fortis*; these are features of model 2. However, *G. magnirostris* ate the three smallest seed types too rarely to be timed, and *G. fortis* consumed the two smallest ones too rarely, so there may be an efficiency difference among the species with these small types of seeds in accordance with model 1, but not revealed by observations because the critical feeding behaviors are very rare. Possibly the truth lies somewhere between the alternatives depicted in Figure 40. To pursue the interspecific comparisons further it would be necessary to scale efficiencies to a common maximum, since species differ conspicuously in their maxima, and to substitute seed biomass for seed number in the reward rate used as a measure of efficiency (e.g. as in Fig. 35). As an example, I have used data from Schluter's (1982a) study on Pinta to calculate *relative* efficiencies for the three ground finch species feeding on the smallest seed type taken by them all, namely, *Paspalum galapageia*, a species of grass. The relative efficiencies are 31% for *G. fuliginosa*, 14% for *G. fortis*, and 6% for *G. magnirostris*. In other words, they are unequal, and inversely related to bill size, as expected from this modified form of model 1. To repeat, and arguing by analogy with this interspecific variation, I suggest that model 1 may best represent the feeding efficiency differences among phenotypes in a population.

Tests of the models. Two tests of both models can be made by comparing the performance of birds dealing with moderately large and hard seeds with their expected performance (see Fig. 40). The first expectation is that large phenotypes in a population are more efficient than medium phenotypes. This is a feature of both models. The second expectation follows from the fact that a sample of seeds of a given type (species) varies in physical properties. Such variation will be more important to the medium phenotype than to the larger one because over the range of physical properties of the seeds the efficiency of the medium phenotype will vary much more. Therefore the second expectation is a prediction that medium phenotypes will be much more choosy, or selective, in the seeds of this type that they attempt to crack. It should be understood that the argument is presented in terms of two phenotypes only, for simplicity. In reality bill sizes, feeding efficiencies, and physical properties of the seeds all vary continuously.

To test these expected relationships, I made timed observations of banded *G. fortis* individuals feeding on the woody fruits of *Tribulus cistoides* (caltrop) on Daphne in 1977 (Figs. 33 and 34). The difficulty confronted by *G. fortis* individuals in attempting to extract the seeds has already been described. The first expectation is that large birds should be more efficient than smaller birds. Detailed observations agree with it, although they are not as simple as in the examples of feeding on *Bursera* and *Opuntia*, and therefore they are not merely a duplication of the observations that led to the construction of the models. The time taken to crack a mericarp is not correlated with bill size. But, as described previously, there are complicated maneuvers involved in the handling of a mericarp, i.e. cracking the woody tissue, exposing and then extracting the seeds, and consuming them. Total handling time, the sum of all these maneuvers, is shorter for birds with small bills than for those with large bills (Fig. 41). If the seed rewards were constant, this would be the exact opposite to the expected relationship. In fact the rewards are not constant. By persisting longer, a large-billed bird gains two or more seeds during its total handling of the mericarp whereas a small-billed bird gains only one. I have calculated that the larger birds gain the greater reward per unit time spent dealing with one mericarp (Grant 1981b). In this sense they are more efficient.

The second expectation is that small birds should be more selective and large birds less selective in dealing with mericarps. This prediction is matched by observations (Fig. 42), and suggests one possible reason why cracking times were not correlated with bill size; small-billed birds may exploit only those mericarps that they can crack quickly, and such selectivity prevents us from detecting poor performance with large mericarps. Controlled experiments would be needed to deal with the problem. Notwithstanding this complication, individuals with large bills have two advantages over individuals with small bills when exploiting *Tribulus* seeds: they reject fewer mericarps and they extract more seeds from a mericarp. Overall their energy reward per unit time is greater than is the reward gained by smaller birds (Grant 1981b), a result that parallels the difference between *G. magnirostris* and *G. fortis* (p. 122).

The models withstood the two tests. In addition, this study corroborated another feature of the models: larger phenotypes can deal with some seeds that are too large or hard for smaller phenotypes to deal with. Forty-two individuals were observed to exploit *Tribulus*, but six were never observed even to try to crack a mericarp despite being seen to feed on smaller seed types many times in the vicinity of patches of *Tribulus*. The *Tribulus*-exploiters were larger than the non-exploiters in all dimensions, but especially in bill width and depth. This finding has been duplicated by T. D. Price (unpublished data).

Independent observations by P. T. Boag on Daphne Major confirmed the main result of these field studies, that individual *G. fortis* with different bill sizes have somewhat different diets. In 1977 he walked around the island repeatedly, each time noting the first seed type to be consumed by the 320 banded birds he saw. He classified the seed types into three hardness classes, and the birds according to the largest and hardest seed they were ever observed to consume, then tested the bill depths of the birds for heterogeneity with an F test. Highly significant heterogeneity was found ($F = 22.39$, $P < 0.001$); the birds feeding on the smallest and softest seeds had the smallest mean bill depth, those feeding on the largest and hardest seeds had the deepest beaks on average, while those feeding on seeds of intermediate size had bills of intermediate mean size (Table 5). Morphological differences between the sexes, with males being slightly larger on average than the females, contributed only a little to this heterogeneity (Boag and Grant 1984b; see also Chapter 8). Thus *G. fortis* on Daphne is a generalist species comprising individuals that are, to some degree, specialized at dealing with part of the food spectrum exploited by the population as a whole.

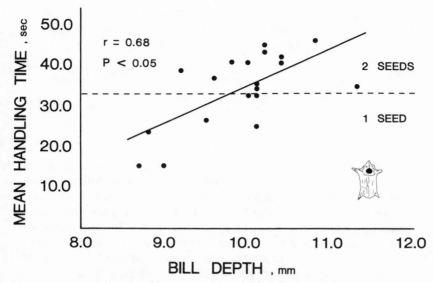

FIG. 41. Mean handling times of *Tribulus* seeds by 19 medium ground finches, *G. fortis*, with different beak sizes, on Daphne. The broken line represents the average amount of time devoted to the cracking of the mericarp and the extraction of one seed; while some single seeds take longer to extract, mean handling times greater than this value probably reflect an extraction of two seeds. Original data from Grant (1981b).

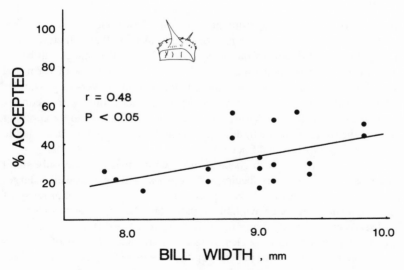

FIG. 42. Proportions of *Tribulus* mericarps which medium ground finches (*G. fortis*) on Daphne attempt to crack, as a function of their bill size. Original data from Grant (1981b).

TABLE 5. Bill depths of medium ground finches, *G. fortis*, feeding on soft, medium, and hard seeds on Daphne Major in 1977. (After Boag and Grant 1984b.)

Seeds	Beak Depth Mean ± Standard Error	Number of Birds
Soft	9.28 ± 0.06	199
Medium	9.73 ± 0.12	39
Hard	9.95 ± 0.08	82

In concentrating on the relationship between bill size and diet I have almost entirely ignored the variation in behavior shown by different finches of approximately the same bill size when feeding on the same food type. It would be wrong to assume that behavioral variation is unimportant. For example, different individuals of *G. fortis* used slightly different maneuvers in attempting to crack a *Tribulus* mericarp and extract the seeds. Peculiarities of individuals stand out after repeated observation. In addition some individuals, but only some, fed parasitically from others, and from *G. magnirostris*, by rushing in to seize a seed or mericarp fragment when a large piece was split or shattered, and then rapidly hopping away to consume it. Thus bill size is an important determinant of diet, but so too is behavior that varies to some extent independently.

SUMMARY

The beak of a bird is an instrument for dealing with food. Darwin's Finch species differ in the size and shape of their beaks, skull architecture, head musculature, and details of the horny palate. These differences can be related to different feeding functions. There are three classes of form and function. Long, pointed bills are suitable for probing flowers (e.g. warbler finch) or soft woody tissues (e.g. woodpecker finch), bills with a convex curvature of both upper and lower mandibles are well designed for applying force at the tip (tree finches), and bills that are deep at the base are suitable for crushing seeds (ground finches).

Among the ground finch species, those with small bills eat only small and/or soft seeds, whereas species with large bills eat small as well as large, and hard, seeds. Thus the larger species have a broader potential range of seed sizes and hardnesses available to them, and they can crack moderately hard seeds more quickly than can smaller species. The difference in diet between any two sympatric species, in terms of size and hardness of seeds, is a function of the difference in beak depths. Proportions of different seeds in the diet of a species are determined by the relative availabilities of the seeds and by the rewards obtained as a function of time spent handling them.

In the wet season and early part of the dry season, species of ground finches tend to have a common diet of soft, easy-to-handle, seeds and fruits, as well as caterpillars and other arthropods (e.g. spiders) when they are available. There are some subtle differences among the species in the foods they deliver to their nestlings. About two weeks after leaving the nest, young birds start to show adult, species-specific, feeding behaviors, but fully adult diets form slowly as the skills needed to deal with different food types are acquired and the bill and head structure matures. As the dry season progresses food abundance declines, as does seed diversity, and diets of sympatric species diverge, with the result that the differences in their diets increasingly reflect their bill differences.

Morphologically well-differentiated populations of the same species, such as *G. difficilis*, also differ in diets, but the correspondence between bill differences and dietary differences is less pronounced than it is in comparisons of different species.

Within a single variable population, adult individuals with different bill sizes differ in their diets in a manner that parallels the variation shown by different species, although dietary differentiation is much weaker within populations of the same species than it is between populations of different species. Individuals in a population of *G. fortis* on Santa Cruz distribute themselves in different habitat patches in the dry season, those with small bills tending to occur most frequently in patches with the richest supply of small seeds. In another population of this species on Daphne Major, only those birds (of either sex) with the largest bills can crack moderately large and hard seeds, and they do so with an efficiency that is bill-size dependent.

Therefore at all levels of comparison there are associations between the size and shape of the beaks of birds and the nature of their food or the substrate where it is found. Interspecific differences in beak sizes and diets are magnified versions of differences between populations of the same species, and between individuals in the same population.

PLATE 43. Tameness: adult male large cactus finch, *Geospiza conirostris*, peering into a camera. Española (*photo by C. P. Hickman, Jr.*).

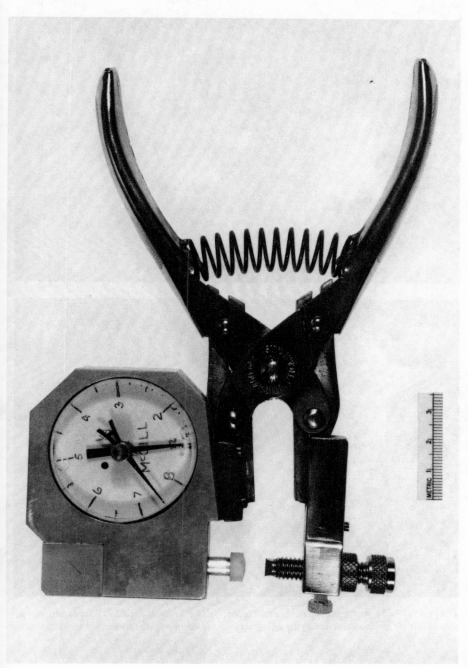

PLATE 44. The McGill seedcracker, a pliers device for measuring the force necessary to crack a seed, in kilograms-force.

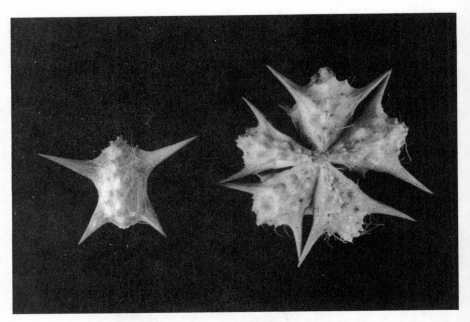

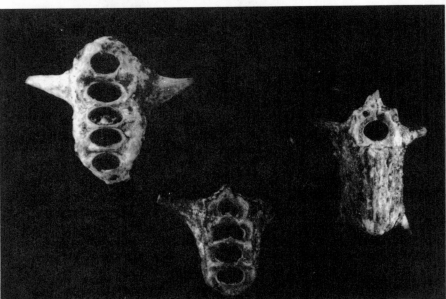

PLATE 45. Woody mericarps comprising a fruit (upper right) of *Tribulus cistoides*. Finches have removed part of the tissue surrounding the cells that contained one seed each (lower). See also Figure 33.

PLATE 46. Flowers (upper) and fruits (lower) of *Croton scouleri* on Genovesa. Caterpillars that feed on the flowers are eaten by finches. The fruits, though large, are soft and are eaten by all finches.

PLATE 47. Fledgling cactus finch, *Geospiza scandens*, at a stage when it is learning to feed by itself. Plaza Sur (*photo by H. Snell*).

PLATE 48. Parkland in the arid zone of Santa Cruz, north shore, at Bahía Borrero. Upper: Dry season. Vegetation comprises *Bursera graveolens* trees, *Opuntia echios*, a few bushes of wild cotton, *Gossypium barbadense*, and several grasses. Lower: Wet season, and with rain approaching. All trees are in leaf. A slightly different view from the one above (note the outline of the hills).

PLATE 49. Abundant small flowers and fruits on the shrub of a euphorb, *Chamaesyce amplexicaulis*. Daphne Major (*photo by P. T. Boag*).

CHAPTER SEVEN

The Importance of Food to Finch Populations

INTRODUCTION

The preceding chapter has shown that species have different diets, as determined in part by their beak sizes and shapes, and that diets change as food supply changes. In this chapter I will shift the focus from beak morphology to feeding ecology, and explore the manner in which populations of finches are affected by the vicissitudes of their food supplies. I shall do this first by describing the typical events of an annual cycle of plant production, in amplification of the general remarks made in Chapter 2, next by doing the same for an annual cycle of the finches, and then by establishing the connection between them.

PLANT PHENOLOGY IN THE ARID ZONE

Most plant production occurs in the wet season, which, in a typical year, extends from January to April or May. The majority of plant species at low elevation are dry-deciduous, that is, they only form leaves in the wet season. Leaves are produced within a week of the first heavy rain on common species of plants, such as *Bursera graveolens*, *Croton scouleri* (Plate 46), and *Lantana peduncularis*, and they are accompanied or quickly followed by flowers. This is also a time of rapid development of arthropod populations, especially caterpillars. By the second month after the first heavy rain, seeds and fruits are maturing. The amount of rain and the length of the wet season influence the quantity of seeds produced.

In the early part of the dry season leaves are shed at a time when some seeds and fruits are still maturing; the caterpillar supply declines rapidly. Some species of plants flower at this time, *Cordia lutea* (Plate 10) and *Waltheria ovata* (Plate 42) being examples, so their major fruiting season occurs after the rains have ceased. In addition, other species of plants do not lose their leaves immediately, but continue flowering and fruiting to an extent probably determined by the amount of rain that has fallen in the wet season; the species of *Chamaesyce* (Plate 49) are examples of this pattern. Nevertheless, the bulk of seed production occurs within the wet season or shortly thereafter, mostly between one month after the onset of the heavy rains and their cessation.

From the end of one wet season to the beginning of the next one the standing crop of seeds gradually declines. Ground finches are the main consumers, but there are a few others. Doves feed on small seeds but also on large seeds, including two types avoided by finches: *Merremia aegyptica* and *Ipomoea linearifolia*, both members of the Convolvulaceae (P. R. Grant and K. T. Grant 1979). Mockingbirds feed extensively on the fleshy fruits of *Cordia lutea* (Plate 10) and *Opuntia* species, but regurgitate the undigested seeds that are then cracked and eaten by finches (P. R. Grant and N. Grant 1979). On Fernandina and Santa Fe native rats (*Nesoryzomys narboroughi* and *Oryzomys bauri*) consume the larger seeds, as do the introduced rats (*Rattus rattus*) on several other islands. Native rats were once present on Santa Cruz, Baltra, San Cristóbal, Isabela, and Santiago, but have become extinct in the last 150 years (Clark 1984). This is a relatively short list of potential competitors. Conspicuously lacking are seed-eating ant species, which are important consumers in arid regions of continents (Brown et al. 1979).

The last phase of the annual cycle is a burst of flowering by *Opuntia* cactus towards the end of the dry season, starting anywhere between mid-October and mid-December. On some islands such as Daphne Major, where the cactus is locally dense, the abundance of flowers (nectar and pollen; Plate 50) is sufficient to alleviate the effects of a diminishing seed supply.

FINCH PHENOLOGY

Finch production closely follows the pattern of plant production, with the period of breeding usually being restricted to the wet season. Like passerine birds in other arid parts of the world, finches respond rapidly to the onset of seasonal rains. Males begin to sing, build display nests, defend their territories against other males, and display to females, and by the end of the week or in the next week the first eggs are laid. A clutch comprises 2–5 eggs, very rarely 6, usually 3 or 4, and is completed in as many days. Incubation begins before the clutch is complete, and continues for another twelve days until hatching. Nestlings stay in the nest for a further two weeks, grow rapidly on a diet of caterpillars, spiders, pollen, fruit, and some seeds, and then fledge. So from the time the first egg is laid until the time the last nestling fledges approximately one month has elapsed. If the plants have not started to lose their leaves, a second breeding attempt will be undertaken, sometimes followed by a third and more. If food conditions are really good, the second clutch of eggs may be laid in a new nest before the first set of nestlings has even fledged. Thus reproduction is rapid (Grant and Grant 1980a, Boag and Grant 1984a). It is followed by an annual molt that is usually concentrated in a 2–3 month period, but which can be inter-

PLATE 50. Flowers of the cactus, *Opuntia helleri*, on Genovesa. Upper: Just opened, central stigmas and surrounding anthers intact. Lower: Exploited, with pollen consumed and stigmas snipped off and removed by a finch, probably to facilitate probing to the base of the style for nectar.

rupted by a resumption of rain and breeding activity, and is sometimes protracted (Snow 1966).

Breeding ceases less concertedly than it begins, and correspondingly it is less obvious why it ceases than why it begins. The best guess is that a decline in the caterpillar supply is the single most important factor; even though caterpillars are not essential for successful breeding, and some clutches are initiated after caterpillar numbers decline (Schluter 1984a), they are probably very important for the energy balance of both parents and offspring.

In the ensuing dry season many finches, mainly young birds, disperse into other areas, even into areas not occupied for breeding. On the large islands some tree finches and *G. difficilis* disperse from upland areas into coastal regions, and other ground finches, especially *G. fuliginosa*, move in the opposite direction, occasionally in large flocks (Schluter 1982b, 1984a). Dispersal between islands occurs also, to a limited extent (Chapter 8). The dry season is a time when survival is at a premium; many birds, mostly young ones, die.

I have just described an annual cycle that stands for all finch species on the Galápagos. It is simplified by omission of the phenology of arthropods (except caterpillars) and of plants in transition and humid zones, since these are not known in detail. All finches breed in response to rain, they all build dome-shaped nests that differ in size in proportion to their own size (Plate 51), they show similar courtship behavior (Chapter 9), they all lay reddish- or brown-spotted white eggs, and they all incubate the eggs and tend the young in similar ways. Differences among them are minor. For example, all ground finch species nest in cactus, as well as in many other bushes (Grant and Grant 1980a), but tree finches and the warbler finch never build nests in cactus. According to Lack (1947), and to our own observations, the cactus finch *G. scandens* nests only in cactus.

What little is known of the Cocos Island finch suggests that it has the same characteristics as the Galápagos finches (Lack 1947), except that clutch sizes appear to be no higher than two (T. W. Sherry and T. K. Werner, pers. comm.). On this island, which lacks a definable dry season, it would be interesting to know how long individual pairs continue to breed before stopping to undertake their molt.

Although similar to each other, the species on the Galápagos are not identical in breeding phenology. A few pairs of *G. scandens* regularly breed before the onset of heavy rain on Daphne Major, which we have attributed to their comprehensive exploitation of the major pre-rains food, pollen and nectar from *Opuntia* cactus (Boag and Grant 1984a). This fits with the observation that *G. conirostris* is the earliest to breed on Genovesa, for it too exploits *Opuntia* fully (Grant and Grant 1980a, 1981). A few pairs of *G.*

PLATE 51. Dome-shaped nests, differing in size and placement but not in shape. Upper left: Nest of large ground finch, *Geospiza magnirostris*, in a tree of *Croton scouleri*. Isabela. Upper right: Nest of small ground finch, *Geospiza fuliginosa*, in *Opuntia echios*. Santa Cruz. Lower: Nest of mangrove finch, *Cactospiza heliobates*, in tree of *Maytenus octagona*. Punta Tortuga, Isabela (*upper photos by H. Snell, lower photo by E. Curio*).

fortis also bred before the onset of heavy rain on Daphne Major in 1984, but in none of the preceding eight years. Tree finches may start to breed earlier, and stop later, than the ground finches, because the habitats they occupy at higher altitudes are wetter (Lack 1947, 1950; see also Snow 1966). For example, by the end of January 1985 the woodpecker finch and the two *Camarhynchus* species on Santa Cruz had raised a brood in the *Scalesia* forest at high elevations (R. Å. and U. M. Norberg, pers. comm.), but the breeding of ground finches was apparently later in the lowlands around Bahía Academía (T. D. Price, pers. comm.). Therefore tree finches may have longer breeding seasons than ground finches, but it is not known if this is a consistent difference between the groups, because none of the tree finch species has been studied in detail.

FINCH POPULATIONS IN RELATION TO FOOD SUPPLY

The rise and fall in food supply is summarized in Figure 43. Typically it rises in January at the beginning of the wet season as a result of the production of flowers, seeds, fruits, and arthropods, chiefly caterpillars. It falls after April at the end of the wet season at a time when only a few new seeds and fruits are being produced, and it rises a little in November as a result of the flowering of *Opuntia*. The demand placed upon the supply is also shown to rise and fall; first it rises as a result of population increase due to breeding in the wet season, then it falls as a result of population decrease due to mortality caused by food shortage.

Consumption by finches is the main reason for the decline in food supply from May to November. The latter part of this period is a time of potential food limitation for finch populations. The length of time during which food limitation occurs varies among islands, among altitudinal sites within islands and from year to year, in parallel with variation in rainfall and in primary and secondary production. Variation in plant species composition among islands adds further complexity. For example, Daphne Major usually has an abundance of cactus flowers in December, but at the Bahía Borrero site on Santa Cruz at the nearest point to Daphne both cactus bushes and flowers are sparse. Thus Figure 43 represents average conditions, especially relevant to low altitude sites in non-extreme years. The heights of the curves and the duration of each phase are annually and spatially variable.

EXTREME CONDITIONS

As discussed in Chapter 2, the rains occasionally fail, and in other years rain is exceptionally heavy and prolonged. These drought and wet years are times of extreme scarcity and of extreme plenty, respectively. We have been

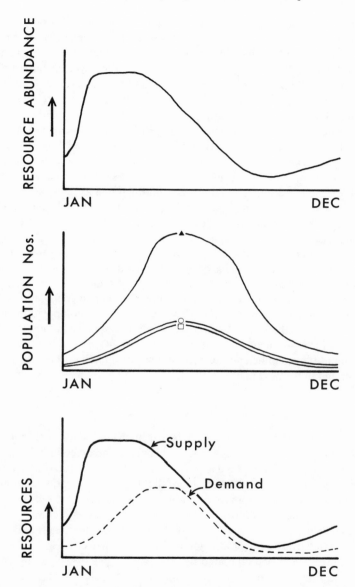

FIG. 43. A standard pattern of seasonal variation in finch population numbers and their food resources, based on observations of *G. magnirostris* (○), *G. conirostris* (□), and *G. difficilis* (▲) on Genovesa, and other finches on other islands. In the lower figure resource abundance represents supply, and finch numbers represent demand. From Grant and Grant (1980a) and Grant (1985b).

fortunate to witness both extremes in the last ten years. In 1977 only 24 mm of rain fell in the normal wet season on Daphne Major. Plant production was extremely reduced, and many perennials died (Plate 8). The medium ground finch, *G. fortis*, progressed through a breeding cycle as far as building nests, but did not lay eggs. The cactus finch, *G. scandens*, did breed but had poor success, and none of the juveniles survived for more than three months (Boag and Grant 1984a).

Five years later an exceptionally strong El Niño event occurred. On this island 1359 mm of rain fell between November 1982 and July 1983, and then the rains ceased abruptly. Plant growth was rampant (Plates 9, 52, and 53), seed production was enormous, and finches bred throughout the 8-month period. One female medium ground finch laid as many as 39 eggs, and the most successful pair fledged 25 young. Not only was the breeding effort of the populations sustained for a long time, it was augmented by recruitment of birds that were born at the beginning of the season! Young of both species born in the first 2–3 months of the breeding season were, themselves, breeding at 80–90 days of age and onwards. Some of these precocious individuals successfully fledged a few chicks (Gibbs et al. 1984).

Extreme conditions thus have a major influence on finch populations. Not only do population numbers oscillate strongly, but age composition varies according to whether, in preceding years, conditions were favorable or unfavorable for reproduction and survival (see Fig. 44). The influences of extreme conditions on finch populations will be considered further later in this chapter and in the next one.

FOOD LIMITATION OF POPULATION SIZES

The question I will now consider is whether finch populations are limited in size by their food supply. Figure 43 and some of the preceding discussion imply that ground finch populations are limited in the approximate period October to December. If they are, there should be a systematic relationship between ground finch biomass or ground finch numbers and seed biomass at this time (Grant 1985b). In preceding months there should be fewer finches than can be supported by the seed supply, because the production of seeds has outstripped the production of finches in the wet season. After December the finches are supported by other food types as well as by seeds. Figure 45 illustrates the expected relationship. It shows the finch carrying capacity for a given biomass of seeds as a line which is crossed twice during the course of a year. It also illustrates in an arbitrary fashion the different maximal biomasses of both finches and seeds in successive years.

To see if such a relationship between finches and their food exists we need quantitative estimates of the sizes of finch populations and their food sup-

PLATE 52. Production in an El Niño year: two small-seed bearing plants that responded strongly to the extensive rains. Upper: *Heliotropium angiospermum*. Daphne Major. Lower: *Eragrostis cilianensis*. Genovesa (*upper photo by P. T. Boag*).

PLATE 53. Production in an El Niño year, and the aftermath. Upper: *Portulaca howellii*, producer of many small seeds, nectar, and pollen consumed by ground finches. Daphne Major. Lower: *Opuntia helleri*, smothered by a vine, *Ipomoea habeliana*. Genovesa.

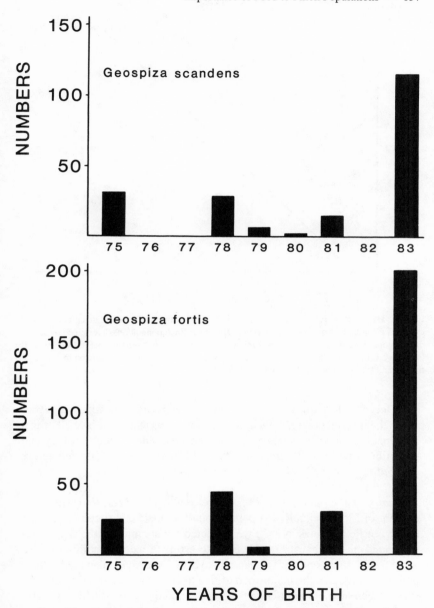

FIG. 44. Age composition of the breeding populations of medium ground finches (*G. fortis*) and cactus finches (*G. scandens*) on Daphne Major in 1984. The populations bear the fingerprints of drought-related mortality and little or no breeding in 1977, 1980, and 1982, and of the enormous production of young in 1983. The "1975" birds were born in 1975 or earlier.

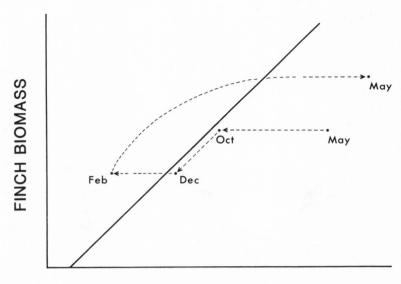

SEED BIOMASS

FIG. 45. Hypothetical relationship between finch biomass and seed biomass in a typical annual cycle. The solid line represents the carrying capacity of finch populations determined by seed biomass alone. Finches are shown to be limited by the seed supply from October to December. Biomasses are shown to be different in successive Mays to illustrate annual variation in both finches and seed production. From Grant (1985b).

ply. Recent field studies have provided these estimates. They have also provided the information on diets used in the previous chapter. Since all of this information plays a crucial role in the understanding of finch ecology and evolution, I will first describe how the field studies were conducted before proceeding with the question of food limitation.

Estimating finch numbers and their food supply and diets. Estimating the amount of food available to a population of animals is difficult. When we began our field studies in 1973 we restricted our attention to the six species of ground finches because it is easier to estimate the availability of their basic foods, seeds and fruits, than it is to estimate the abundance of arthropods that constitute the major food of tree finches. At each of eight sites we surveyed an area of about 23,000 m² and marked the coordinates of a grid of 10 m × 10 m units with flagging (Abbott et al. 1977). We then chose fifty 1-m² quadrats with the aid of a random numbers table, marked the edges of each one with string, counted the number of fruits and seeds attached to each plant species, and counted the number of fruits and seeds on the ground and in the surface layer of the soil in two 0.1-m² areas of the quadrat. Seeds and fruits were identified to species or genus with the help of Wiggins and Porter's (1971) *Flora of the Galápagos Islands*. Representative specimens were measured.

Finch numbers were estimated by mist-netting. Two or three 9-m nets were opened at dawn and closed two to three hours later. They were opened once more in the next two days, and then used in a new position in the study area. We stopped netting when the total net-hours had reached 22 (198 net-meter-hours). Captured finches were weighed, measured, color-banded, and released.

Diets were determined by recording the amount of time spent by each bird dealing with an identified food item, up to a maximum of 300 seconds of total observation per bird, and for as many birds as possible.

Field studies on one or more islands have been conducted every year since 1973, but a few small changes in methods have been made since 1976 (Smith et al. 1978, Grant and Grant 1980a, Schluter 1982a, 1982b, Schluter and Grant 1982, 1984a, Boag and Grant 1984a, 1984b). Several additional islands were visited, and on some of them censusing and sampling were conducted on more than one study area: two on Santiago, two on Marchena, three on Española, four on Fernandina, and six on Pinta. The six on Pinta were approximately evenly spaced along an altitudinal transect from near the coast to just below the summit. Twelve-meter nets replaced the 9-m nets, and they have been used for 16.5 net hours (198 net-meter-hours) most frequently. Captures from longer or shorter periods have been adjusted to this total. Visual censuses were also adopted on some islands. Numbered metal bands as well as color bands were used for subsequent identification of individual finches on Daphne, Santa Cruz, Pinta, and Genovesa. Seeds on the ground were counted in two 0.0625-m² areas instead of two 0.1-m² areas. To increase the efficiency of seed detection soil samples were removed in plastic bags and spread on white enamel trays. Kernels of almost all seed types were extracted, dried, and weighed. Finally the study area on Daphne was enlarged and extended to include part of the outer slope (Plate 16), and diets were recorded in units of numbers, one food item per bird, instead of units of time spent on each food type.

All methods of estimation have their problems. The two main problems with estimating food supply in the way we did it are, first, sample sizes are small and confidence limits on the estimates are correspondingly large, and second, standing crops of seeds may not be a reliable index to the amount of food produced and hence available to the birds over a period of time if different seed types are produced at different rates and/or they are depleted by the finches at different rates. With more time we would have increased the number of samples by a factor of ten and used seed traps to measure production. The main problem with estimating finch numbers by netting is that on the large islands, large but loosely aggregated flocks of finches form soon after the breeding season, and being very mobile the finches are prone to being caught in large numbers. Finch numbers are overestimated when this happens. Recorded diets may be distorted by the over-representation of food types that are taken during a conspicuous foraging activity, such as caterpillar-gleaning from the leaves of *Bursera* trees, so an effort was made to guard against this bias.

Estimates of ground finch numbers, food supply, and diets have been obtained by these methods on all the large islands except San Cristóbal, Floreana, and Pinzón, and on some of the small ones. Information on diets is used mainly in the previous chapter and the next one.

Variation among islands. If finch populations on different islands are limited by food at the same time, there should be a correspondence between total biomass of finch populations on different islands and seed biomass on those islands. Figure 46 shows that there is a correspondence between total ground finch biomass and seed biomass on seven different islands in the months October to December. The linear correlation between the two biomasses is positive and highly significant ($r = 0.96$, $P < 0.01$). In this and subsequent analyses I have used only those seeds whose proportional representation in the diet of a finch species is at least 50% greater than their proportion in the environment (Abbott et al. 1977). By restricting attention to these "preferred" seeds I exclude a few types that are eaten rarely.

A curvilinear relationship is suggested by the data. However, the second term of a polynomial regression does not statistically explain a significant amount of the residual variation in finch biomass ($P = 0.15$), and the Y intercept is only reduced to 190.7. Nevertheless, a curvilinear relationship might be expected on logical grounds if at low seed densities it is disproportionately time-consuming for finches to find the seeds, and if mortality rates do not remain constant in relation to seed biomass but are relatively high at the low seed densities. Indeed, given the particular association between seed and finch biomasses over the upper range of seed biomasses, a curvilinear relationship is a biological necessity. Assuming that the sampled food truly reflects the supply on each island, the linear relationship shown in the figure has the impossible feature that 238 g of finch biomass are supported by 0 mg/m² of seed biomass! The line should curve downwards near the origin in such a way that when seed biomass declines to a low but positive value, finch biomass declines to zero; the figure shows this as a broken line. Expressed another way, if a finch population went extinct through food shortage, there would still be a few widely scattered seeds (the starvation threshold) at the time of death of the last finch. We have no idea how many seeds there would be.

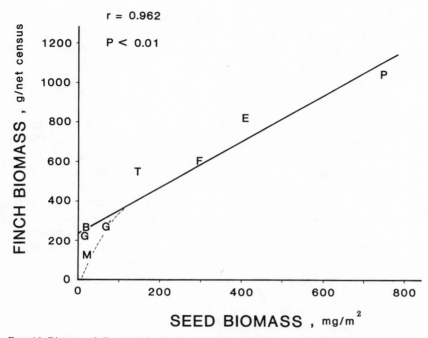

FIG. 46. Biomass of all ground finches in relation to seed biomass in the late dry season on seven islands: Marchena (M), Genovesa (G, G'), B. Borrero on Santa Cruz (B), Tortuga (T), Fernandina (F), Española (E), and Pinta (P). A least squares regression line is fitted to all points except G'. The broken line shows a more biologically reasonable relationship at low values (see text). From Grant (1985b).

Since energy requirements are related to body size, and average body size varies among the islands, it would be more realistic to use an adjusted biomass value for finches. The adjustment, yielding an energy-demanding or "consuming" biomass, can be made by multiplying each individual finch weight by the exponent (0.724) in the interspecific equation which relates standard metabolic rate to body weight (Lasiewski and Dawson 1967). This refinement is unnecessary here, however; the correlation among islands between finch standing crop biomass and total consuming biomass is 0.99.

The analysis represented in Figure 46 assumes that seeds are the only foods available to finches at this time, and that all species consume the seeds. These assumptions are only approximately correct, yet the correlation is strong. All ground finch species eat seeds, but, as discussed in the previous chapter, not all species eat all seed types. They also feed on other food types, even at this time of the year, although these other types are minority items in the diet except for arthropods in the diet of *G. difficilis* on Pinta (Schluter 1982a). For example, nectar, pollen, and arthropods were consumed by finches at B. Borrero (Santa Cruz), Genovesa, Pinta, and Marchena. These extra foods are responsible for finch biomass being greater than that supported by seeds alone if the carrying capacity determined by seeds is a downcurving line reaching the seed axis at a starvation threshold greater than 0 (Fig. 46).

In the above analysis some judgments had to be made in deciding the length of the potential food-limiting period since, as discussed earlier, this period may vary among islands and among years. I excluded data from Cerro Ballena on southeastern Isabela (August 1975) and Santiago (September 1981) because the wet season ended late in those years, in May-June, and thus the data were collected only 2–4 months after the rains had ceased. At the far end of the dry season I excluded data from B. Academía on the south side of Santa Cruz and from Daphne (December 1973), and from Santa Fe (early January 1974), because of the abundance of *Opuntia* flowers at those sites. Pollen and nectar feeding from *Opuntia* flowers was a major feeding activity at these three sites. What happens to the correlation if the length of the potential food-limiting period is expanded to include these data? If the late dry season data, either separately or together, are included in the analysis, the correlation is surprisingly unaffected; with data from all three sites included the correlation coefficient is exactly the same ($r = 0.96$). If, instead, the early dry season data from Isabela and Santiago are included, the correlation is strongly affected ($r = 0.49, P > 0.1$). But if both sets are included the correlation is significant ($r = 0.93, P < 0.01$).

According to Figures 45 and 46 finch biomass should be below the seed-determined carrying capacity in the first few months after the breeding season. Data are available from six islands in the period April to September to test this. In agreement with expectation, all points for finch biomass fall below the extrapolated carrying capacity line (Fig. 47). With one exception (Genovesa) seed biomass is greater at *all* of the six early dry season sites than at *all* of the seven late dry season sites. Although the early dry season seed biomass on Genovesa is low, it is much greater than the late dry season seed biomass there.

There is no necessary relationship between finch biomass and seed biomass expected in the early dry season, although one does in fact exist for the seven sites studied in April to July ($r = 0.86, P < 0.05$). It is weaker than the late dry season relationship. Thus 92.5% of the inter-island variation in finch biomass in the late dry season is statistically explained by variation in seed biomass at that time, whereas only 74.6% of variation in finch biomass in the early dry season is explained by seed biomass variation.

Figure 47 also shows two complications discussed earlier. The first is the unreliability of finch estimates when large flocks are formed. Large flocks occurred at B. Borrero on Santa Cruz and on Española on one occasion, and

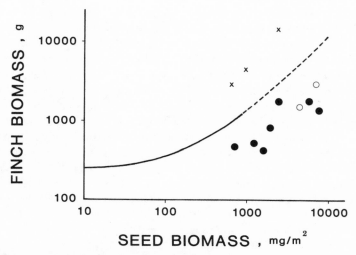

FIG. 47. Ground finch biomass in relation to seed biomass on several islands in the early part of the dry season (●), April to September, and in the late dry season (○) at a time of *Opuntia* flower abundance, in December. The solid line is the regression line (carrying capacity) taken from Figure 46, and extrapolated beyond the original range with a broken line; it is curved as a result of a change in scale. The three crosses indicate finch estimates that are unreliably elevated as a result of flocking at B. Borrero on Santa Cruz (April–May 1973 and July 1975) and Española (May–June 1973).

finch biomasses at these sites appear to be larger than the carrying capacity (see p. 163). The other is the relatively low finch biomass when major alternative food sources are available—in this case pollen and nectar in *Opuntia* flowers at B. Academía (Santa Cruz) and Daphne in December. Alternative food sources can result in finch biomass being either below or above the carrying capacity, depending upon when the alternatives become available. If they become available before the carrying capacity is reached it may never be reached that year, while finch mortality and seed consumption both decrease. On the other hand, if carrying capacity is reached before the alternative food source becomes available, then finches may continue to deplete the seeds, although at a reduced rate, and this will result in finch biomass being sustained at a level higher than the seed-determined carrying capacity.

The Daphne and B. Academía data appear to be examples of the first pattern, with finch biomasses below carrying capacity, and the Santa Fe data are an example of the second pattern of excessive finch biomasses. The Santa Fe point is not shown in Figure 47 because it is off the low end of the scale. The 2 mg/m² of seed biomass is predicted to support 241 g of finch biomass by the relationship in Figure 46, but in fact the net census yielded an estimate of 401 g of finch biomass. The lateness of the study (early January) suggests that pollen and nectar feeding had been important for some time, because *Opuntia* flowers were moderately common already by late December on nearby Santa Cruz. Moreover, if the study site on Santa Fe had a higher than average density of cactus it may have also had a higher than average density of finches. Such is the problem of using small study sites to estimate finch-food relationships on the island, or the arid zone, as a whole.

We were fortunate to have one independent check on the suggestion that *Opuntia* flowering occurred before the carrying capacity was reached on Daphne. A large number of finches had been color-banded in April 1973, and resightings in December enabled us to estimate survival rates. In this year they were very high: 85% for *G. fortis* and 90% for *G. scandens* (Grant et al. 1975). On a return visit in June–July 1975, birds were seen that had been missed in December 1973. The revised estimates of survival in 1973 are 90% for *G. fortis* and 95% for *G. scandens*. Therefore it is unlikely that the seed-determined carrying capacity was reached in 1973 on this island, despite augmentation of finch biomass through immigration of several *G. magnirostris* and *G. fuliginosa*. In July 1975 on Daphne the finch biomass was again below carrying capacity, as expected during this early time in the dry season.

Variation within islands in space. By the same reasoning that guided the discussion in the previous section, total biomass of finch populations may vary in relation to seed biomass in different places on the same island. To examine this possibility I have compared total finch biomass with seed biomass at the six sites at different altitudes on Pinta studied by Dolph Schluter in November 1979. The data are plotted in Figure 48. The correlation between the biomasses is highly significant ($r = 0.97, P < 0.01$).

The evidence is consistent with a hypothesis of food limitation but does not demonstrate it. On the island of Pinta, as elsewhere, birds might distribute themselves more commonly in the food-rich sites than in the poorer sites, thus giving rise to the observed correlation even when population sizes are not limited by food (Schluter 1982b). Birds do not redistribute themselves under changing food conditions *among* islands, however, at least not to any important degree, yet the same relationship between consumers and resources is seen *among* islands as *within* the island of Pinta; the inter-island regression line from Figure 46 (recalculated without the Pinta data) provides an adequate fit to the intra-island scatter of points in Figure 48.

Variation within islands in time. The direct way to investigate food limitation is to follow populations of known size through time. The only available

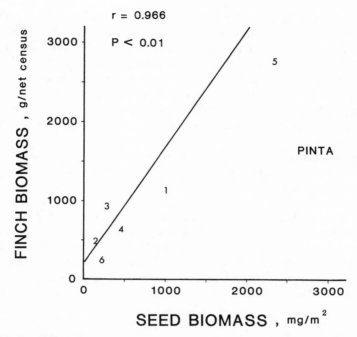

FIG. 48. Ground finch biomass as a function of seed biomass on Pinta. The numbers refer to points on an altitudinal transect, from low (1) to high (6) elevations. The line has been estimated by least squares regression from the points in Figure 46, excluding the data from Pinta (P) and the second survey on Genovesa (G′). The fit to the six points from Pinta is quite good, although the slope is a little high. From Grant (1985b).

data that permit this come from the island of Daphne Major. I have compared total finch biomass with seed biomass on the island before and during a drought in 1977. Finch populations were large at the end of the breeding season in 1976, but thereafter they declined until early 1978. Between these two times *Opuntia* pollen and nectar became available twice, in the late dry seasons of 1976 and 1977. As mentioned earlier in this chapter, *G. scandens* (only) reproduced, in early 1977, and all fledglings died within three months. To a rough approximation then, it may be said that the period May 1976 to January 1978 was one long dry season as a result of the nearly complete failure of the rains in early 1977.

Although immigration of some *G. fuliginosa* and *G. magnirostris* occurred during this period, there was a strong correlation between total finch biomass and seed biomass ($r = 0.92$, df 3, $P < 0.05$). Each of the resident species treated separately showed a significant correlation; for *G. fortis* $r = 0.91$ ($P < 0.05$) and for *G. scandens* $r = 0.95$ ($P < 0.05$). The correlations are thus very strong, although since the observations along the time sequence are not strictly independent it would be preferable to use a repeated measures test of the associations (e.g. see Davidson et al. 1985). The correlation with total finch biomass is illustrated in Figure 49, with an emphasis on the time course. The most rapid reduction in both finch and seed biomasses seems to have occurred in the last phase. Many plants died in the drought, with *Heliotropium angiospermum* all but disappearing (Grant and Grant 1980b, Boag and Grant 1984a), and even those which usually produce some seeds after the wet season (e.g. *Chamaesyce* species) scarcely reproduced at all.

The correlation between finch biomass and seed biomass is consistent with the hypothesis that food limitation occurred throughout most of this extended dry period.

Finch numbers in relation to seed biomass. So far I have only considered the biomass of finches, but the food limitation hypothesis implies that individual species are limited in numbers by their particular food supplies. Therefore there should be a positive relationship between the numbers of a given species and the biomass of seeds known to be consumed by that population. The dry season data will now be reexamined.

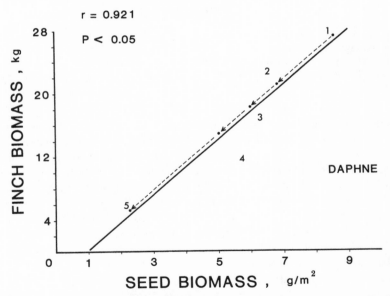

FIG. 49. The decline in ground finch biomass on Daphne Major as the seed biomass declined from mid-1976 (1) to the end of 1977 (5). The line of best fit (solid line) was calculated by least squares regression; the arrows and broken line are added to emphasize the time course. From Grant (1985b).

Three species have been studied in the late dry season on a barely sufficient number of islands for analysis: *G. magnirostris* (7), *G. fortis* (5), and *G. fuliginosa* (8). Correlations between finch numbers (or biomass) and food supply are positive and statistically significant ($P < 0.05$) in each case (Schluter and Grant 1984b).

Along the altitudinal gradient on Pinta, *G. fuliginosa* numbers were positively correlated with food supply (Table 6). Correlation coefficients are positive for the remaining three species, but none is significantly different from zero. The sample sizes are small, however. The correlation coefficients exceed 0.5 in all cases except for *G. difficilis*, but even here the coefficient exceeds 0.4, and is statistically significant, when August data for the six sites are added to the November data used in Table 6 (Schluter and Grant 1982). I therefore suspect that finches of all species are truly distributed in approximate proportion to variations in their food supplies, but that inaccuracies in the data obscured some of these relationships.

TABLE 6. Correlation coefficients for associations in the late dry season between finch numbers and the biomass of food consumed by each species among six sites on Pinta: seeds for the first three species, and invertebrates for *G. difficilis*. An asterisk = $P < 0.05$.

	Coefficients
G. fuliginosa	0.883*
G. fortis	0.707
G. magnirostris	0.622
G. difficilis	0.442

Over the long dry period on Daphne from May 1976 to early 1978 (Boag and Grant 1981), *G. fortis* numbers declined precipitously in parallel with a decline in their food supply ($r = 0.91$, $P < 0.05$). *G. scandens* numbers declined in parallel with their food supply too, but the correlation is only marginally significant ($r = 0.83$, $0.1 > P > 0.05$; Spearman's rank correlation coefficient $r_s = 1.0$, $P = 0.05$). Over a slightly longer period of time, starting in January 1976 and ending in June 1978, the relationships are stronger; for *G. fortis* $r = 0.86$ ($P = 0.003$) and for *G. scandens* $r = 0.78$ ($P = 0.014$). The relationship is weaker for *G. scandens* than for *G. fortis* probably because *G. scandens* feeds to a greater extent on other food types at most times of the year. These include extra-floral nectar from *Opuntia*, and beetle larvae and pupae excavated from rotting *Opuntia* pads. The abundance of these foods was not estimated, and therefore they did not enter the calculations.

THE FREQUENCY OF FOOD LIMITATION

Short of experimental manipulations of finch numbers or food supply on the appropriate scale and with appropriate controls, these correlations provide as strong a case as can be made that food is sometimes limiting. We know that food is not limiting in every year. It is not limiting in years when the rains are extensive, seed production is large, and the supply of seeds is not severely depleted by the time that *Opuntia* flowering begins, as happened in 1973 and in 1983 on Daphne. The important question is how often does food limitation occur? Is it so rare as to be of trivial significance to finch populations, or is it sufficiently frequent that individual finches are likely to experience conditions of hardship once or more in their lifetime?

The combined data from all of the study sites in all of the years lead me to believe that food limitation occurs in most years (Grant 1985b). Long-term studies are required to establish its frequency. The only data bearing directly on this issue come, once again, from Daphne. Figure 50 shows how population sizes changed drastically on this island over a ten-year period. It would appear from the figure that populations might have been limited by food prior to 1977, but thereafter they were not, as indicated by their sustained low levels for several years. Such reasoning implies an inability of the populations to increase rapidly through reproduction. This is partly correct; given the amount of reproduction that subsequently occurred, populations could not have reached pre-drought levels in less than two years, even without mortality. But the rates of increase of populations, and hence their recovery times, were governed by food supply in two ways. In the breeding season the supply of caterpillars and pollen restricted the period of reproduction to 1–4 months in each of the years 1978–1982 (Boag and Grant 1984b, Millington and Grant 1984). In the nonbreeding season the reduction in seed biomass caused high mortality, especially among young birds (Price and Grant 1984).

The high mortality appears paradoxical in view of the low population density, but the paradox disappears when it is realized that food supply was also low. Many seed-bearing plants died in the drought of 1977, especially *Heliotropium angiospermum*, and they were not immediately replaced. In fact *Heliotropium* (and finches) only became common again in the extraordinarily wet year of 1983. The seed-determined carrying capacity for finches was depressed for several years after the drought of 1977 (Fig. 50), and this affected the populations because small seeds are crucial to the survival of birds in their first year (Price and Grant 1984). We will understand the determinants of survival better when the fluctuations in finch numbers since 1982 have been examined in relation to the food supply.

I conclude that finch populations are limited in the dry season of many,

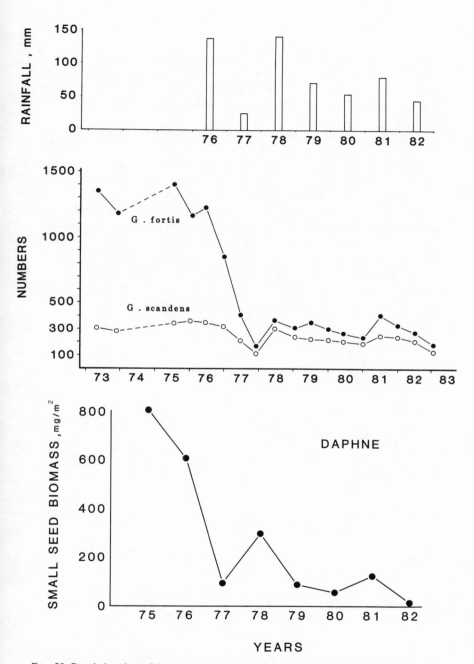

FIG. 50. Population sizes of the two resident species of ground finches on Daphne Major, censused twice a year except in 1974. Populations crashed in the drought of 1977. The biomass of small seeds, on which young finches are dependent for survival, crashed at the same time. From Grant (1985b).

perhaps most, years. Droughts have occurred recently once every 10 years (Grant and Boag 1980, Grant 1985b; see also Fig. 9), and less severe dry conditions accompanied by high juvenile mortality have occurred on Daphne recently once every 2–3 years (Price, Grant, Gibbs, and Boag 1984).

To put the frequency of food limitation in perspective we need to compare it with the generation length of finches. The study on Daphne shows that generation length, that is the mean period between the birth of parents and the birth of their offspring, is rather long for small passerine birds, on the order of 3–5 years. We also know that finches can live for more than 10 years; one *G. scandens* individual and one *G. fortis* individual, banded as adult males on Daphne in 1973, were still alive in 1984. The steady decline in survival with age (Fig. 51) enables us to estimate that as many as 10 adult *G. scandens* out of 100 of comparable age that survive the first few months are alive at age 10 years, and perhaps one or two may live to 15 years or more. Therefore the effects of food limitation may be experienced as frequently as once or more each generation, and strong effects may be experienced once a decade.

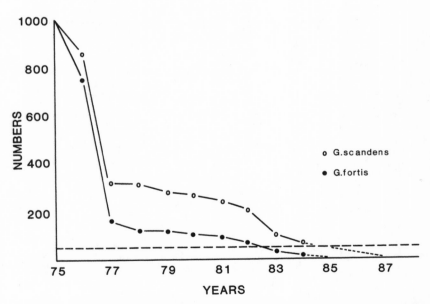

FIG. 51. Survivorship of two species of ground finches on Daphne Major. The birds were adults in 1976, but of unknown age. Note the heavy mortality in 1977; this was followed by a gradual decline in numbers. The broken line marks 5% of starting numbers, and the dotted lines estimate the remaining course to extinction of the two groups.

OTHER FACTORS LIMITING FINCH POPULATIONS

In reaching the above conclusion I have ignored other potentially limiting factors such as nest sites, predation, and disease. To judge from the abundance of places in bushes and cactus where nests could be placed, nest sites are never in short supply, nor is the material to build the nests ever limiting. The effects of predators on finch populations seem to be generally small where they have been studied in detail (Grant and Grant 1980a, Boag and Grant 1984a). On Genovesa short-eared owls prey upon nestlings and fledglings of finches, particularly those of the largest species *G. magnirostris*. They thus reduce breeding success of the finches. Their effects are augmented to a small but unknown extent by the consumption of finch eggs and nestlings by mockingbirds on this island. We have observed a lava heron trying to kill a *G. conirostris* fledgling, and suspect that night herons occasionally do the same. But analysis of pellets shows that owls feed predominantly on storm petrels (*Oceanodroma* spp.) and herons feed predominantly on arthropods on Genovesa. Thus finches appear to be minor elements in the diets of these species, and of mockingbirds, and the predators appear to have a minor influence on finch populations that is almost restricted to the breeding season. On Daphne there are no predators on eggs and nestlings. Predation by owls and egrets falls largely on the fledglings, and thus curbs the rate of increase of the two finch populations but without having a major influence (Boag and Grant 1984a). It presumably affects the time at which finch carrying capacity is reached (see Fig. 45).

The exact influence of predation is difficult to estimate because, despite our intensive searches all over Daphne for pellets in the small caves and crevices that owls use as roosts, we must miss some of the finch remains. The task of determining the influence of disease on finch populations is much more difficult because birds may die without showing external symptoms. Blood parasites have been sought by a parasitologist, Wallace Harmon (pers. comm.), in 94 finches of five species on Santa Cruz and Santiago, but have never been found. This is significant because *Plasmodium* parasites causing avian malaria are known to affect continental finches frequently (Manwell 1955) and have had devastating effects on Hawaiian honeycreeper finches (Warner 1968). But the signs of pox, or pox-like symptoms (Plate 54), are observed periodically in Darwin's Finch populations. In the study on Daphne the incidence among resident finches has varied from 0 to 6%. It has been higher, up to 20%, in the immigrant *G. fortis* and *G. fuliginosa*. The highest incidence was recorded at the end of the very long breeding season in 1983. Many birds that have the symptoms disappear, and so presumably die, but this happens at a time when other birds that do not have the symptoms also disappear and presumably die, and some birds recover from the disease and subsequently breed. Given the usually

PLATE 54. Symptoms of pox on the foot and leg of an immigrant small ground finch, *Geospiza fuliginosa*, on Daphne Major.

low frequency of occurrence and the less than maximal mortality that could be associated with it, I doubt if this particular disease has a major limiting influence on finch populations.

Predation, disease, and other factors may interact with finch food supply in some unknown way to determine finch numbers, but I cannot think of a simple biological explanation that would render spurious the demonstrated correlations between finches and their food supply.

INTERSPECIFIC COMPETITION FOR FOOD

Competition between species can be defined as a negative effect of one species upon the population size of another arising from their joint exploitation of environmental resources. Darwin's Finch species compete for food whenever they exploit the same food resources at a time when the food supply limits population sizes. Experiments are needed to demonstrate competition conclusively, but it seems very likely that it occurs between some pairs of species in most years in view of the evidence for frequent food limitation of species whose diets overlap (Grant 1985b). For example, on Daphne Major populations of *G. fortis* and *G. scandens* were food-limited in 1977, and their diets overlapped (Fig. 52); therefore they were in competition for food. Their diets diverged during the drought and the overlap diminished; as a result, direct competition was possibly reduced but certainly not eliminated because there was dietary overlap at all times. This argument would be incorrect if the seeds jointly exploited by the two species had no influence on the survival of members of either. This seems highly unlikely, especially as the feeding niche of *G. scandens* is contained (included) entirely within the feeding niche of *G. fortis* (Grant and Grant 1980b, Boag and Grant 1984a).

SUMMARY

This chapter reviews the major phenological events of a typical annual cycle in providing a background to the relationship between finch populations and their food supply. The latter part of the dry season, October–December, is identified as a period of potential food limitation for finches. Actual food limitation at this time is indicated by a strong positive correlation between variation in total ground finch biomass among seven islands and variation in food biomass. Food (seed) biomass was lower on all islands at this time than in the early part of the dry season. On the island of Pinta total finch biomass in the late dry season was positively correlated with food biomass among six sites at different elevations.

On some islands and in some years food is not limiting in the dry season.

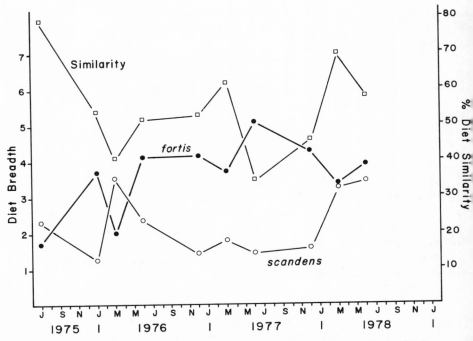

FIG. 52. Changes in diet breadths of *G. fortis* and *G. scandens*, and in the similarity of their diets, on Daphne Major. In the drought of 1977 their diets were at their most dissimilar; at this time the diet of *G. fortis* was unusually broad, and the diet of *G. scandens* was narrow. From Boag and Grant (1984b) and Grant (1985b).

In 1973, survival from April to December of the two species of finches on Daphne, *G. scandens* and *G. fortis*, was 90% or higher. Ten years later so much rain fell in association with an El Niño event, and so many seeds were produced, that it is doubtful if birds died solely from starvation in the ensuing dry season. By contrast the biomasses of these two species fell precipitously in parallel with a decline in food supply during an extended dry period in 1977. Populations of individual species such as these are probably limited in most years by the supply of the food they consume. In terms of generation length, which is on the order of 3–5 years, food limitation probably occurs at least once a generation. Since the diets of *G. scandens* and *G. fortis* overlap at times of food limitation, the species are in competition for food, although competition is reduced as a result of the divergence of diets in the latter part of the dry season. Predation and disease affect finch populations also, but do not appear to contribute in a major way to the limitation of population sizes.

CHAPTER EIGHT

Population Variation and Natural Selection

INTRODUCTION

Finch populations differ in the degree to which they vary in continuously varying traits such as bill size or body size. This is illustrated in Figure 15, which shows differences in the variation of bill depth between populations of the same species; for example, the population of *G. fortis* on Daphne varies less than does the population of this species on Santa Cruz. The figure also shows differences in variation between species; *G. fuliginosa* varies less than do *G. fortis* and *G. magnirostris*. These features were pointed out in Chapter 4 without discussion. I return to them here and attempt to explain them in the light of the information on feeding habits given in the previous two chapters.

RELATIVE VARIATION

Variation is partly a function of size. In general, the larger the mean the larger the variance, so some correction for the effects of scale needs to be made to see whether populations with large means are proportionately or disproportionately more variable than those with small means. The effects of size can be removed by transforming the original measurements to logarithms and then calculating the variance, or by calculating the coefficient of variation (100 standard deviations divided by the mean) from the original measurements as I have done (Table 7). Coefficients provide measures of relative variation that can be compared between species of such disparate sizes as the small and the large ground finches (*G. fuliginosa* and *G. magnirostris*).

Values listed in Table 7 demonstrate substantial differences in relative variation in beak depth between populations of the same and different species. Many of the differences between coefficients for the ground finch samples are statistically significant, as revealed by F tests of ratios of squared coefficients (Grant et al. 1985). Population variation is more uniform, and low, among the tree finches, the Cocos finch and, according to my unpublished measurements, the warbler finch.

Species of passerine birds elsewhere have coefficients in the range of 3.0 to 5.0 (Grant 1967, 1979b, Grant et al. 1985). Judged against this standard,

TABLE 7. Coefficients of variation for bill depth of males, calculated from data in Lack (1947); see also Bowman (1961) and Grant et al. (1985). Bill depth of *Certhidea olivacea* was not measured by Lack. Samples from north and south Isabela were treated by him as separate populations in four species: *G. fuliginosa*, *G. fortis*, *C. parvulus*, and *C. pallida*. In all other cases each island is represented by one population. The population of large ground finches from Darwin has been omitted (see text below).

	Minimum	Maximum	Average	No. Populations
G. fuliginosa	2.60	5.73	4.54	14
G. difficilis	3.40	4.16	3.70	6
G. scandens	3.20	4.96	4.29	9
G. fortis	3.38	10.30	7.04	12
G. conirostris	6.91	7.39	7.23	3
G. magnirostris	4.34	9.17	6.54	7
P. crassirostris	2.76	4.42	3.70	6
C. psittacula	3.05	5.98	4.79	5
C. pauper	4.89	4.89	4.89	1
C. parvulus	3.73	5.41	4.46	6
C. pallida	2.31	4.94	3.11	5
C. heliobates	3.41	3.41	3.41	1
P. inornata	4.03	4.03	4.03	1

all populations of tree finches exhibit typical variation (Table 7), with minor exceptions being two populations of *C. psittacula* (Santa Cruz and Santiago) and one population of *C. parvulus* (N. Isabela). Variation in populations of *G. scandens* and *G. difficilis* is also typical of birds elsewhere, but the three largest species of ground finches are unusually variable. Most populations of *G. fortis* and *G. magnirostris*, and all populations of *G. conirostris*, have coefficients greater than 5.0. *G. fortis* populations span the remarkable range of 3.38 (Marchena) to 10.30 (Santa Cruz). The largest coefficient of all is not shown in the table. The large ground finch on Darwin has a coefficient of 13.13 (Bowman 1961); owing to uncertainties (Chapter 3) about its identity (one species or two?), I have omitted it from the table.

Thus the values in the table, and the statistical analyses, confirm the impression gained from looking at Figure 15 that there are differences in intrinsic variation among some populations of ground finches.

THEORETICAL BACKGROUND

Why do some populations vary so much? As Lack (1947, p. 94) expressed it: "It would be of great interest to determine the critical factors controlling the variability of each species, and to know why some species are so much more variable than others." This question is best approached by considering those factors tending to increase the variation in a population, and the countervailing forces tending to reduce it.

Variation in quantitative traits is governed by the action of many genes, each with small additive effect upon the expression of a trait, and by environmental influences. Nonadditive genetic effects (dominance, epistasis) are generally assumed to be small for such traits. Genetic variation is translated into phenotypic variation each generation through recombination of alleles. Figure 53 shows how new alleles enter a population, and increase the genetic variation, in three ways: by mutation; by introgression of genes through the immigration, and subsequent breeding, of conspecifics from another population; and by hybridization with either a sympatric or an immigrant allopatric species. Genetic variation is depleted by random (drift) or systematic (selection) processes. From these considerations it can be seen

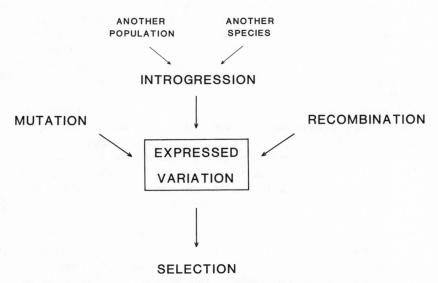

FIG. 53. A model for the maintenance of genetic variation in a population. Additions, through mutation and introgression, are balanced by losses through selection. Based on Lande (1976a), whose model also shows that recombination exposes the hidden variation which is stored in negatively correlated and linked combinations of alleles.

that one population could be more variable genetically than another for one, or both, of two reasons: the rate of genetic input through mutation and/or introgression is higher, or the rate of genetic output (depletion) through selection, or drift, is lower (Grant and Price 1981).

The ecological circumstances giving rise to selection pressures on variation in resource-utilization traits such as bill depth were first explored in detail by Van Valen (1965). His scheme, simplified in Figure 54, is useful for visualizing how different populations are subject to the variation-depleting effects of stabilizing selection to different degrees according to the way in which they exploit the environment. In model 1 a large segment of a resource spectrum (e.g. food size) is available to, and exploited by, a species, whereas in model 2 the exploited segment is narrower. Model 2 situations might arise in species-rich communities, such as occur in continental regions, whereas model 1 situations are likely to arise in species-poor communities such as on islands, providing the resources are sufficiently varied. The two versions of model 1 show alternative ways in which individuals with different phenotypes respond to a diverse environment: in the first case all members of a population are generalists and ecologically similar; in the second case all individuals are specialists and different. Where individuals differ in their use of the environment (1b) selection pressures on them differ. In contrast, where individuals are similar, that is, they all exploit the environment in a similar generalized (1a) or specialized (2) manner, selection pressures on them are uniform. A large variation in resource-utilization traits is only expected for a 1b population because selection pressures on the individuals are not uniform and the constraints of stabilizing selection are relatively weak.

The limitation of the Van Valen scheme is that it stops short of specifying the different conditions under which models 1a and 1b are expected to apply (Grant et al. 1976). Several, but not all, attempts to test the basic ideas embodied in the models have foundered for this reason (Grant and Price 1981). Nevertheless, it is unequivocal in setting forth a causal relationship between morphological variation and eco-behavioral variation. In populations which vary strongly in morphology, such as some of the populations of Darwin's Finches, individuals should differ from each other to some degree in the way they exploit the environment, and to a greater extent than do individuals in morphologically less variable but ecologically similar populations.

Populations of *G. fortis* conform to these expectations. On Daphne Major, *G. fortis* is a generalist species comprising individuals that are to some extent specialized at dealing with part of the food spectrum exploited by the population as a whole (Chapter 6). The population of this species at Bahía Academía on the south side of Santa Cruz is significantly more variable in bill depth, and exploits a greater food diversity, than the population at B.

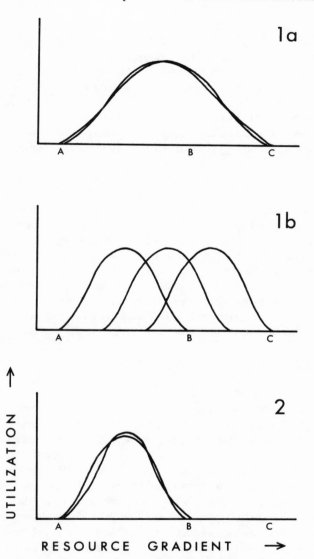

FIG. 54. Three models for the way in which resources are exploited by members of a species. A broad range of resources (A–C) is exploited by populations 1a and 1b; in the first individuals are generalists, in the second they are specialists. A narrow range of resources (A–B) is exploited by all members of the specialist species 2. Adapted from arguments in Van Valen (1965). The overlaps among individuals in 1b are shown to be symmetrical, for simplicity; in fact asymmetrical overlaps, as in the feeding efficency models in Figure 40, are more realistic.

Borrero on the north coast of Santa Cruz and the population on Daphne Major (Abbott et al. 1977). The three larger *Geospiza* species have significantly more diverse diets as well as significantly greater variation in bill depth than do the three smaller species (Abbott et al. 1977). Members of one population of the larger species, *G. conirostris* on Genovesa, differ from each other in their somewhat specialized diets (Grant and Grant 1979, B. R. Grant 1985). In contrast, dietary differences have not been found among members of a smaller species, *G. difficilis*, on the same island. Thus some populations belong to model 1b of the Van Valen scheme.

FIELD STUDIES

How well do these theoretical ideas concerning genetic enhancement and depletion apply to Darwin's Finches? From field studies we need to know if populations exchange genes at a high enough frequency to have a perceptible effect upon variation, and we need to know if selection operates in the postulated way. But we also need to know if the phenotypic variation we observe and measure is strongly influenced by genetic variation, for the theoretical ideas embodied explicitly in Figure 53 and implicitly in Figure 54 require that causal connection. I shall treat these subjects in the following order: genetic variation, selection, and introgression.

GENETIC VARIATION

According to standard quantitative genetics theory (e.g. Falconer 1981), the phenotypic variance (V_P) of a trait such as bill depth can be partitioned into two components that represent genetic (V_G) and environmental (V_E) effects. The genetic variance in turn is made up of additive, dominance, and interaction (epistasis) variance: $V_G = V_A + V_D + V_I$. The additive genetic variance is the predominant component of V_G (Falconer 1981), and is important because its magnitude, together with the strength of selection, or the selection differential (s), determines the evolutionary response (R) to selection acting solely on the trait under consideration: $R = h^2 s$. Additive genetic variance enters this equation as part of the heritability (h^2). The heritability of a trait in the narrow sense is V_A/V_P or the proportion of the phenotypic variance of a trait which is attributable to the additive effects of genes.

Heritability can be estimated in several ways. In an uncontrolled field study the most satisfactory technique is to regress the average of the values of the offspring, when fully grown, on the average for their parents. The slope of the regression provides an estimate, with attendant confidence limits, of the heritability, that is, the strength of the genetic resemblance between offspring and their parents. The estimates and confidence limits should theoretically lie between 0 and 1. Heritabilities have been estimated in this way for three populations of ground finches.

In 1976 Peter Boag and I color-banded nestlings in many nests of *G. fortis* on Daphne Major when they were eight days old. Boag did the same again in 1978. Almost all nestlings fledged, and many were later recaptured and measured when fully grown, or almost so. Measurements of birds captured and measured more than once were found to be highly repeatable. The offspring-midparent regression technique was applied to the measurements of families, i.e. to the offspring, usually one per family, and to the father and the mother. The resulting heritability estimates are given in Table 8. Figure 55 illustrates the relationship between offspring and midparent bill depth.

TABLE 8. Heritabilities of morphological traits, calculated from offspring-midparent regressions and shown with one standard error. *N* refers to number of families, followed by number of offspring. Statistical significance of the regressions is indicated by one ($P < 0.05$), two ($P < 0.01$), or three ($P < 0.001$) asterisks. Sources are Grant (1983a) for *G. conirostris*, and Boag (1983) for the other two species, except for *G. scandens* bill heritabilities which are from Price, Grant, and Boag (1984).

	G. fortis	*G. scandens*	*G. conirostris*
Weight (g)	0.91 ± 0.09***	0.58 ± 0.39	1.09 ± 0.27***
Wing (mm)	0.84 ± 0.14***	0.12 ± 0.26	0.69 ± 0.21**
Tarsus (mm)	0.71 ± 0.10***	0.92 ± 0.23***	0.78 ± 0.23**
Bill length (mm)	0.65 ± 0.15***	0.58 ± 0.24*	1.08 ± 0.19***
Bill depth (mm)	0.79 ± 0.09***	0.80 ± 0.32*	0.69 ± 0.09***
Bill width (mm)	0.90 ± 0.10***	0.56 ± 0.25*	0.77 ± 0.25**
PC I	0.75 ± 0.12***	0.20 ± 0.39	0.79 ± 0.15***
PC II	0.36 ± 0.10**	1.07 ± 0.20***	0.92 ± 0.17***
PC III	0.47 ± 0.15**	0.64 ± 0.22**	0.52 ± 0.21*
N_1, N_2	39,82	16,29	20,24

Heritabilities are uniformly high for the six traits in *G. fortis* (Table 8) and there is no difference in the slopes of regressions in different years (e.g. Fig. 55). The first three components from a principal components analysis of all morphological dimensions are also significantly heritable. Subsequent work confirmed the high heritabilities in 1981 (Price, Grant, and Boag 1984). It has shown heritabilities to be high in the population of *G. conirostris* on Genovesa, but a little lower in the less variable population of *G. scandens* on Daphne (Table 8).

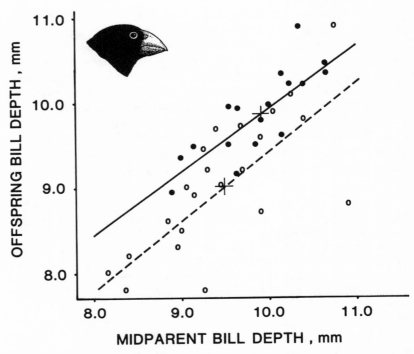

FIG. 55. The relationship between the beak depth of offspring and their parents in the medium ground finch (*G. fortis*) population on Daphne Major. The slope of the relationship is the heritability. In two years the slopes were nearly the same; 0.82 in 1976 (○) and 0.74 in 1978 (●). Offspring reached larger sizes in 1978 than in 1976. Crosses indicate bivariate means. From Boag (1983).

Heritabilities estimated in this way could be inflated by genotype-environment correlations. For example, a distortion could be produced by large parents raising offspring in the most food-rich territories. The offspring would then reach large size as adults partly for genetic reasons and partly for environmental reasons (good rearing conditions). Being large they might stand the best chance of acquiring good territories, and so their offspring in turn would grow to large size for the same reasons. The perpetuation of such a process would result in a distortion of the heritability estimates, and an overestimation of the degree of genetic resemblance; the parents and offspring would resemble each other partly because they experienced similar environments during growth.

Boag (1983) explored such a possibility, but the search for these effects was unsuccessful. There was a tendency for birds nesting on the inner slopes of Daphne to be larger than elsewhere, but their offspring did not reach a larger size, relative to their parents, than did young elsewhere, and breeding success was not higher on the inner slopes than elsewhere (Boag and Grant 1978, Boag 1983). Moreover, birds do not tend to breed in the parts of this island where they were born. A similar study of *G. conirostris* on Genovesa concluded that if genotype-environment correlations occur, their effects are likely to be small (Grant 1983a).

An alternative approach to the problem of genotype-environment correlations is to redistribute different genotypes randomly among environments in controlled experiments. Any possible correlations would be broken by this manipulation, except for those arising in the egg, and the resulting offspring sizes, when regressed on midparent sizes, would yield better estimates of heritabilities. The experiment can be performed best by exchanging eggs among nests. We planned such an experiment in 1977 (Boag and Grant 1978), but we were unable to carry it out owing to the drought that year (Chapter 7). In the following year this experiment was performed with an island population of song sparrows, *Melospiza melodia*, in Canada. Heritabilities of bill size and body size had previously been determined to be fairly high (Smith and Zach 1979). The clutch-exchange experiment produced no detectable effect of the foster-parent rearing environment on the final size of the offspring (Smith and Dhondt 1980): the heritability estimates of control and experimental birds did not differ. Consistent results have since been obtained in a study of blue tits, *Parus caeruleus*, in Belgium (Dhondt 1982). If these studies with song sparrows and blue tits are a guide to what might be expected with Darwin's Finches, they suggest that our estimates of heritabilities are not seriously distorted.

Thus there is a large amount of additive genetic variance for bill and body size traits in the populations of Darwin's Finches studied. The connection between genetic and phenotypic variation required by the theoretical framework outlined in the introduction to this chapter is indeed present.

NATURAL SELECTION

I shall now describe what is known about the forces of selection acting on modern populations of Darwin's Finches. But first I must make explicit what I mean by natural selection. Probably the most widely held meaning of natural selection is the differential perpetuation of genotypes (Mayr 1963, p. 183) or genes. For general discussion of natural selection this is quite sufficient, but for purposes of measurement and analysis it is unsatisfactory, and I have therefore followed instead the working definition used by many geneticists which makes a distinction between selection on phenotypes and its genetic consequences. In the words of Haldane (1954, p. 480): "Natural . . . selection acts on phenotypes. It is ineffective unless it favours one genotype at the expense of another. But it may occur without doing so."

Selection is restricted here to events occurring within a generation. The consequences in the following generation are a separate matter, contingent upon genetic variation and its transmission from one generation to the next. The consequences are referred to as the evolutionary response to selection. Other authors (e.g. Endler 1986) prefer to reserve the term phenotypic selection for changes in phenotype occurring within a generation, and only refer to them as natural selection if they are accompanied by genetic change.

The essence of selection is that certain individuals in a population do better than others in part because they possess traits, or expressions of traits, not possessed by other individuals. Statistical analysis can identify those traits which are important for success. I shall refer to traits so identified as the targets of selection. Again, this usage is not universal. For example, Mayr (1983) defines the target of selection as the whole individual and not a single gene or trait. Used in this sense its scope is too restricted for my purposes, and it conflicts with the quantitative genetic framework in which a discussion of continuous variation naturally fits (e.g. see Falconer 1981, p. 26).

Directional selection in G. fortis on Daphne Major. Forces of selection acting in this population have principally directional, and not stabilizing, effects on the size of traits. The directions of selection change, however, with the result that small size and large size are alternately favored. Over a long period of time the net effect of these oscillations in directional selection may be roughly equivalent to a weak form of stabilizing selection.

The outstanding episode of natural selection that we have been fortunate enough to witness occurred during the 1977 drought described in the previous chapter. The size of the *G. fortis* population on Daphne Major declined from mid-1976 until early 1978 (when the rains returned and breeding resumed) from about 1200 individuals in mid-1976 to about 180 by the end of 1977. Seed abundance declined in parallel with a decline in finch abundance (Chapter 7), from approximately 10 g/m^2 to 3 g/m^2. During this long period of declining numbers, survival was not random with respect to phenotype: large birds survived better than small ones. Therefore natural selection occurred.

Using O'Donald's modification of Haldane's method of estimating the intensity of selection (O'Donald 1970, Haldane 1954), we calculated that this was the most intense selection on metric traits documented for a natural population of animals up to that time (Boag and Grant 1981)!

Figure 56 represents these events in terms of an index of overall size. The index is the first component from a principal components analysis of seven morphological variables. The index is stable at the beginning, starts to rise in early 1977, and ceases to rise in early 1978, remaining stable thereafter at a value 5 to 6% higher than at the beginning. Mean values are shown for males, for females, and for these two groups combined with birds of undetermined sex who were mainly born in 1976. Since breeding did not take place in 1977, the measured traits did not grow, and all birds had been measured prior to 1977, the rise in average PC I scores can only have resulted from one process, the differential loss of small birds.

Loss could be due to mortality or to emigration. Detecting emigration is a needle-in-the-haystack problem, and the only emigration we know of with certainty involved three juvenile cactus finches (*G. scandens*) that were banded on Daphne Major and observed on Daphne Minor, about 8 km east of Daphne Major, in 1979 (Grant et al. 1980). There were no signs that birds emigrated temporarily, taking the form, for example, of disappearance throughout 1977 and reappearance on the island in 1978 or subsequently. On the other hand *in situ* mortality was detected (Plate 55). Thirty-eight banded *G. fortis* individuals were found dead on Daphne Major in 1977 and early 1978. In their measurements they were statistically indistinguishable from the remaining banded birds that disappeared but were not found, and they were significantly smaller than the survivors. Therefore a part, and probably a large part, of the selective disappearance of small birds during the drought of 1977 can be attributed to mortality on the island.

The reason why small birds suffered most lies in the changing nature of the food supply. Following the large production of seeds in the wet season in 1976 all birds fed on the abundant, small, and soft seeds of grasses and herbs such as *Portulaca, Chamaesyce*, and *Heliotropium*. When the supply of these seeds, which are easy for finches to deal with, declined, birds increasingly turned to the larger and harder seeds, principally *Opuntia* and *Tribulus*. Those with the largest bills are the most efficient at dealing with these seeds, as discussed in Chapter 6; birds with smaller bills are either relatively inefficient or incapable of dealing with large and hard seeds at all. Therefore birds died through starvation (Plate 55). Small birds of both sexes and all ages disappeared from the population throughout 1977 when the average size and hardness of the seeds remained high because the supply of small seeds had been depleted (Fig. 56). The average size-hardness index of the seeds only fell again in early 1978 when a normal wet season resulted in the production of large numbers of small seeds again.

Thus strong directional selection occurred in both sexes. Variances remained unchanged, so the forces of selection did not act in a stabilizing manner.

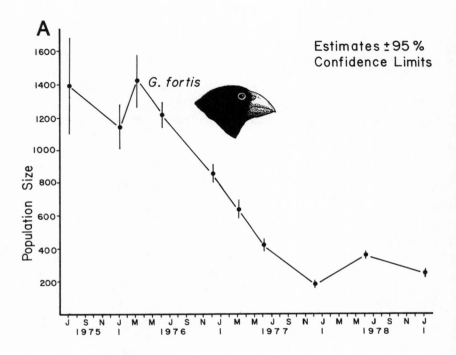

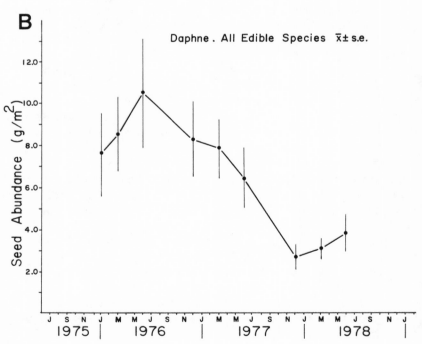

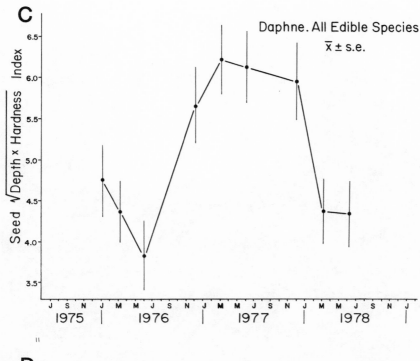

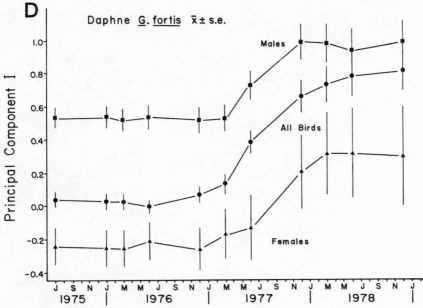

FIG. 56. Changes in finch numbers, morphology, seed abundance and average seed size and hardness on Daphne Major. Means and standard errors are shown. Principal component I is a measure of overall size; the analysis was conducted on all birds measured in 1975 and 1976, and the changes in the mean scores are the result of loss of birds from the initial sample. From Boag and Grant (1981).

PLATE 55. Medium ground finch, *Geospiza fortis*, on Daphne Major: casualty of the 1977 drought. Note the extremely worn plumage (*photo by P. T. Boag*).

The targets of selection. The foregoing presentation of facts and interpretations implies that bill size was the crucial variable that determined success or failure under environmental stress. This is consistent with the feeding efficiency models and the results of subsequent tests (Chapter 6). Yet in Figure 56 I have shown an index of overall size, not just bill size, as the morphological index that changed under directional selection. The question arises, what was the target or targets of selection? Was it bill size, for reasons developed earlier, or was it body size or some other trait with which bill size is correlated? The question is difficult to answer because all the traits we have measured are positively correlated with each other (Chapter 4), so disentangling their effects on survival is not easy. For example, correlations between PC I scores on the one hand and measurements of weight, wing length, tarsus length, bill length, bill depth, and bill width on the other, are all positive and strong, with coefficients falling in the range 0.60 to 0.94 (Boag and Grant 1981). Any one of these substituted for PC I means in Figure 56 would give the same pattern of increase as shown by PC I.

Disentanglement can be achieved using a partial regression analysis (Lande and Arnold 1983). It is a device for finding the relative importance of the different morphological traits to survival. A measure of relative fitness of individuals is regressed on the values of each of their morphological traits. For this analysis the absolute fitness of an individual is scored as 0 if it disappeared and 1 if it survived, and these scores are then converted to relative fitness values by dividing them by the mean absolute fitness. The partial regression coefficient for each trait estimates the strength (by its magnitude) and direction (by its sign) of selection acting directly on that trait, i.e. independent of any indirect effects caused by selection on other traits with which that trait is correlated. To eliminate problems of scaling, values are converted to standard units prior to the analysis by dividing them by the standard deviation of the trait before selection.

Table 9 shows the results of regression analysis applied to three equal segments of the selection period, mid-1976 to the end of 1977. One body size measure (weight) and three bill dimensions are analyzed. The partial regression coefficients (β) constitute the *selection gradient*, or the direct effect of selection on each trait; and the *selection differential* (s), measured as the difference between the mean value of a trait before and after selection in standard deviation units, represents the total or net effect of selection, i.e. the direct effects and the indirect effects arising from character correlations.

The relative magnitudes of the partial regression coefficients in the analysis reveal that selection acted directly on both body size and bill depth, occasionally strongly and consistently positively, i.e. favoring large size. In addition, and surprisingly, selection acted directly and negatively on beak width, i.e. a narrower beak width for a given beak depth was consistently favored, but weakly. Beak length was never subjected to direct selection.

TABLE 9. Natural selection in the population of medium ground finches (*G. fortis*) on Daphne Major. *s* = standardized selection differential, and β (with standard error) = standardized selection gradient. An asterisk indicates statistical significance (*P* < 0.05). (After Price, Grant, Gibbs, and Boag 1984.)

	June–December 1976		December 1976–June 1977		June–December 1977	
	s	β + SE	*s*	β + SE	*s*	β + SE
Weight (g)	0.05*	0.03 ± 0.04	0.28*	0.23 ± 0.08*	0.28*	0.29 ± 0.12*
Beak length (mm)	0.06*	−0.01 ± 0.05	0.21*	−0.17 ± 0.10	0.22*	−0.04 ± 0.14
Beak depth (mm)	0.08*	0.19 ± 0.07*	0.30*	0.43 ± 0.13*	0.23*	0.19 ± 0.19
Beak width (mm)	0.04	−0.15 ± 0.06*	0.24*	−0.19 ± 0.12	0.21*	−0.14 ± 0.17
Sample surviving	640		442		194	
Proportion surviving	0.69		0.44		0.50	

One can conclude from these results that bill depth was indeed the direct target of selection, but it was not the only target. Independently, birds of large body size but relatively narrow bills were favored. Large birds may have been favored for metabolic reasons, and they may have gained an added advantage over small birds by being dominant at food sources. Birds with relatively narrow bills may have been favored because a narrow yet deep bill was the best instrument for performing the difficult task of tearing, twisting, and biting the mericarps of *Tribulus* to expose the seeds (see Chapter 6).

We have no data that bear on the interpretation of selection for large body size, but feeding observations of the survivors tell us something about the beak characteristics of those that fed on *Tribulus* seeds. I have analyzed the data by substituting feeding or not feeding on *Tribulus* seeds (Chapter 6) for survival or nonsurvival as measures of fitness in an analysis of differentials and gradients. I refer to these as trophic differentials and gradients, following B. R. Grant (1985). Trophic differentials for beak depth and beak width were strongly positive and significant ($P < 0.01$ in each case), a result expected from previous treatment of the data in Chapter 6 (also Grant 1981b). Coefficients in the trophic gradient were not significant, but the signs were the same as in the selection gradient: for beak depth, $\beta = 0.22 \pm 0.16$; for beak width, $\beta = -0.04 \pm 0.13$. Therefore overall beak size made the difference between feeding and not feeding on *Tribulus* seeds, with perhaps a small tendency for beak shape to be involved as well. Selection for deep beaks is more satisfactorily accounted for than is selection for relatively narrow beaks. An alternative explanation for the latter is that some birds with deep but relatively narrow beaks survived because they were efficient at cracking the other type of large and hard seed, *Opuntia*. We do not have enough feeding observations of the survivors to make a comparable check on this possibility.

Note that the analysis of selection gradients uncovered the direct effects (β) of selection that were, in some cases, not even hinted at by the total effects (s) of selection. The direct effect of selection to decrease bill width was more than compensated by the indirect effects arising from positive correlations with bill depth and body size that were selected to increase. Bill length increased in each six-month period as a result of similar character correlations and indirect effects, and not because it was a target of selection. This makes functional sense because the seed cracking ability of a bird is primarily a function of the depth and width of the bill (Chapter 6), and only secondarily a function of bill length.

These results are contingent upon correct identification of the relevant variables. They could be spurious if selection had acted on some other unanalyzed trait with which all of the included traits are correlated. This does not seem likely, as addition or deletion of variables had no major effect on the estimates of β values for the traits body weight and bill depth (Price, Grant, Gibbs, and Boag 1984). These two traits appear to have been the major targets of selection.

Weaker selection events occurred in the relatively dry periods of mid-1979 to the end of 1980 and mid-1981 to the end of 1982, and in both cases the targets of selection were the same as in 1976–1977 (Price, Grant, Gibbs, and Boag 1984).

Evolutionary response to selection. As mentioned earlier, the predicted response to selection is given by the product of the heritability of a trait and the selection differential when selection acts solely on that trait. This relationship cannot be used to estimate the predicted response when selection acts on several traits, as it did on Daphne. Nevertheless the evolutionary response to selection is expected to be large when heritabilities are uniformly high, as here, and it was observed to be large. The offspring of the 1978 parents were 0.31 standard deviations larger, on average, than the 1976 midparent values. Therefore a measurable evolutionary shift occurred in the *G. fortis* population in response to the forces of directional selection in 1977. Average beak depth increased in the next generation by about 4%.

SEXUAL SELECTION

Hidden in the response to selection is another component of selection: sexual selection. Males are larger than females on average by about 5%. Partly as a consequence of the size-selective mortality, the sex ratio of the population became unbalanced in favor of males. At the most extreme point just prior to the breeding season in 1978, males outnumbered females by approximately 5 or 6 to 1. Not a single pair of birds that had bred in 1976 was known to have survived, as a pair, to 1978, so when breeding commenced in 1978 all pairs were newly formed. It is highly probable that pairing was nonrandom with respect to size at this time, because in the next three years, when there were enough banded birds as pairs to make the appropriate comparisons, Trevor Price found clear evidence of nonrandom pairing.

In 1979 those males with mates were significantly larger in bill depth and bill width than those without (Price 1984a); those without mates occupied territories, behaved normally, and were apparently just as motivated to obtain mates as those who succeeded. This pattern of nonrandom mating (pairing) was probably present in 1978 too because the known pairs were the same except for three or four cases of re-pairings. In 1980 some males acquired mates, born in 1978, but these newly mated males were not significantly larger than the unsuccessful males. In 1981, however, another group of males acquired mates and these were significantly larger than those who did not, in all dimensions except weight. In both 1979 and 1981 the sexual selection differentials were statistically significant for bill depth and bill width (Price 1984b). Throughout this period the sex ratio was male-biased, between 2:1 and 3:1, and all females of two years and older were mated.

At the beginning of the breeding season males are largely confined to their territories, where they sing repeatedly and display to visiting females, whereas females visit many territories before pairing (monogamously) with a male. These patterns of behavior, and some theoretical arguments concerning the unequal investment of energy into reproduction, at least in the early stages, suggest that female choice is largely responsible for the observed nonrandom pairing, and hence for the sexual selection (Price 1984a).

An unresolved issue is why females chose particular males. They might have done so because those males were large, because they tended to be blacker in their plumage, or because they had larger territories than the males not chosen. All three variables were found to correlate significantly with mating success. Price (1984a) used regression analysis to disentangle their statistical effects upon mating success, and found each one made a significant contribution when effects of the other two were held constant. A broader unresolved issue is the importance of these factors and their ultimate significance; for example, does territory size convey important information about resources, quality of the male as a parent and/or as a provider of genes, or all three? Price (1984a) tentatively concluded that territory size was most important as a sign of the parental quality of the male.

Despite these uncertainties concerning interpretation, the consequence of sexual selection was clear; it augmented the effects of natural (survival) selection during the drought of 1977.

COUNTERVAILING SELECTION

If the forces of selection have an overall stabilizing or normalizing effect on the variation of a population, but occasionally a strongly directional effect, when does selection favor birds with small bills, and under what circumstances?

If feeding efficiency model 1 (Fig. 40) is correct, small-billed birds have an efficiency advantage over large-billed birds when feeding on the smallest and softest seeds. This would translate into a fitness advantage if small seeds predominated but the overall seed supply was low. Alternatively there could be an advantage to birds of small body size, for metabolic and energetic reasons, giving as a consequence an advantage to birds with small bills, since bill size and body size are so strongly correlated.

We have identified two times in the life cycle when being small appears to be advantageous.

Survival in the first year of life. The first period occurs in the first year. In 1976 many young birds were measured at ages two–four months, at which time they were close to being fully grown (Chapter 5). Between this age, and age 7–10 months, their survival was nonrandom; survivors were significantly smaller in bill depth and bill width than those who disappeared (Price and Grant 1984). Survival of young birds over the same period in 1981 was also nonrandom; again, survivors were significantly smaller in bill width. In 1978, the year of lowest mortality among the three years, survival was random with respect to size; and in the three remaining years between 1976 and 1981, very few birds survived in 1979, none survived in 1980, and none were even born in 1977.

It thus appears that in some years small-billed birds are favored by selection in their first year of life. This might happen because small-billed birds are most efficient at handling small seeds at a time when the supply of small seeds is diminishing due to consumption, and when all young birds are feeding on small seeds because none of them have yet acquired the skills, and the fully adult muscle masses and ossified skulls, to deal with large and hard seeds (Chapter 6). For example, even at the age of one year, *G. scandens* are less efficient at cracking the seeds of *Opuntia echios* than are older birds (Millington and Grant 1984). A second possibility is that small-billed birds (*G. fortis*) are favored because, being also small in body size, they have lower daily energy requirements than do larger birds and can meet them more easily on the common diet of small seeds. Surviving young in 1981 had significantly lower weights at age 2–4 months, and significantly narrower beaks, than did those that died.

Consistent with the selection argument, birds born in 1976 and in 1981 and surviving to the following January in each case were significantly smaller in all bill dimensions than the group of adults from which they were descended. In contrast, there was no difference between the juveniles surviving to January and the adults in 1978–1980, years in which there was no evidence of size selection (above). The evidence for selection is thus internally consistent, but it is not conclusive because of the possible complication of incomplete growth. A trait may grow on a trajectory that approaches different asymptotes in different years, or it may take longer to complete in some years than in others, with the result that some birds take longer to reach final size than others for environmental and not for genetic reasons. These complications cannot be ruled out with certainty because the potential final size of dead young can never be determined!

Initial reproduction. The second circumstance under which small size is advantageous is the time of first reproduction. In some years conditions for breeding are sufficiently suitable that some one-year old females breed, but not so suitable that they all do. This happened on Daphne in 1979 when 20 females born in 1978 bred, and 31 did not do so despite the availability of unmated, territorial, males. The females which did breed were significantly smaller in the three bill dimensions and in weight than those which did not. An analysis of selection showed that all selection differentials were statistically significant, although the only significant target in the gradient was bill length (Price 1984b). Selection had no evolutionary consequences in this case, however, since none of the offspring of the early-breeding young females survived.

There is no evidence of a comparable reproductive advantage to small young males; on the contrary large males breed first, not because they are the first to come into reproductive condition, apparently, but because they are chosen by females. Nor is there any evidence of selection on metric traits arising from correlations with other components of reproduction. For example, small and large birds do not differ in their dates of egg laying, in their clutch sizes, or in their fledging success (Price 1984b).

A SUMMARY OF SELECTION PRESSURES

Figure 57 represents the four selection pressures identified as acting on the two sexes of *G. fortis* on Daphne. Quantitative values for the various components over a whole life cycle cannot be combined because there are too many errors of estimation, and the study on Daphne has been too short. We have observed stronger selection for large size than for small size, but, in view of the short period of study, the directions of selection are more informative than are their strengths. At any one time the selection pressures on different sexes and age groups may or may not be of the same type and in the same direction. In 1977, for example, when survival was at a pre-

FIG. 57. Opposing selective forces on distributions of male and female sizes in the population of medium ground finches (*G. fortis*) on Daphne Major. The two sexes are subject to some dissimilar forces of selection and some similar ones. The opposing selective forces may balance, approximately. For the purpose of clear illustration the distributions are shown separately; in fact, for all traits male and female distributions overlap, males being on average slightly larger than females. From Price (1984b).

mium, selection favored large overall size in both sexes. But in 1979 during the breeding season when mortality was low, small females were favored by reproductive selection whereas large males were favored by sexual selection.

Selection thus fluctuates in direction within a generation, and possibly between generations as well, driven by the unpredictable and large fluctuations in rainfall (Grant et al. 1976). The effects of oscillating directional selection, integrated over a long period of time such as a decade, are perhaps the same as weak, continuous, stabilizing selection. This does nothing to preserve variation in a population, but it is relatively ineffective at reducing variation. In this restricted sense a large amount of variation is maintained by weak stabilizing selection.

These observations are inconsistent with an alternative model for the maintenance of variation proposed by Soulé and Stewart (1970). They suggested that large variation in general could be the result of a breakdown in canalization under a regime of sustained directional selection. This hypothetical mechanism does not apply to *G. fortis*. Directions and strengths of selection on the Daphne population fluctuate, with the net effect being apparently stabilizing.

SEXUAL DIMORPHISM

An additional source of variation in a population is sexual dimorphism, i.e. the degree to which the sexes differ in mean sizes. In theory a population can become more variable under a regime of essentially disruptive selection in which different mean sizes of the sexes are favored because the sexes are thereby able to exploit the resources differently and compete with each other less (Slatkin 1984). It is a variant of the adaptive variation model 1b in Figure 54, in which the specialists are the different sexes (Rothstein 1973).

In fact there is little evidence that the sexes feed differently. In the breeding season they each feed to approximately the same extent, for themselves and for their offspring, chiefly on caterpillars, pollen, and nectar (Price 1984b, Boag and Grant 1984a), and sometimes on seeds (Downhower 1976), spiders (Grant and Grant 1980a), and other arthropods. In the non-breeding season on Daphne, *G. fortis* males, being larger than females, tend to feed more on large and hard seeds. During the drought of 1977, six females, never seen to feed on large seeds, nevertheless survived by exploiting only small seeds. These were very small birds. They constituted a small second peak in the population frequency distribution of survivors. A significant disruptive component to the overall selection on the population was statistically detected (Schluter et al. 1985), but its effect, in the face of strong net directional selection on both sexes (Boag and Grant 1981), was small, and the sexual dimorphism after selection was virtually the same as it was before selection.

Sexual differences in the means of morphological traits actually vary very little among populations of the same species (Price 1984b, Grant et al. 1985). The variances, or coefficient of variation, for a given trait in one sex differ much more among populations (Table 7). This applies equally to the two sexes, and is the reason why I restricted attention to variation in one sex (males) in the table. The lack of large variation in sexual dimorphism could mean that what we see is the ancestral condition, unaltered to any large degree during the several speciation events.

The rate at which changes in sexual dimorphism could occur is not only determined by selection but by the genetic correlation between the sexes for a given trait, i.e. the degree to which the sexes share the genes for that trait. The correlation can be calculated by comparing the slopes of the separate regressions of male offspring and female offspring value on the values for one of the parents. For this purpose it is assumed that phenotypic variances of the two sexes are similar, the heritabilities for the two sexes are similar, and the contributions of sex-linked genes to variation in body size of the offspring are small enough to be ignored (Lande 1980a). These assumptions are probably met satisfactorily. For example, with regard to the last one, the sex chromosomes are relatively small components of the genome (Jo 1983; see also Fig. 58). Price (1984b) used measurements of *G. fortis* offspring of each sex and just the female parents to calculate genetic correlations, and found them to be close to 1.0 for all traits (but within broad confidence limits). This means that forces of selection would have to be strong, or persist for a very long time, or both, to materially change the dimorphism because selection on one sex has correlated effects on the other sex.

Thus resource partitioning by the sexes does not appear to be a significant factor in the maintenance of current levels of dimorphism. Rather, the scheme in Figure 57 depicts the relevant pressures. These pressures include reproductive selection favoring small females, as suggested by Downhower (1976) and demonstrated by Price (1984a). Whatever the origin of the dimorphism, current levels appear to be maintained as a result of overall stabilizing selection on each of the sexes, and genetic inertia. The genetic correlations between the sexes are close to 1.0, and as a consequence the difference between the sexes remains unaltered, even though the sexes are subjected to different selection pressures, occasionally of opposite sign.

Genetic Drift

Allelic variation could be lost as a result of random processes. This would be mostly likely to happen in small and somewhat inbred populations. On

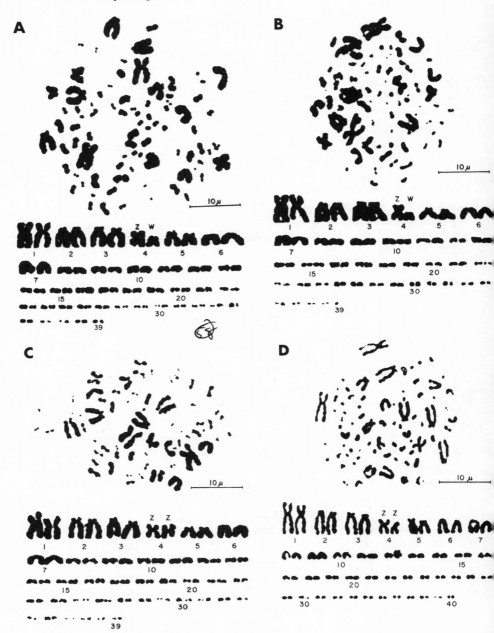

FIG. 58. The chromosomes of four species of finches: A, *Platyspiza crassirostris*; B, *Camarhynchus psittacula*; C, *Camarhynchus pauper*; D, *Camarhynchus parvulus*. Sex chromosomes are Z and *W*. From Jo (1983).

Daphne Major, for example, the *G. fortis* breeding population reached a low point at the beginning of 1979 when only 32 adult females remained. But while genetic drift is a real possibility at such low numbers, the population did not stay at this small size for very long (Chapter 7). Its importance is likely to be overshadowed by the demonstrably strong and frequent selection on this population.

ENHANCEMENT OF GENETIC VARIATION

From factors tending to deplete genetic variation I now turn to factors tending to restore and increase it. The two sources of allelic novelty are mutation and introgression.

It is conceivable that mutation rates are unusually high in Darwin's Finches, and higher in some species than in others, but nothing is actually known about mutation in Darwin's Finches. For a long time it was thought that hybridization, as a mechanism of introgression, was exceedingly rare. Lack (1947) made a special effort to search for evidence of hybridization but failed to find any, nor was it observed by previous workers in the field (e.g. Gifford 1919). Orr (1945) failed to induce interspecific matings in captivity, and the only successful interbreeding ever recorded in captivity, between a male *G. scandens* from Santa Cruz and a female *G. difficilis* from Wolf, ended in the death of the offspring two days after hatching (Bowman 1983). These failures undermined the interpretation of four strange specimens of tree finches as intergeneric hybrids (Stresemann 1936), an interpretation which nevertheless is supported by their morphological intermediacy between *Certhidea olivacea* and *Camarhynchus parvulus* in three cases, and between *C. olivacea* and *Cactospiza pallida* in the fourth (Grant et al. 1985; see also Bowman 1983, pp. 315–316). Lack (1947; also Bowman 1961) inferred that hybridization was possibly responsible for the intermediate or "freak" appearance of several specimens in museum collections (Table 10; see also Plate 56), but was forced to conclude "Clearly hybridization between species is rare, if not absent" (Lack 1947, p. 95).

Hybridization was only established unequivocally when the reproductive fates of banded birds were followed for several years (Grant and Price 1981, Grant and Grant 1982, Boag and Grant 1984b).

Hybridization. Each year two species of finches that do not regularly breed on Daphne Major, *G. fuliginosa* and *G. magnirostris*, immigrate after the breeding season (Grant et al. 1975). The source, identified on the basis of measurements of the birds and in a few cases by song (Chapter 9), is probably Santa Cruz. As many as 150 *G. fuliginosa* were estimated to be on Daphne in December 1973, for example, and the same number was again

PLATE 56. A possible hybrid (*G. fortis* × *G. scandens?*) on Santa Cruz. Note its resemblance to the large cactus finch, *Geospiza coniros-tris*, on Genovesa, in Plates 33 and 35 (*photo by M. P. Harris*).

TABLE 10. Hybrids. Shown above the diagonal are specimens of intermediate appearance between two species and therefore possibly hybrids. From Lack (1945, Table 16); in a later publication Lack (1947, Table XXXII) reduced the number from 65 to 26. Crosses below the diagonal show the pairs of species that are known, from field studies, to hybridize.

	A	B	C	D	E	F	G	H	I	J	K	L	M
Geospiza													
A *magnirostris*		39	—	—	—	—	—	—	—	—	—	—	—
B *fortis*			8	1	4	—	—	—	—	—	—	—	—
C *fuliginosa*	X	X		—	—	—	—	—	—	—	—	—	—
D *difficilis*					—	—	—	—	—	—	—	—	—
E *scandens*	X	X				—	—	—	—	—	—	—	—
F *conirostris*	X			X			—	—	—	—	—	—	—
Camarhynchus													
G *psittacula*								—	9	—	—	—	—
H *pauper*									—	—	—	—	—
I *parvulus*										—	—	—	3
Cactospiza													
J *pallida*											—	—	1
K *heliobates*												—	—
Platyspiza													
L *crassirostris*													—
Certhidea													
M *olivacea*													

present in August 1983. Usually there are fewer, and *G. magnirostris* numbers have never been above 50. Typically, numbers of both species decline sharply just before or at the beginning of the following breeding season. Although dead birds of both species have been found on the island we assume that most immigrants return to their island of origin. Very few stay to breed, and, surprisingly, those that do more often pair with a member of another species than with a member of their own.

Most *G. fuliginosa* pair with *G. fortis*, and some offspring have been successfully raised to independence (Table 11). One *G. fuliginosa* female even bred successfully with a male *G. magnirostris* in 1984! Seven instances of hybridization between resident *G. fortis* and *G. scandens* are known, with five giving rise to fledged young (e.g. see Plate 57). Thus interspecific gene flow is one form of introgression of genes into the *G. fortis* population, providing that the hybrids are recruited to the breeding population of *G. fortis* (see below).

TABLE 11. The occurrence and fate of hybridization of ground finch species on Daphne Major.

	Parents			Reproductive Success		
	♂	♀	Pairs	Eggs	Chicks	Fledglings
1976	fortis	fuliginosa	5	11	6	5
1977	—	—				
1978	fortis	fuliginosa	1	10	9	7
	fortis	scandens	1	5	0	0
1979	fortis	fuliginosa	1	6	3	2
1980	fortis	fuliginosa	1	3	0	0
	fortis	scandens	1	6	6	0
1981	fortis	fuliginosa	3	14	7	7
1982	—	—				
1983	fortis	fuliginosa	2	32	17	13
	fuliginosa	fortis	4	18	0	0
	scandens	fortis	2	31	16	12
	magnirostris	fortis	2	6	0	0
1984	fortis	fuliginosa	3	10	10	9
	fuliginosa	fortis	1	?	?	2
	scandens	fortis	2	7	4	4
	fortis	scandens	2	8	4	4
	magnirostris	fuliginosa	1	4	4	4

PLATE 57. Two known hybrids on Daphne Major. The parents are: Upper: *Geospiza fortis* and *Geospiza scandens*. Lower: *Geospiza fortis* and *Geospiza fuliginosa*. The upper one, whose mother was a cactus finch (*G. scandens*), has a longer beak than the lower one, whose mother was a small ground finch (*G. fuliginosa*).

Some calculations, taken from Grant and Price (1981), show how hybridization can elevate the variance of a trait such as a bill depth in the population of *G. fortis* on Daphne, given the observed frequency of interbreeding. The steps involved are as follows. Since *G. fuliginosa* usually pair with average *G. fortis* in terms of size (e.g. see Fig. 59), we assume a pairing between mean phenotypes (by assumption also the mean genotypes) of *G. fuliginosa* (6.95 mm) and *G. fortis* (9.39 mm). We calculated that a single hybrid offspring at the expected mean position of 8.17 mm will increase the genotypic variance in the *G. fortis* populations by 1.77% if the deviation of the hybrid individual from the *G. fortis* mean is entirely due to genetic factors. The increase in genetic variance is multiplied by the heritability (0.82) to give the increase in phenotypic variance, 1.45%, for every 100 *G. fortis* individuals. These calculations should be multiplied by a factor of about four, as the frequency of hybridization between these species has been about 4% over the period 1976–1984. The increase in phenotypic variance due to hybridization should be thus 5.8%; in contrast the effect of hybridization on the mean is negligible (~0.5%).

Two out of 55 male offspring surviving from 1978, their year of birth, until 1979 were hybrids. Their presence increased the variance in the juvenile male component of the population by 6.5%, which is in rough agreement with the calculation above. This correspondence provides empirical support for the theoretical arguments at the beginning of this chapter.

The demonstrated hybridization and morphological intermediacy of the hybrids (Fig. 59) raises the question of how many birds that we identify as *G. fortis* are really hybrids. This question has been addressed (Boag and Grant 1984b) by using multiple discriminant function analysis to find the best equation for separating four groups of birds: *G. fuliginosa* from B. Borrero on the north of Santa Cruz, *G. fortis* from the same locality (Fig. 59), *G. fortis* from Daphne, and the known hybrids from Daphne in 1976 and 1978. The equation was then used to classify all individuals in the Daphne population in 1976. The result was an assignment of 6.3% of birds identified by us as *G. fortis* to the category of hybrids. Since the group of known hybrids used in the analysis is so small, the estimate of the frequency of hybrids is not precise. It would be best to say that approximately 5% of the *G. fortis* population in 1976 could have been F_1 hybrids.

The final effect of hybridization is difficult to estimate because it depends, among other things, on the probability that hybrids will survive to breed, the number of genetic loci involved, and the breaking up of groups of genes (linkage disequilibrium). Survival probability is low, but so is it low for nonhybrid offspring. A bird thought to be a hybrid on the basis of measurements bred successfully in 1976, but it was not until 1983 that we first established the breeding of a hybrid. Altogether, of the 50 *fortis* × *fu-*

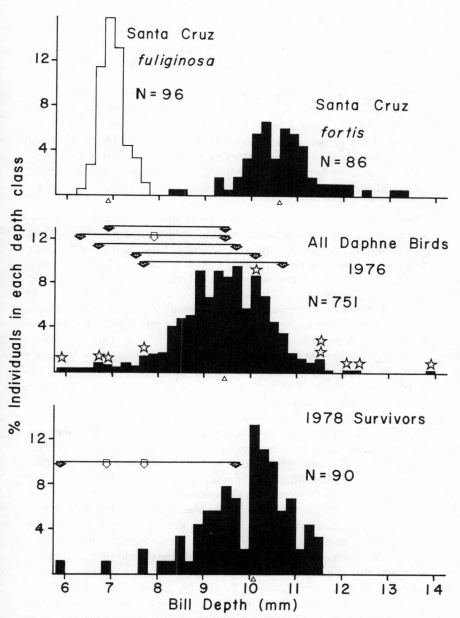

FIG. 59. Frequency distributions of beak depth at B. Borrero, Santa Cruz, and on Daphne Major. Small triangles below the horizontal axes indicate group means. The horizontal lines with terminal arrows represent matings between *G. fortis* and *G. fuliginosa*, and the open arrows indicate three measured offspring. The stars pinpoint 10 birds known to have immigrated to Daphne from Santa Cruz in May 1977. From Boag and Grant (1984b).

liginosa hybrids known to have fledged on Daphne, 6 have survived to breed. They have bred with each other (one case) or with *G. fortis*, and with the latter they have fledged young.

These facts demonstrate gene flow from an allopatric *G. fuliginosa* population into the population of *G. fortis* resident on Daphne. Sparse as the data are, they do not indicate any block to genetic exchange between the species in any particular combination of parents (Tables 11 and 12). And they do not indicate a major disadvantage to the hybrids; note that the long-term frequency of hybridization ($\sim$4%) is about the same as the frequency of hybrids in the population in 1976 ($\sim$5%).

Conspecific gene flow. Conspecific *G. fortis* and *G. scandens* also immigrate occasionally. Their measurements, and subtle features of plumage and bill coloration, suggest that the source is Santa Cruz, as it is for the other two species (Boag and Grant 1984b). *G. fortis* are larger on Santa Cruz than on Daphne, and *G. scandens* are smaller and have more needle-like bills on Santa Cruz than on Daphne. Immigration of conspecifics is probably more frequent than we have recorded, because morphological frequency distributions overlap, especially those of the two *G. fortis* populations, and not all immigrants on Daphne would be recognized as such. Only five immigrant *G. fortis* have been known to stay to breed, and they have bred successfully with local *G. fortis*, whereas no *G. scandens* immigrants have been known to breed. This is the second form of gene flow into the population of *G. fortis* on Daphne.

The total effect of gene flow on variation. Calculations of introgression rates by direct observation are hampered by their extremely low frequency. Presumably the effects of hybridization are weakly augmented by gene flow from the moderately different *G. fortis* populations on other islands. The augmentation is weak because the incidence of gene flow appears to be rare, and because the *G. fortis* populations are phenotypically similar. Nevertheless, the calculations are sufficient to show that the increase in genotypic variance due to hybridization alone is one to two orders of magnitude greater than that expected ($\leq 0.1\%$) from mutation (Lande 1976a). This justifies ignoring mutation, which has not been measured in Darwin's Finches anyway. It has been suggested by Turelli (1984, 1985) that mutation rates are generally too low to offset, by themselves, losses due to selection, and to account for observed levels of additive genetic variance maintained in natural populations. But mutation may not be a trivial factor if the finches are like *Drosophila* flies, for then gene flow into a population could increase mutation rates, through the unmasking of controls over mutator gene activity (Woodruff and Thompson 1980), and thereby contribute indirectly (as well as directly) to an increase in genetic variance.

TABLE 12. The breeding of hybrids and immigrant species of ground finches on Daphne Major in the years 1976–1984; see Table 11 for the breeding of immigrant species with resident species.

Parents			Reproductive Success		
♂	♀	Pairs	Eggs	Chicks	Fledglings
HYBRIDS					
1983 *fortis/fuliginosa*	*fortis/fuliginosa*	1	12	1	0
fortis/fuliginosa	*fortis*	1	6	0	0
fortis	*fortis/fuliginosa*	1	8	5	1
1984 *fortis/fuliginosa*	*fortis*	1	4	3	3
fortis	*fortis/fuliginosa*	1	3	3	3
IMMIGRANTS					
1981 *fuliginosa*	*fuliginosa*	1	4	4	4
1983 *fuliginosa*	*fuliginosa*	3	19	10	8
magnirostris	*magnirostris*	3	24	19	17
1984 *fuliginosa*	*fuliginosa*	2	6	3	3
magnirostris	*magnirostris*	2	5	5	3

One reason why *G. fortis* is more variable than *G. scandens* is that it receives a higher rate of genetic input from conspecifics and heterospecifics. It is possible too that the rare gene exchange between these two species is asymmetrical, with *G. fortis* receiving more than its fair share, although none of the twenty known hybrids has bred so far.

VARIATION IN RELATION TO ABUNDANCE

Fisher (1937) argued on theoretical grounds that small populations should be less variable than larger ones, but Lack (1947) could find no relation between variation in Darwin's Finches and population size as indexed by island size, and simple geographical trends in population variation are generally weak or absent (Grant et al. 1985). Furthermore, small species of ground finches are more abundant yet less variable than large species.

Nevertheless, variation in one species is sometimes correlated inversely with the abundance of another species. Four populations of *G. fortis* are more variable on islands where *G. magnirostris* is absent or rare than on four others where it occurs commonly (Grant 1967). This could be the result of occasional hybridization with *G. magnirostris* that are immigrants or residents, for if members of the rare species cannot find conspecific mates they may seek heterospecific ones (Chapter 9). The island where *G. fortis* is exceptionally variable is Santa Cruz. *G. magnirostris* is relatively rare, and it too is very variable. *G. fortis* is more variable on the south of the island, around Bahía Academía, than on the north at B. Borrero (Boag and Grant 1984b), and *G. magnirostris* is present on the south side but apparently does not breed on the north side (Abbott et al. 1977). G. L. Stebbins (in Bowman 1961) suggested the possibility of hybridization between the two species, and later Snow (1966) made the same suggestion. Another possible explanation for the large variation in *G. fortis* is immigration of conspecifics from one of the southern islands, and subsequent interbreeding (Grant et al. 1976). Yet another possibility is disruptive selection on the *G. fortis* population (Ford et al. 1973) in an ecologically diverse environment (Abbott et al. 1977). Field observations do not contradict any of these possibilities, but the populations on Santa Cruz need to be studied in the breeding season to see if interbreeding occurs.

Thus the inverse relationship between variation in one species and abundance of another is interpretable in terms of the reproductive and ecological factors that increase and decrease variation as modeled in Figure 53.

OTHER SPECIES

This chapter has focused on *G. fortis* because most is known about it, but no one species can be considered typical of Darwin's Finches as a whole. Other species differ, in both morphological and ecological variation. Three have been studied in detail, and I shall conclude the chapter by briefly outlining the features that are relevant to their morphological variation.

1. *Geospiza scandens*, a specialist. *G. scandens* is the only other resident species of finch on Daphne Major. As the name cactus finch implies, it is proficient at exploiting parts of *Opuntia* cactus for food: pollen, nectar and seeds, additional nectar from extra floral nectaries at the base of clusters of spines, and insect larvae and pupae in rotting pads (Plates 4, 12, 41, 50, 58, and 59). Its beak shape reflects these abilities, being proportionally much longer than the bill of *G. fortis* and thus more suitable for probing large flowers (Plates 40 and 41). The differences between the species are summarized as follows (from Boag and Grant 1984a, p. 484): "*G. fortis* is a small, opportunistic breeder, with a large clutch size, an occasionally large but variable population size, a broad diet and a seasonally varying pattern of territory occupation. *G. scandens* is larger, aggressively dominant to *fortis*, and occupies an included niche; it produces smaller clutches, breeds more regularly and with more uniform success, and as a result suffers less extreme population changes; and it exhibits highly specialized feeding behavior, particularly in times of food shortage, and has larger more tightly structured territories based on a reliable and defendable resource [*Opuntia*]

used all year long (see also Millington and Grant 1983)." *G. scandens* feeds on all of the food types exploited by *G. fortis* except the large and hard seeds of *Bursera* and *Tribulus*.

While *G. fortis* fits model 1b in the Van Valen scheme represented in Figure 54, *G. scandens* fits model 2. No between-phenotype component of niche width, in the terminology of Roughgarden (1972), has been detected in this species. Males and females feed in the same way on the same food types (Boag and Grant 1984b). *G. scandens* responds to a changing food supply in a different way from *G. fortis*, for as both the number of available seed species and flower abundance decline the diet of *G. scandens* narrows while the diet of *G. fortis* broadens (Fig. 52).

As expected for a model 2 species, variances of bill traits are low in this species, and much lower than variances of *G. fortis* (Grant and Price 1981; also Table 7). There is heritable variation in the bill dimensions of *G. scandens*, but apparently less than in *G. fortis* (Boag 1983, Price, Grant, and Boag 1984; also Table 8). Precise estimation of heritabilities is hampered by a larger measurement error associated with a lower phenotypic variance.

Corresponding to its different feeding niche, *G. scandens* is subject to different regimes of selection. Like *G. fortis*, the population suffered heavy mortality (66%) in the drought of 1977, but whereas selection was directional on *G. fortis* it was stabilizing on *G. scandens*. Trait means of each sex remained the same but variances decreased, significantly so for some traits in both sexes (Grant and Price 1981, Boag and Grant 1984b). Figure 60 shows the contrast between the effects on the two species in terms of fitness functions. An earlier episode of selection, which was also stabilizing but weaker, occurred in 1974 (Grant et al. 1976).

After the drought in 1977 the same opportunity for sexual selection arose, because males survived better than females and by the same factor as in *G. fortis* (6:1). This observation raises the possibility, incidentally, that differential survival of the sexes in *G. fortis* may not be attributable solely to selection for large size, since *G. scandens* was not subjected to selection for large size; perhaps some females of both species emigrated when feeding conditions deteriorated. Despite the opportunity for strong sexual selection to occur in *G. scandens*, it occurred only weakly. The males which acquired mates in 1979 were significantly larger, as measured by wing length, than those which did not (Price 1984b). There was no directional sexual selection in 1980 (as in *G. fortis*). In 1981 the males which acquired mates were significantly larger in bill depth and bill width than those which did not. These results were significant only at the 5% level. There is an additional but weak indication of stabilizing sexual selection on bill length of males ($P = 0.06$, two-tailed Levene test). These slight tendencies for large males to be favored under sexual selection contrast with strong tendencies in *G. fortis*.

Individuals of *G. scandens*, like *G. fortis*, feed extensively on small

PLATE 58. Cactus finch, *Geospiza scandens*, cracking a seed of *Opuntia echios*. More than a year is usually required to develop maximal efficiency at cracking seeds. Santa Cruz (photo by C. P. Hickman, Jr.).

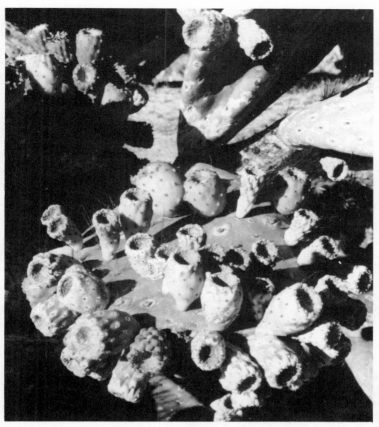

Left: Large cactus finch, *G. conirostris*, cracking a seed of *Opuntia helleri* on Genovesa (see Plate 59). Right: A variety of fruits on the pad of *Opuntia helleri*, on Genovesa. Several are shriveled and infertile, as a result of cactus finch damage to the stigmas of the flowers (see Plate 50).

PLATE 59. Effects of large cactus finch, *Geospiza conirostris*, feeding on *Opuntia helleri*. Genovesa. Upper: Excavated fruit. When the seeds are soft, cactus finches remove them from the fruit, eat the surrounding arils, and discard them, sometimes wiping them off their bills on to the pad. Later, when they harden, the seeds are cracked open by large cactus finches, but more frequently by large ground finches, *Geospiza magnirostris*. Lower: Partly excavated rotten pad beneath bush. Fly and beetle larvae and pupae are extracted and consumed by cactus finches.

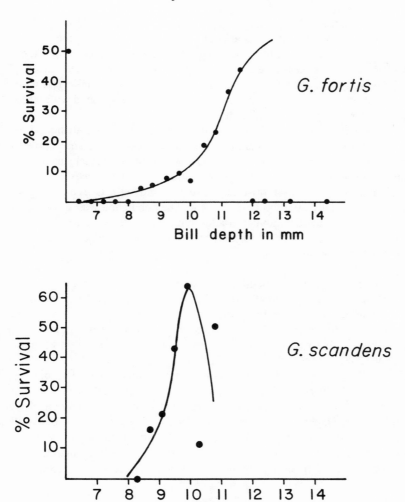

FIG. 60. Fitness functions for the two species of ground finches on Daphne Major in terms of survival from 1976 to 1978. Curves are drawn by eye. *G. fuliginosa*, represented by one survivor out of two in its size class, and immigrant *G. fortis* and *G. magnirostris* with beak depths of 12.0 mm or greater, are also shown in the *G. fortis* diagram. From Boag and Grant (1984b).

seeds in their first year. They suffer heavier mortality than *G. fortis*, perhaps because they are larger (Boag and Grant 1984a). The mortality does not appear to be size selective, but with lower variances and higher proportional contributions of measurement error, the task of detecting small selection differentials if they exist is that much more difficult in the (smaller) *G. scandens* samples. This may be the reason why selection associated with reproduction has not been detected in this species either.

A summary of the forces of selection acting on frequency distributions of *G. scandens* is shown in Figure 61. The largest single difference between this figure and the comparable one for *G. fortis* (Fig. 57) is that survival selection is stabilizing on *G. scandens* and directional on *G. fortis*.

In addition to selection, drift may have occurred in the *G. scandens* population, especially after the 1977 drought. The population was at its lowest in early 1979 when the number of adult females had fallen to 23. The population did not remain small for long, however, and in the next three years the number of breeding females doubled.

2. *G. conirostris*, an intermediate species. This species occurs on two main islands lacking resident populations of *G. fortis* and *G. scandens*: Genovesa and Española. Morphological variation is substantial in both populations (Table 7, Plate 33, Fig. 62). Fieldwork on Genovesa throws light on the question of why this population displays a high level of heritable variation (Grant 1983a, Grant and Grant 1983, B. R. Grant 1985).

In bill shape and body size *G. conirostris* on this island resembles *G. scandens* elsewhere, but it is larger, and has a proportionally deeper bill, in which respect it resembles *G. fortis*. The diet is similar to the diet of *G. scandens* on Daphne (Grant and Grant 1982). *G. conirostris* exploits all parts of *Opuntia*, and in the dry season this is the major feeding activity, but unlike *G. scandens* it feeds on *Bursera* seeds after cracking the hard "stones" (Plate 10), as *G. fortis* does on Daphne, and it strips the bark from

FIG. 61. Opposing selective forces on distributions of male and female sizes in the population of cactus finches (*G. scandens*) on Daphne Major. Stabilizing selection appears to be more prevalent than it is on *G. fortis* (see Figs. 57 and 60).

FIG. 62. Variation in beak size and shape among members of the population of large cactus finches, *G. conirostris*, on Genovesa. Drawn from photographs.

dead *Bursera* branches (Plate 60) to feed on cryptic arthropods, as tree finches (*Camarhynchus* and *Cactospiza*) do on other islands (e.g. see Plate 30). Thus the diet is less restricted than is the diet of *G. scandens* and about as broad as the diet of *G. fortis* (Grant and Grant 1982).

Part of the reason for the high level of variation is hybridization with the two sympatric species on the island, *G. magnirostris* and *G. difficilis*. In the period 1978–1984 we recorded one *G. conirostris* × *G. magnirostris* pair and one *G. conirostris* × *G. difficilis* pair (males given first). Offspring fledged from the pair involving a *G. magnirostris* parent but not from the other. Two other pairings involved possible hybrid females as their dimensions were close to the midpoint between averages for *G. conirostris* and *G. difficilis*. Both of these pairs successfully raised offspring to independence, and one of the offspring bred successfully with a *G. conirostris*. Finally, two other pairings involved possible hybrid males, one who sang a *G. magnirostris* song and was paired with a *G. magnirostris* female, and the other who sang a *G. conirostris* song and was paired with a *G. conirostris* female. Both pairs bred successfully. Collectively these observations suggest infrequent gene flow into the *G. conirostris* population from both sympatric congeners. Gene exchange with *G. difficilis* could be predominantly unidirectional, since *G. difficilis* varies relatively little (Table 7), whereas mutual exchange with *G. magnirostris* is suggested by the high level of variation

PLATE 60. Bark of *Bursera graveolens* stripped by a large cactus finch, *Geospiza conirostris*, while searching for insects and termites. Genovesa. Upper: General appearance of branch and trunk, showing extent of activity. Lower: Close-up, showing tunnels and grooves in the wood.

in this species. Field data to test these possibilities are still lacking, even after seven years of intensive study.

There is no evidence for the other potential sources of alien genes, namely immigrants. One *G. scandens* and eight *G. fortis* were observed and/or captured on the island in the period 1978–1984, but none were known to stay to breed. No immigrant *G. conirostris* from Española or Gardner (Fig. 1) has ever been identified on Genovesa, even though they probably could be recognized by their unusual beak shapes.

The forces of selection on this population have been less easy to study than those on the congeneric populations on Daphne because sample sizes have been smaller, the island is much larger, and the study only began in 1978, after the 1977 drought. If the population size in 1976 was as large as in 1982, which seems reasonable in view of the generally favorable conditions prevailing in the archipelago in 1972–1976, then the population suffered about 70% mortality in the drought of 1977 (B. R. Grant 1985). If the morphological characteristics of the population were the same in 1976 and 1982, then the population was subjected to selection, not directional or stabilizing but disruptive selection. This deduction follows from the observation that variances were greater in the post-drought (1978) sample than in the 1982 sample. For bill depth of females the difference was statistically significant (B. R. Grant 1985).

There are two other indications of disruptive selection. First, variances of bill dimensions of young birds in their first dry season increased between ages 3–4 months, when approximately fully grown, and about 12 months, by as much as a factor of three (B. R. Grant 1985). The constancy of means and the (nonsignificant) increases in variances reflect a tendency for disruptive selection to occur. Second, disruptive selection has been inferred from the significant difference in bill length in 1978 between the two song groups of males, those singing song type *A* and those singing song type *B* (song variation is discussed in the next chapter).

In each of these cases a link with feeding ecology can be established. There are four feeding modes in the dry season, and they appear to be different enough to set up different selection pressures on bill size and shape among individuals that tend to specialize. The four activities are (1) barkstripping to obtain arthropods (Plate 60), (2) cracking the seeds of *Opuntia helleri* (Plate 58), (3) extracting seeds from ripe *Opuntia* fruits to obtain the surrounding arils (Plate 59), and (4) tearing open rotting *Opuntia* pads to obtain arthropods (Plate 59). Birds that stripped bark had significantly deeper beaks than those which did not. Birds that have been observed to crack the moderately large and hard seeds of *Opuntia* had significantly larger beaks than those not observed to do so. Birds observed to open *Opuntia* fruits had significantly longer bills than those observed to feed on *Opun-*

tia arils at already opened fruits but never observed to open the fruits themselves (B. R. Grant 1985). This last result is germane to the apparent disruptive selection on the male component of the population in 1977. In the dry season of 1978 only the song A males, with the longer average beak, were observed opening *Opuntia* fruits (Grant and Grant 1979). Unfortunately it is not known how or why long bills predominated among the *A* males and short bills predominated among the *B* males, since the study only began in 1978; this intriguing difference is discussed further in Chapter 10.

Whatever the explanation for causes, these results show that *G. conirostris* conforms to model 1b in Figure 54. The effects of disruptive selection are to maintain a broad variation in a population when the variants exploit either partly or completely different niches or habitat patches, and not to erode the variation. In this respect, as well as in its feeding ecology, it appears to be different from *G. fortis* and *G. scandens* on Daphne. How might the limits to variation be set? Stabilizing selection during a stressful season is a possible answer but it has not been observed, nor has sexual selection on bill dimensions. Limits to variation may be set by fluctuating directional selection, but the only indications of directional selection have been in reproductive performance. Among males, those who gained a territory (for the first time) were larger, and had longer and deeper beaks, than those who failed to do so. Among females, long-billed birds lived the longest and produced the most clutches and the most offspring that later bred (B. R. Grant 1985). Their advantage derived from breeding earlier in the season than the others (Grant and Grant 1983). A possible cause of their earlier breeding is their ability to feed on the nectar and pollen in *Opuntia* flowers without being aggressively displaced, and as a result being in a greater state of readiness to breed at the onset of rain. Exploitation of these resources has been implicated as a factor responsible for the earlier breeding of *G. conirostris* than of *G. magnirostris* and *G. difficilis* on Genovesa (Grant and Grant 1980a), and for the usually earlier breeding of *G. scandens* than of *G. fortis* on Daphne (Boag and Grant 1984a).

In summary, the maintenance of a diversity of bill sizes and shapes in this population is due to a set of factors similar, but not identical, to those operating on the *G. fortis* population on Daphne. Variation is enhanced by a low frequency of hybridization. It is depleted at a slow rate by different types of selection: selection acting in opposite directions, and a tendency for disruptive selection to occur under dry conditions. Selection acts in opposite directions on positively correlated bill dimensions, on the same dimension at different stages of the life cycle, and on the same dimension in the two sexes (B. R. Grant 1985).

3. *Pinaroloxias inornata*, an unvarying generalist. This is the sole Darwin's Finch species on the very isolated Cocos Island (Fig. 1). Here, in the ab-

sence of introgression of any sort, population variation is minimal (Table 7). The niche, in constrast, is extremely broad (Slud 1967, Smith and Sweatman 1976). Finches search in different parts of the tropical rain forest from the ground to the canopy on leaves, flowers, seed heads, and beneath bark, for arthropods, nectar, small seeds, and fruit (Plate 26). These are elements of the feeding repertoires of *Geospiza difficilis, G. fuliginosa, Certhidea olivacea*, and tree finches on the Galápagos. Detailed observations of banded individuals by T. W. Sherry and T. K. Werner (pers. comm.) have revealed that the broad niche of the population is the product of individuals specializing on different feeding tasks. This species has no single counterpart on the Galápagos in the degree to which the diet of the population is broad and the diets of individuals are narrow and consistently different from each other.

Feeding characteristics are too varied in type to be represented on a single resource axis as in Figure 54. Nevertheless, from the feeding observations made by T. W. Sherry and T. K. Werner it would appear at first glance that the species conforms to model 1b in Figure 54; yet the population varies very little. Individuals are behavioral, but not morphological, specialists; they are not, as in 1b, restricted to exploiting a segment of the population's resource spectrum for morphological reasons. Every individual's beak can be used, apparently, for all of the feeding tasks displayed by the population as a whole, even though it may not be best suited for any one of the tasks, or more than one or two of the tasks, as was described for *G. difficilis* on p. 131. The species therefore conforms more to model 1a, with behavioral specialization and dietary restriction apparently, but not necessarily, under environmental control (e.g. see Van Valen 1965). Why the beak of a generalist is actually used in a specialized manner is not clear. Specialization is presumably fostered by a year-round supply of a large variety of food types in different microhabitats, and has possibly been favored by natural selection under conditions of intraspecific competition for food. It would be interesting to know whether feeding specializations run in families or not, how they develop as the individuals mature, whether different food availabilities influence the choice of a foraging specialization, and whether these availabilities vary very much, predictably or unpredictably.

In terms of the theoretical scheme in Figure 53 it appears that the low morphological variation is more attributable to the absence of genetic input through introgression than to intense stabilizing selection.

SUMMARY

This chapter raises the question of why some populations of Darwin's Finches are so variable, and provides a theoretical framework for answering it. Variation in a trait such as beak depth is determined by the resolution or balance of two opposing sets of forces. Introgression of genes from conspecific and heterospecific populations, together with mutation, tends to increase variation, and the forces of selection, and possibly random drift, deplete it. Stabilizing selection is relatively weak on populations made up of individuals that specialize to some extent on parts of a broad spectrum of food resources for morphological reasons. Some populations vary more than others because they differ in the point of balance between introgression and selection.

Results of field studies of ground finch populations are consistent with this framework in demonstrating the presence of a large amount of additive genetic variance (high heritabilities) for bill and body size traits, the repeated occurrence of natural and sexual selection, and interbreeding between species and between immigrant and resident members of populations of the same species.

Heritabilities, estimated by the offspring-midparent regression technique, are uniformly high for six traits in populations of *G. fortis* and *G. conirostris*, and a little lower in a population of *G. scandens*. On Daphne Major, the resident *G. fortis* hybridizes with resident *G. scandens* and immigrant *G. fuliginosa*. Although hybridization is rare, it has important effects on population variation because the hybrids are viable and back-cross to *G. fortis*. *G. fortis* and *G. scandens* individuals also immigrate from elsewhere, apparently rarely, and their effects on the variation of the two resident populations seem to be relatively small. Thus a high level of genetic variation is present in the population of *G. fortis* because it receives new alleles each generation from heterospecific sources, and because selection on this variation, although occasionally intense, acts in different directions on different age and sex groups, and to different degrees in this temporally varying and moderately heterogeneous environment.

The forces of both natural and sexual selection operate repeatedly on observed variation. During a drought on Daphne Major, *G. fortis* was subject to exceptionally intense directional selection favoring individuals large in body size and bill depth. The largest individuals survived best because only they could crack and deal efficiently with the large and hard fruits and seeds remaining in relative abundance after the large stock of small and soft seeds had been depleted. The effects of natural selection were augmented by sexual selection; females chose as mates the largest among the available males at a time when the sex ratio was heavily male-biased. The net result of natural and sexual selection was a small evolutionary increase in the mean size of all traits in the *G. fortis* population in the next generation. Similar but weaker directional selection occurred in two other years. Selection in the opposite direction occurs at two points in the life cycle, in some years only. During their first year of life small birds survive better than large birds, per-

haps because they need less energy to maintain themselves, and small females breed at an earlier age than large females. Integrated over a whole life
cycle these effects of selection in opposite directions may roughly balance
and be equivalent to weak, overall, stabilizing selection. In contrast, *G.
scandens*, a feeding specialist that exhibits lower variation in all dimensions, is subject more to stabilizing and less to directional selection, and hybridizes less often. The variable *G. conirostris* on Genovesa is an intermediate species in feeding ecology. It may occasionally be subject to
disruptive selection.

The Cocos finch, *Pinaroloxias inornata*, is apparently unique in displaying strong, individual, feeding specializations that are unrelated to bill size.
They are presumably fostered by a year-round supply of a large variety of
food types, and may have been favored by natural selection under conditions of intraspecific competition for food. Behavioral specializations are
less important on the Galápagos possibly because finches there experience
a lower variety of food types and microhabitats at any one time, and less
constancy in the diversity of foods.

CHAPTER NINE

Species-Recognition and Mate Choice

INTRODUCTION

Lack (1940a, 1940b, 1945) drew attention to the difficulty of understanding how species of Darwin's Finches remain distinct. The problem, which was recognized more than a hundred years ago (Salvin 1876), is that morphological ranges of sympatric congeneric species abut, so that, for example, the largest *G. fortis* on Santa Cruz is more similar to the smallest *G. magnirostris* individual on the same island than it is to a typical (average) member of its own population (e.g. see Fig. 15). The same problem exists for *G. fortis* and *G. fuliginosa*, which are another pair of size-neighbors. An extra twist to the problem is given by geographical variation. For example, on the island of Rábida where *G. magnirostris* is at its smallest average size, the small members of the population are almost indistinguishable in visible traits from the largest members of *G. fortis* on another nearby island (Isabela) where the species reaches its maximum size.

If discrimination is difficult for finches, it is worse for us! A list of 85 museum specimens of questionable identity (Table 16 in Lack 1945) testifies to the unusual difficulties faced by taxonomists wishing to classify Darwin's Finches (see also Table 10). So bewildering was the diversity of species and subspecies recognized fifty years ago (37), and the intergradations among them, that Swarth (1934) went so far as to suggest that all forms of *Geospiza* to *Certhidea* might almost be classed as one species. This cry of desperation was used by Lowe (1936) to substantiate a theory that most species owed their existence to hybridization, and that interbreeding continued rampantly in the absence of ecological differences among the species. It required a strong counterargument from Stresemann (1936) and a breeding study by Lack (1945) to curb this incorrect speculation. The breeding study sought evidence for hybridization on San Cristóbal and Santa Cruz (Lack 1945), as well as in captivity (Orr 1945), and failed to find any. This brought the problem of species-recognition into sharp focus. It is the subject of this chapter.

THE POSSIBLE CUES USED IN SPECIES-RECOGNITION

The following brief description of courtship and mating will help us to identify the cues used by finches in recognizing members of their own species;

it is based upon the observations of Lack (1945), Orr (1945), and Ratcliffe and Grant (1983a).

Males stay on territory, sing frequently, build several display nests and court conspecific females that enter the territory, while females visit several territories and are correspondingly courted by several males. Females vocalize but do not sing. Males initiate courtship, actively pursuing and displaying to females. If a female appears in the vicinity of one of his display nests he flies towards her, singing both in flight and when perched, and often carrying nest material in his bill at the same time. The wings are drooped and quiver, and he sways from side to side, between repeated fluttering flights from one perch to another. He then flies to the nearest nest, which may even belong to another species, and attempts to attract the female by repeated enterings and exitings. Another feature of courtship is the sex chase, which is a weaving and undulating flight usually initiated by the male, with the male usually but not always following the female. This chase may be an assessment of the size of the territory by the female, or an attempt by the male to keep the female on territory, or else is some expression of his vigor, or even a test of hers. Most courtship sequences end with the departure of the female, at any stage during the procedure, and resume with the next entry of a female on the territory. Eventually these various behaviors culminate in pair-formation and mating. The female adopts a horizontal posture, lifts her head and remains still, and the male mounts and copulates with her.

Finches clearly discriminate between conspecific and heterospecific members of the opposite sex because hybridization is so rare (Chapter 8), and the courtship of females by males, described above, is selective. For example, during observations of well over 500 courtship sequences on Daphne Major, Plaza Sur, Pinta, and Santa Cruz, Ratcliffe (1981) observed *Geospiza* males approach and begin to court females of another species on only 26 occasions (i.e. $< 5\%$). This is far fewer than would be expected by chance, given the frequency of occurrence of *Geospiza* species on those islands (Abbott et al. 1977, Schluter 1982b), the overlapping territories of the different species, and their almost simultaneous pairing and breeding (Grant and Grant 1980a, Boag and Grant 1984a) which is entrained by the first heavy rains following a long dry season.

Probably both males and females exercise discrimination. The males do so first when deciding whether to initiate courtship or not. After courtship initiation either sex may cause it to cease, although usually it is the female that does so. It can be argued that both sexes should be discriminating because they are typically monogamous, and polygamous exceedingly rarely, and each member of a pair contributes extensively to parental care.

How, then, do they discriminate? It would be easy for tree finches to exclude ground finches as potential mates, and vice versa, just by observing their plumage. But within each group, plumage is too similar to convey information about species identity, and some other cue or cues must be used. The description of courtship above suggests three possible sets of cues. Members of either sex, but perhaps especially the females, could respond to diagnostic postures and movements employed by members of the opposite sex in courtship. The morphological features by which we distinguish species, namely, the size and shape of the beak and body size, could be used. And females could use male song as a cue to species identity.

Lack (1945, 1947) was the first person to observe the finches with the mate-choice and species-recognition question in mind. He observed that aggressive and courtship behaviors of many of the species appeared to be identical. Likewise Orr (1945) could find no differences in the postures or displays of male *G. magnirostris*, *G. fortis*, *G. fuliginosa*, and *G. scandens* in captivity. If differences do exist they must be subtle, and probably quantitative, for they have not been detected by subsequent observers either. This reduces the list of potential cues from three to two: morphological features and song. I will consider each in turn.

MORPHOLOGICAL CUES

Lack (1945, 1947) gave special attention to beak size and shape as cues to the identity of potential mates. This traces back to Snodgrass's (1902) failure to find a correlation among species between beak size and diet; "If this is true," wrote Snodgrass (1902, p. 381), "then we must look elsewhere for an explanation of the variation in the *Geospiza* bill." Lack found an explanation in a reproductive context; bill differences enable the species to avoid confusion and choose conspecific mates. He repeatedly observed that resident males flew to the front of an intruder, and then ignored it if it was a member of another species. "It was observed so often that it became clear that in Darwin's finches the beak is the chief character used in species recognition" (Lack 1947, p. 53).

Preliminary experiments. Lack (1945) tested the discriminating ability of finches with experiments involving stuffed female specimens as models. Models were placed near the nests of 14 wild *G. fuliginosa* males on Santa Cruz. First a *G. fuliginosa* model was presented, then a *G. fortis* model was substituted for the *G. fuliginosa* model, and finally this was replaced by the original *G. fuliginosa* model. The behavior of the nest-owner was noted at each presentation. Seven males responded generally strongly and sexually to the first model (*G. fuliginosa*), more weakly to the *G. fortis* model, and then somewhat more strongly to the *G. fuliginosa* model when presented again. The specimens were stuffed in such a way that their body sizes were similar, so they differed only or predominantly in beak size and shape. The results are consistent with the idea developed by Lack (1945) that *G. fuliginosa* males can distinguish between females of *G. fuliginosa* and *G. fortis* solely on the basis of the appearance of the bill. Unfortunately, the males varied greatly in their responses, and a statistical analysis of the data failed to show significant discrimination (Ratcliffe 1981, Ratcliffe and Grant 1983a).

Orr (1945) performed similar experiments, but with captive birds. Four individuals or pairs of *G. scandens* were tested with specimens of conspecifics, with congeneric heterospecifics, and with two North American species as controls. Again the results were consistent with a discrimination hypothesis but were inadequate for a proper test. Male *G. scandens* displayed to, and attempted to copulate with, the *G. scandens* model at first presentation in each case. Heterospecific models presented subsequently elicited the same or weaker responses (*G. fortis* and *G. magnirostris* models), or no responses (a *G. fuliginosa* model). The results are difficult to interpret however, because even the two control models elicited attempted copulations.

Definitive experiments. The experiments of Lack and Orr suffered from inadequate sample sizes and from a large variation in the responses of birds that was probably caused, or at least enhanced, by repeated testing of the same individuals. These procedural defects were remedied in an extensive series of experiments carried out on Daphne Major, Plaza Sur, and Pinta with *G. fortis*, *G. fuliginosa*, *G. difficilis*, and *G. scandens* (Ratcliffe 1981, Ratcliffe and Grant 1983a). Two stuffed specimens were set up one meter apart and about five meters from a nest, or in the center of a territory if a nest was lacking. The models were attached either to *Opuntia* cactus pads or to a stick lashed horizontally to a portable tripod. The simultaneous presentation of two models was designed to give responding birds the opportunity to reveal a discrimination, if possessed, through a difference in the attention given to the models. The models were typical representatives of their populations.

In tests of the ability of males to discriminate, conspecific and heterospecific female models were positioned in a copulation-soliciting posture: the bill subtended an angle of about 45 degrees to the horizontal, and wings and tail were spread slightly. This is so strong a stimulus that in preliminary trials a male would swoop down on to a model before we had positioned it, forcing us to cover the models with cloth before starting an experiment.

Results of the experiments were very clear (Table 13). A characteristic sequence of behavior is illustrated in Figure 63. Males showed a strong discrimination by directing more sexual attention to the conspecific model than to the heterospecific one (e.g. see Plate 61). In separate tests on Daphne Ma-

TABLE 13. Discrimination by *Geospiza* males between local female models of the same and of a sympatric species. Statistical significance of the difference in responses is indicated by one ($P < 0.05$), two ($P < 0.01$), or three ($P < 0.005$) asterisks. (After Ratcliffe and Grant 1983a.)

Species Tested	Number of Tests	Models	Mean responses (± SE)			
			Time at Model	Number of Displays	Number of Mountings	Number of Copulations
DAPHNE						
G. scandens	10	*scandens*	43.4 ± 9.2	2.1 ± 0.9	4.2 ± 0.9	1.7 ± 0.6
		fortis	22.3 ± 6.6	1.3 ± 0.6	1.9 ± 0.5	0.7 ± 0.3
			*	*	**	
G. fortis	13	*fortis*	31.3 ± 7.3	1.6 ± 0.5	2.8 ± 0.7	1.2 ± 0.3
		scandens	14.1 ± 3.7	0.2 ± 0.2	1.4 ± 0.3	0.7 ± 0.1
			*	**	*	
PINTA						
G. fuliginosa	21	*fuliginosa*	44.8 ± 9.3	1.6 ± 0.3	3.8 ± 0.6	2.5 ± 0.5
		difficilis	17.3 ± 4.9	0.5 ± 0.2	1.5 ± 0.4	0.8 ± 0.2
			***	***	***	***
G. difficilis	14	*difficilis*	31.1 ± 8.6	1.4 ± 0.5	3.9 ± 0.3	2.1 ± 0.4
		fuliginosa	10.0 ± 3.6	0.5 ± 0.3	0.9 ± 0.4	0.4 ± 0.2
			**	**	***	***

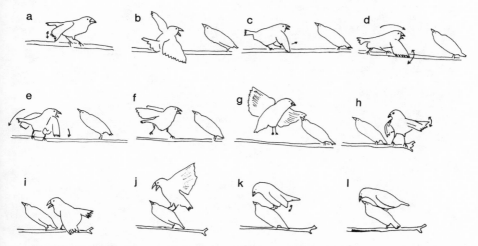

FIG. 63. Characteristic sequence of behavior shown by a male ground finch in species discrimination experiments. The responding bird lands on a rod (*a*) that supports two finch models, and orients towards one of them (*b*). After a series of wing-vibrating displays (*c–e*), it flies over the model (*f–h*), mounts it, copulates and/or pecks it (*i–l*). Drawn from super-8 film of *G. difficilis* taken by the author on Genovesa. From Ratcliffe (1981).

jor, *G. fortis* and *G. scandens* males visited both *G. fortis* and *G. scandens* models but spent longer at the conspecific model, displayed more frequently to it, and mounted and copulated with it more often. On Pinta, in a parallel set of experiments, *G. fuliginosa* and *G. difficilis* males showed a similar and strong discrimination (Table 13).

In many of the tests, responses to the heterospecific model appeared to result from a carryover of behavior initially directed towards the conspecific model. The displays directed towards the heterospecific model often collapsed in mid-performance, as noted also by Lack (1945) in his experiments.

Reciprocal experiments, which test the ability of females to discriminate sexually between conspecific and heterospecific models in male plumage, are difficult to perform because their mates respond to the models aggressively (Ratcliffe and Grant 1983a). In doing so, however, the males discriminate! On Daphne Major, *G. scandens* males ($N = 22$) approached a male *G. scandens* model more often than a *G. fortis* model, spent more time at it, and delivered more pecks to it ($P < 0.05$ in each case). On Plaza Sur, *G. fortis* males ($N = 8$) showed similar discrimination between *G. fortis* and *G. fuliginosa* models in male plumage.

Despite interference from their males, some females responded to the models in the last two sets of experiments. The data are inadequate for sep-

PLATE 61. Male ground finches courting conspecific female models in species-discrimination experiments. Pinta. Upper: *Geospiza fuliginosa*. Lower: *Geospiza difficilis* (*photos by P. T. Boag*).

arate analysis, but when combined ($N = 13$) the responding females showed a statistically significant discrimination similar to that shown by the males. Responses were aggressive, not sexual. The models were in a sitting posture, with wings closed and bill horizontal.

With hindsight it is easy to prescribe the experiments that should have been done to test for female sexual discrimination; remove the female's mate and position the male model in courtship posture with wings and tail spread, preferably swaying from side to side. The importance of the posture of the model is shown by the results of a set of experiments on Daphne Major in which male discrimination was tested with female models in the same posture as that of the male models, namely, sitting upright, instead of in the copulation-soliciting posture. In separate tests, both *G. scandens* and *G. fortis* males showed discrimination, as in the tests with models in soliciting posture, but it was much weaker.

Thus females discriminated, but the relevance to sexual behavior is questionable since the responses were aggressive. The same discrimination was also shown in the first set of experiments described, those in which male discrimination was the object of the test and the female models were in soliciting posture. In some of these experiments the male's mate responded as well, though not necessarily at the same time and in his presence. Sample sizes for each species were too small to be treated separately, so for each island they were combined. On Daphne Major and Pinta, as well as in a separate series of experiments on Plaza Sur, females of the four species tested attacked the conspecific female model significantly more frequently than the heterospecific model ($P < 0.05$ in each case).

In summary, experiments have shown that males of the four *Geospiza* species tested make a sexual discrimination between female models on the basis of visual cues alone; they make an aggressive discrimination between male models; and females make an aggressive discrimination between male models and between female models. In all cases the bias of the discrimination is towards conspecifics.

Bill size or body size as the visual cue. In all these experiments the specimens were stuffed to resemble the natural body size of the bird when alive, unlike in Lack's experiments. In an attempt to see if the bill and head, or the body, provide the most important cues used in discrimination, Laurene Ratcliffe performed a final set of experiments on Daphne Major with the heads of male models removed and reattached with glue (controls) or exchanged with others and glued (experimentals). The series of experiments was incomplete, owing to shortage of time, but the results show that *G. scandens* respond most strongly when both head and body are "correct." The conspecific head (*G. scandens*) on a heterospecific body (*G. fortis*) elicited significantly weaker responses than did the conspecific head on a conspecific body (Ratcliffe and Grant 1983a). Other tests, using a pair of experimental models, or one control and one experimental, elicited no discrimination, nor did a parallel set of tests with G. *fuliginosa* and G. *difficilis* on Pinta (Ratcliffe and Grant 1983a). Since the experimental models are highly artificial, one can only conclude that there is no evidence from these experiments for the primacy of one part in species discrimination.

On Pinta, G. *difficilis* differ from all G. *fuliginosa* in both bill shape and body size. In contrast, frequency distributions of the body weights of G. *scandens* and G. *fortis* on Daphne Major overlap to a small degree, whereas bill length distributions do not, which suggests that on this island bill size (or shape) is more important than body size. On Santa Cruz the frequency distributions of body sizes of these two species overlap even more, whereas beak frequency distributions are more nearly distinct (Fig. 16). Bill size and shape are unambiguous signals of species identity whereas body size is not always so.

SONG

Both Lack (1945, 1947) and Orr (1945) dismissed song as being important in species recognition on the grounds that there is too much variation among individuals of the same species, and too much similarity between individuals of different species, for it to provide an unambiguous signal of species identity. With the benefit of sonagraphic display of tape-recorded songs of all fourteen species, Bowman (1979, 1983) reached the opposite conclusion.

I shall discuss song under three headings: a qualitative description of song, its variation, and the contexts in which it is produced; a quantitative description; and experiments designed to test the ability of ground finches to discriminate between the songs of conspecifics and heterospecifics.

Qualitative description. Bowman (1979, 1983) recorded songs of most populations of all fourteen species, and demonstrated a very great variation in song structure. To reduce this complexity to order, he adopted the approach of a comparative anatomist. He classified song components into basic types, sought homologies, and then reconstructed transformations from one pattern to another on the basis of structural similarities and identified intermediates, i.e. gains, losses, and alterations of components in the transformation of one song into another.

There are two classes of song: a whistle or "hiss" and a territory-advertising song. The whistle is usually sung in or close to the nest, and it appears to function in courtship and pair-bonding. It has a very high frequency that

WHISTLE SONG

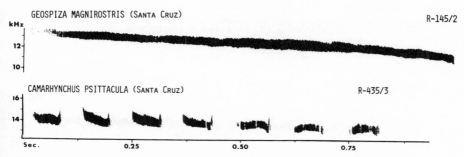

FIG. 64. Whistle songs of *G. magnirostris* and *Camarhynchus psittacula*. From Bowman (1983).

descends, and is produced as a single note or, in the case of *Camarhynchus* species (only), as a repeated note (Fig. 64). It is unlikely to play a role in species recognition because sympatric species produce whistles of very similar frequencies (Bowman 1983). However, whistles do vary somewhat in length among species, and in pattern among the tree finches (Bowman 1983), so whether or not finches discriminate between conspecific and heterospecific whistles needs to be established with experiments.

In contrast, the advertising songs vary greatly. Bowman (1979, 1983) identified five temporal regions in the advertising songs of all fourteen species, and used them to classify songs into two types, basic and derived (Fig. 65). The mono- or disyllabic basic song is so called because all species possess it. The polysyllabic derived song, implicitly a phylogenetically derived song, has been recorded from all species except *Platyspiza crassirostris* and *Pinaroloxias inornata*. Not all populations of the remaining species possess both song types. Upon these two song themes there is much variation, both among individuals of a population and among populations of a species on different islands (Figs. 65 and 66). Bowman (1983) has richly illustrated this variation with more than 800 individual songs.

Sonagraphic analysis has confirmed the extreme similarity of the songs of some sympatric species noted by Lack (1945). For example, Lack (1945, p. 29) wrote, "Some individuals [of *G. fortis*] were indistinguishable from *magnirostris, fuliginosa, scandens,* and occasionally from *Cactospiza.*" Bowman's (1983) analyses allowed the following amplification: "Many individuals of *Geospiza fuliginosa* have songs so similar to those of other species of Darwin's finches that field identification by means of song alone is often impossible. . . . Among species of ground-finches, the problem of field identification by means of song alone is greatest wherever *Geospiza*

BASIC SONG

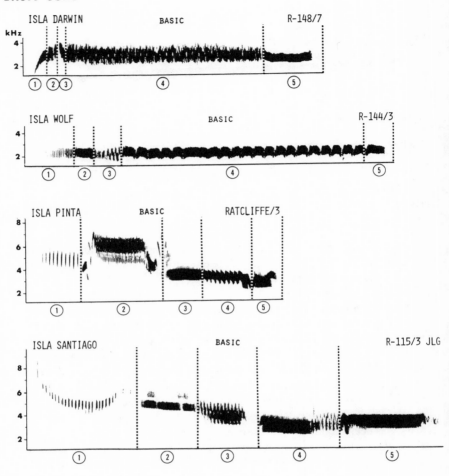

DERIVED SONG

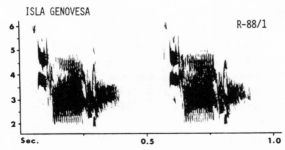

FIG. 65. Advertising songs of *G. difficilis*. Regions of the basic song which are presumed to be homologous are numbered and delimited by vertical lines. From Bowman (1983).

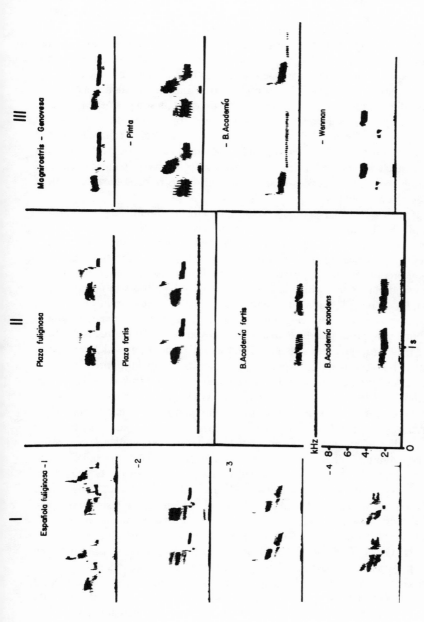

FIG. 66. Similar advertising songs of pairs of species (panel II) contrasted with a variety of songs of the same species on the same island (panel I; Española) and on different islands (panel III). In panel II, the upper two songs were recorded on Plaza Sur and the lower two were recorded on Santa Cruz. From Ratcliffe (1981).

fuliginosa and *G. fortis* occur together . . ." (p. 249). An illustration of this similarity is provided in Figure 66. No wonder Lack (1945, 1947) dismissed song as an important cue to species identity.

Nevertheless, despite these similarities, song could still be used as a species cue if parts of the repertoire, or quantitative features of an individual song such as amplitude or frequency modulation, are unique to the members of a species that sing it. For example, some individual finches have been recorded singing both basic and derived types of song. Possibly one is unique to the species and the other is similar to, or even the same as, the song of a sympatric species. The derived song is the better candidate for the possibly unique song for, according to Bowman (1983, p. 262), "Song divergence has involved mainly the evolution of "derived" patterns, many of which show convergences in structure with songs of ecological equivalent species on mainlands."

If each species has a unique song, all members should sing it, and some or all should sing the other song as well. There is debate about how many songs individuals usually sing. Bowman (1983) has argued that "bilingual" singing is typical of most geospizine species, and that the two songs are sung predominantly in different contexts, in part associated with differing motivational states. This view is supported by his observations: "For example, on Isla Floreana, on one occasion a male *Camarhynchus pauper* on territory sang derived song antiphonally with a distant conspecific male. Upon the approach of the distant bird to within a few feet of the nest site of the new bird, the latter switched to the special basic song, which now dominated his vocalizing, as it pursued the intruder through the forest canopy" (p. 272). Parallel observations were made of *G. conirostris* on Española. Ratcliffe (1981) also recorded a male of this species on Española singing the two types, but it was the only male to sing both types in her sample of recordings of ~650 *Geospiza* males from many islands in the archipelago.

Studies of banded birds are needed to establish just how common bilingual singing is. The only such studies conducted, of *G. fortis* and *G. scandens* on Daphne Major (Ratcliffe 1981) and of *G. conirostris*, *G. magnirostris*, and *G. difficilis* on Genovesa (Grant and Grant 1980a, 1983), have shown that fewer than 5% of the adult males of a population sing more than one song. Five such birds are known out of a total of about 150 male *G. conirostris* observed on Genovesa over periods ranging from one to six years (Grant and Grant 1983, B. R. Grant 1984): no relationship between song type and context could be determined through observation of any of these birds. The remainder sang either basic or derived song. Therefore it cannot be generally true that male finches declare their species identity with one unique song from a repertoire of two. Possibly tree finches (except *Platyspiza crassirostris*) and the warbler finch differ from grounds finches in usually singing two advertising songs, but typically ground finches, at least, have a repertoire of one song that remains fixed for life (Bowman 1983, B. R. Grant 1984).

Therefore, for the songs of ground finches to convey information about species identity they must differ in quantitative features, such as frequencies and note duration.

Quantitative description. Like Bowman (1979, 1983), Grant and Grant (1979, 1983) and Ratcliffe (1981) recognized two song types, but gave them neutral labels, *A* and *B*. These correspond to derived and basic songs respectively. Figure 67 illustrates the two types for each of the the six *Geospiza* species. A song is composed of one or more notes or figures separated by a longer period of silence. Ratcliffe (1981) measured seven standard features of songs from sonagrams made from tape-recordings; three are illustrated in Figure 68. Principal components analysis of the measurements of songs from six populations confirmed the reality of the two types recognized on the basis of figure or note length and the number of figures per song. Only 3% of 232 songs classified by eye were misclassified in a quantitative analysis. Altogether the songs of 617 *Geospiza* males were recorded and examined, and of these only 13 birds had songs not readily categorized by type. It must be admitted that there is some arbitrariness in the grouping of variants into two types (e.g. see Fig. 67), but the important point is that the variety of song types in a population is low.

In quantitative features do the songs of sympatric species, when matched for type, differ? The answer is yes, generally. At most localities, significant heterogeneity among species was demonstrated by analyses of variance (Ratcliffe 1981). Thus songs of species that look similar on sonagrams, and sound similar to the human ear, nevertheless differ subtly. But there are exceptions, for example the songs of *G. fuliginosa* and *G. fortis* on each of the islands Pinta, Santiago, and Isabela were found to be statistically indistinguishable in every measure (see also Fig. 67).

Both principal components analysis and stepwise discriminant function analysis have shown large degrees of similarity in the songs of the six *Geospiza* species, Moreover, in at least half of the sampled populations that exhibited both types of song, the two types were as discrete in quantitative measures as were the songs of diifferent species (Ratcliffe 1981). This means that if females use song as a cue to the identity of potential mates, they must be responsive to two different conspecific songs and not responsive to other, equally different, heterospecific songs, except perhaps on the small islands where only one song type is present in a population (Ratcliffe 1981). Evidence that they are responsive to both conspecific songs comes from the field study of *G. conirostris* on Genovesa (Grant and Grant 1983,

SONG TYPE A   SONG TYPE B

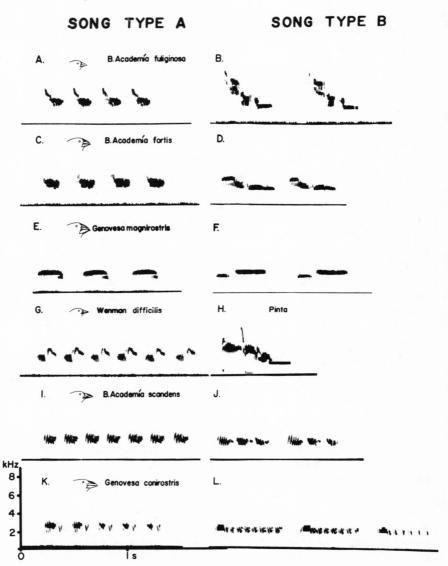

FIG. 67. Two song types of the six ground finch species. From Ratcliffe (1981).

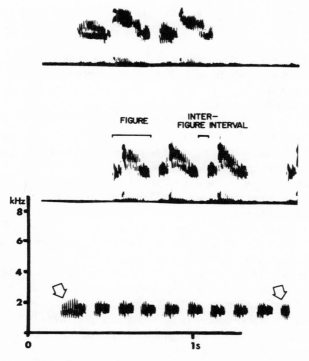

FIG. 68. Measurements of some song characteristics: figure length, interval between figures, and total song length. From Ratcliffe (1981).

B. R. Grant 1984). Whereas males sing only the song type sung by their fathers, females breeding for the first time are as likely to pair with males singing the song *not* sung by their fathers as with males that do sing the same song as their fathers. This song transmission and pairing pattern has been observed also in the *G. fortis* population on Daphne Major (Millington and Price, 1985).

On the basis of this descriptive information, the case for species discrimination by song is not strong. However, it is possible that finches respond to features of song that are neither perceived nor measured by us. In Lack's (1945) words: "[although] song may not be of fundamental importance in keeping these species apart . . . the possibility remains that, while songs overlap to human ears, the birds detect differences." Bowman (1983, p. 292) went further: "For species of Darwin's finches, there appears to be a margin of safety, derived from frequency and quality discontinuities within songs, sufficiently large to permit the birds to distinguish successfully between each other's vocal signals in all situations, despite human difficulties in doing likewise.'' The only way to find out if this is true is to perform experiments.

Playback experiments. The abilities of *G. fortis, G. fuliginosa, G. difficilis,* and *G. scandens* to discriminate between conspecific and heterospecific song were tested with playback of tape-recorded song on Daphne Major, Plaza Sur, and Pinta (Ratcliffe and Grant 1985). Songs were broadcast from a speaker placed in the center of a territory and the behavior of the tested bird was registered before, during, and after playback. In some series of experiments each bird was tested with one playback lasting for three minutes. In other series of experiments, bouts of conspecific and heterospecific song were broadcast, separated by one minute of silence, to each bird. The order in which songs were played was random. Birds were never tested with recordings of their own song.

These experiments failed to test the reproductive responses of females to the songs of males. Only three females responded in 481 tests! But the experiments succeeded in eliciting aggressive responses from the males in 93% of the tests. Therefore the results should be viewed in the general context of discrimination and not in the specific context of female preference for songs of particular males.

As a first step it was necessary to determine experimentally if a male responded differently to type *A* and type *B* conspecific songs, i.e. the same as its own song (homotypic) and the opposite one (heterotypic). *G. conirostris* individuals on Genovesa were found to respond equally to homotypic and heterotypic songs, regardless of whether they themselves sang *A* or *B* songs; later this result was duplicated in another set of tests that used the same experimental protocols (Grant and Grant 1983). On Daphne Major, *G. fortis* individuals that sang *A* or *B* songs responded equally to playback of the common song, song *B*; they were not tested with song *A*. In addition to these experimental results, observations of intraspecific territorial chases involving all six *Geospiza* species on several islands gave no indication of discrimination between homotypic and heterotypic song. With this information, the species discrimination experiments were designed and carried out with playback of just song *B*, except in one case explained below.

On Plaza Sur, where the songs of *G. fortis* and *G. fuliginosa* are similar in syllable structure and timing (Fig. 67), *G. fortis* ($N = 23$) responded significantly more strongly to playback of local *G. fortis* song than to playback of local *G. fuliginosa* song (Table 14). Males in neighboring territories were also attracted to the playback, but only *G. fuliginosa* ($N = 8$) approached the speaker when *G. fuliginosa* song was played, and only *G. fortis* ($N = 7$) approached at the playback of *G. fortis* song. This asymmetry is distinctly nonrandom (Fisher's Exact Test, $P < 0.005$). Responses of *G. fuliginosa*

TABLE 14. Discrimination by *Geospiza* males between local conspecific and heterospecific song. Statistical significance of the difference in responses is indicated by one ($P < 0.05$) or two ($P < 0.01$) asterisks. (After Ratcliffe and Grant 1985.)

Species Tested	Number of Tests	Song Playback	Mean responses ($\pm$ SE)			
			Number of Flights over Speaker	Number of Other Flights	Closest Approach to Speaker (m)	Number of Approaches to within 3m
PLAZA SUR						
G. fortis	12	*fortis*	0.9 ± 0.4	11.2 ± 1.5	0.8 ± 0.3	6.9 ± 1.2
	11	*fuliginosa*	0.5 ± 0.3	6.5 ± 1.7*	3.2 ± 0.7*	2.3 ± 0.8*
PINTA						
G. fuliginosa	10	*fuliginosa*	1.8 ± 0.6	10.6 ± 1.8	1.2 ± 0.4	6.3 ± 1.4
	10	*difficilis*	0.5 ± 0.4*	6.5 ± 1.3*	4.5 ± 1.4*	1.2 ± 0.7
G. difficilis	10	*difficilis*	0.7 ± 0.3	9.1 ± 1.6	1.7 ± 0.4	6.5 ± 1.9
	10	*fuliginosa*	0.1 ± 0.1	4.2 ± 0.9*	2.9 ± 0.5**	2.2 ± 0.7*
DAPHNE						
G. scandens	31	*scandens*	1.6 ± 0.4	7.8 ± 0.8	0.7 ± 0.1	5.9 ± 0.7
	8	*fortis*	1.1 ± 0.4	6.1 ± 1.2	1.0 ± 0.5	4.4 ± 1.0
G. fortis	41	*fortis*	3.0 ± 0.6	7.0 ± 0.8	0.8 ± 0.2	6.6 ± 0.7
	8	*scandens*	0.1 ± 0.1*	3.5 ± 1.0*	3.9 ± 1.3	1.8 ± 0.7*
	13	*fuliginosa*	3.6 ± 1.3	6.3 ± 1.1	0.5 ± 0.1	6.6 ± 1.2

males were not tested directly by playing back recorded conspecific and heterospecific song to them on their territories.

To see if the result of this experiment was either generalizable or peculiar to this pair of species, the experiment was repeated with different pairs of species on other islands. The same results were obtained with *G. difficilis* and *G. fuliginosa* on Pinta (Table 14). Each responded more strongly to conspecific song than to heterospecific song, and playback attracted only conspecific neighbors.

The third and final series of tests, on Daphne Major, produced almost the same results (Table 14). Here song *A* of *G. scandens* was used in the experiments because only song *A* and a variant is sung by *G. scandens* on this island. *G. fortis* responded significantly more strongly to conspecific song than to heterospecific song. In 25 tests *G. fortis* neighbors were attracted to the playback of song, but only to *G. fortis* song. *G. scandens* responded differently to the playback of the two songs, although the overall quantitative analysis showed little evidence of discrimination. To *G. scandens* song they responded by approaching the speaker closely and frequently but silently, whereas in response to *G. fortis* song they made fewer close approaches but increased their singing rate. *G. scandens* neighbors ($N = 23$) were attracted to *G. scandens* song, but also to *G. fortis* song in four tests. The reason for the response of this species to heterospecific song seems to lie in their defense of *Opuntia* cactus bushes against all finches. *G. scandens* is not interspecifically territorial, but it does chase *G. fortis* individuals regularly and almost as frequently as other *G. scandens* individuals (Ratcliffe and Grant 1985, Boag and Grant 1984a).

The conclusion from these experiments is that *Geospiza* males know the difference between the songs sung by other members of the population, be they types *A* or *B*, and the songs sung by other species in the same environment even when those songs are structurally similar to their own. This leaves unanswered the question of whether females make the same discrimination in a reproductive context. One piece of information suggests that they do, although it is highly indirect because it comes from an experiment in which only conspecific song was used. Songs *A* and *B* were played in random order to *G. conirostris* on Genovesa in the early and late parts of a breeding season (Grant and Grant 1983). In the first set of tests males showed no discrimination, and prevented females from showing any response by chasing them away from the speaker. In the second set of tests females were alone on the territories in 12 tests, and they showed a discrimination: 9 of the 10 females that showed a differential response responded more strongly to playback of heterotypic song, i.e. the one not sung by their mates, than to homotypic song. If they distinguish between conspecific songs *A* and *B*, for whatever reason, it is reasonable to suppose that they also distinguish between conspecific and heterospecific song. Just as in the previously described experiments, other sympatric species (*G. magnirostris* and *G. difficilis*) with territories overlapping the *G. conirostris* territories did not respond to the playback of *G. conirostris* song, of either type.

The particular components of song that signal the species identity of the singer are not known from any of these experiments. To identify them it would be necessary to test finch responses to manipulated song, as has been done with related North American species (e.g. Emlen 1971, Searcy et al. 1982).

SONG AND BILL MORPHOLOGY AS SPECIES CUES

Past disagreement over the importance of song and morphological cues used in species recognition can be summarized as follows: "Lack (1945) was of the opinion that because of song overlap among geospizines, song could not be of fundamental importance in keeping the species apart. However, all the evidence now available from field and laboratory studies suggest that normal positive assortative (selective) mating in species of Darwin's finches is mediated predominantly, if not entirely, by advertising song . . . structural differences [thought by Lack to be of prime importance] seem not to be overly significant in species recognition" (Bowman 1983, pp. 316–317).

In fact, the current evidence suggests that these views are unnecessarily in opposition, and that finches use not one but both cues in recognizing members of their own species. Song acts as a long-range (auditory) signal, and appearance, especially bill morphology, acts as a short-range (visual) signal, each transmitting information about the identity of the sender.

As described at the beginning of this chapter, courtship is initiated by males and is selective. Since females do not "sing" their identity, males must use some feature of the appearance of the female in taking the decision to court. Experimental evidence confirms that they have the ability to discriminate between conspecific and heterospecific females when those females are stationary and silent.

Females, unlike males, have both song and morphological cues available to use in taking their decisions. Their first decision, to enter a territory, could be determined by the song of the male. Their subsequent decisions, to stay and be courted or to leave, could be influenced by both cues. Here the experimental evidence falls short of the ideal, for whereas discrimination by females between conspecific and heterospecific male appearance has been shown, the responses were aggressive and not sexual, and discrimination by females between conspecific and heterospecific song has not been demonstrated. There is suggestive evidence that song is not essential for the females to take their decisions. Female *G. magnirostris*, *G. fuliginosa*, and

G. scandens mated in captivity with conspecifics, even though most of the young males, born in captivity, did not develop a fully formed, and presumably diagnostic, song (Orr 1945). Details of mating opportunities were not described, so it is not clear how often females had to make a discriminatory choice.

If both song and morphology are important to females, this should be demonstrable in experiments employing a combination of models and song playback. The appropriate experiments were carried out by Laurene Ratcliffe, but once again only the males responded. The results shed light on female sexual behavior only to the extent that females would make the same discrimination, in a sexual context, that males make in an aggressive context.

Stuffed specimens in male plumage were mounted above speakers placed 10 m apart on Daphne Major (Ratcliffe and Grant 1985). Song was broadcast through one speaker for 3 minutes and then, after a 3-minute silence, song was broadcast through the other speaker for 3 minutes. The models and the songs were of local *G. fortis* or of *G. fuliginosa* from Santa Cruz. When both models were *G. fortis*, the one that ''sang'' conspecific song elicited significantly more approaches from *G. fortis* males than the other which ''sang'' *G. fuliginosa* song. When both models ''sang'' the same *G. fortis* song but one of the models was *G. fuliginosa*, there was no discrimination. These results demonstrate the importance of song in species recognition. However, when the data from the second experiment were combined with the results of a third in which a large *G. fortis* model from Santa Cruz was substituted for the *G. fuliginosa* model, an interesting difference was revealed; in the absence of a song difference between the models, the Daphne *G. fortis* model received more approaches in a significant proportion of the 19 tests than did the other model. Hence it appears that close-range visual cues enhanced the initial attractiveness of the long-range acoustical signal.

IMPRINTING

Learning is probably involved in the relatively fine discriminations that reproductive individuals sometimes make, but nothing is known of its role in the development of responses to the signals that proclaim the species-identity of an individual. A little is known about the role of learning in the development of one of the signals, namely, song.

Sons copy their father's song, even in the fine details of the structure (B. R. Grant 1984, Millington and Price 1985; e.g. see Fig. 69). The precise replication of a father's song by his sons raises the question of whether the characteristics of song are learned by listening to the father (i.e. imprinting), or whether they would be produced by the sons anyway in the absence

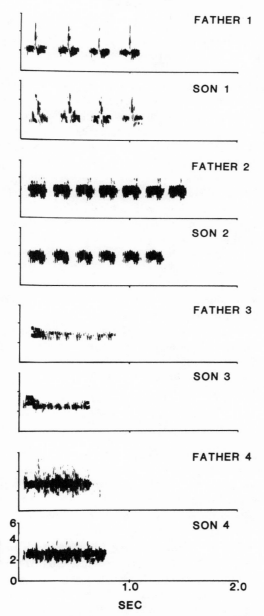

FIG. 69. Precise copies of the song of a father by his son; *G. conirostris* on Genovesa. The upper four songs are type *A*, the lower four are type *B* (see text). From B. R. Grant (1984).

of father's song. Some experimental evidence from the laboratory suggests that imprinting does occur. Bowman (1983) reported the results of nine unreplicated experiments in which young birds of various ages were subjected to playback of tape-recorded song, either conspecific or heterospecific, and then tape-recorded when they themselves sang several months later. Some of the subjects eventually sang a song that more closely resembled the heterospecific song they heard in their youth than typical conspecific song. Others that had been tutored at a later age never sang a proper song. In the aggregate these experiments suggest that the characteristics of adult song are learned when the young are between about 7 and 40 days old. This period corresponds to their second week as nestlings, when their fathers are singing at a high rate (Downhower 1978, Ratcliffe and Grant 1985), and the subsequent month as fledglings on their natal territory when they are largely fed by their fathers (Grant and Grant 1980a). Song frequency is at a peak just prior to the time of fledging, when the father sings loudly close to the nest, and continues during the period when fledglings are fed by their fathers.

If sons learn their father's song, it is logical to suppose that daughters learn it too, even though they do not sing it subsequently. Similarly, if both sons and daughters learn song from their fathers, it is logical to suppose that they learn morphological cues as well, not only from the father but also from the mother since both parents feed them in the nest, and sometimes, but apparently not always, both parents feed each of the fledglings. These suppositions need to be tested experimentally, as they are supported by only fragmentary evidence.

The Learning of Heterotypic Song

A question that arises from the imprinting hypothesis is how females acquire their positive responses to heterotypic song when choosing mates. One possible answer is that females learn the essential features of both types of song just by listening to their fathers. This seems unlikely in view of the large quantitative differences between the two types of conspecific song, which are comparable to the differences between some conspecific and heterospecific songs (p. 235). But there may be crucial elements, not measured by us but common to both types of conspecific song, which are learned by listening to only one. An alternative possibility is that the main features of conspecific song are learned from the father, and heterotypic song is then identified as conspecific in association with visual perception of the morphological features of the singer who resembles the father to some degree. Experiments could discriminate between these alternatives, but they have not been done. Until such tests are carried out we can only use natural observations to infer the process of learning.

Some features of the *G. conirostris* population on Genovesa (Grant and Grant 1979, 1983, B. R. Grant 1984) are explicable in terms of the second alternative: specifically, that young females learn the heterotypic song from a territorial neighbor, in association with its morphology, while they are on their natal territories. Not all territory owners have neighbors that sing a heterotypic song. Some have homotypic neighbors (one or more), some have heterotypic neighbors, and some have both kinds (Fig. 70). Those males with at least one heterotypic neighbor acquire mates earlier than those without a heterotypic neighbor, suggesting that one factor in female choice of a male is the song type of the neighbor. This early pairing advantage is translated into a reproductive advantage, because early breeders produce the largest number of broods and fledge the most offspring in a breeding season. This reproductive advantage is translated in turn into a recruitment advantage, for significantly more offspring born on territories with heterotypic neighbors acquire a mate in subsequent years than do those born on territories without a heterotypic neighbor. Furthermore, among the males, significantly more of those born on territories with a heterotypic neighbor obtained territories with heterotypic neighbors, and significantly more obtained a mate.

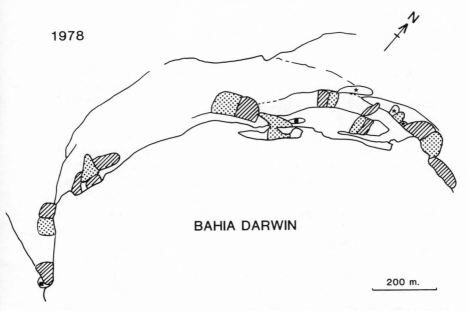

FIG. 70. Territories of *G. conirostris* males around B. Darwin, Genovesa. No two mated males with the same song type have adjacent territories. Symbols: stippled, mated song *A*; lines, mated song *B*; star, unmated song *A*; square, unmated song *B*. Continous and broken lines indicate major and minor faults respectively in the lava. From Grant and Grant (1979, 1983).

Females born on territories without a heterotypic neighbor may be less likely to learn heterotypic song and suffer, as a consequence, a restricted choice of mates when they breed. A testable prediction of this hypothesis is that they pair predominantly with males singing the same song as their fathers sing. Not enough data have been obtained so far to make the test.

Males have less flexibility than females. With rare exceptions, once they acquire a territory they retain it for life. Nevertheless, they too show signs of attempting to breed next to a heterotypic singer, both in their initial settling pattern and in their responses to new neighbors. In some but not all years of the study there was a significant tendency for adjacent breeding males to sing different songs (Fig. 70). This pattern may reflect more on female choice than on male settlement, since no such tendency was shown by the unmated males. A differential responsiveness to new neighbors can be inferred from the results of the song playback experiments described earlier. Neighbors were attracted to the playback, but only when it was the same song as their own. In half of the instances the territory owner in whose territory the playback was broadcast possessed the same song. Thus the intruding neighbor discriminated between the owner's song and playback of the same type. We have interpreted this as an indication of the neighbor's intention to prevent a new homotypic singer from settling next to him (Grant and Grant 1983).

MISIMPRINTING

The learning process can be perturbed, naturally if accidentally. The father of one of the five bilingual males in the *G. conirostris* population on Genovesa sang only one song type (*A*) in every year from 1978 to 1982; his bilingual son was born in 1978. If birds do not possess a pre-programmed song but rather learn their song by listening to others, as Bowman's experiments suggest, how did the bilingual son acquire the heterotypic song? Possibly he learned it from the male on a neighboring territory who sang only type *B* song. It is less likely that he learned it from a neighbor when he bred because there is no evidence of males acquiring any of the fine structure of their songs from such neighbors (B. R. Grant 1984). The son of another of the bilingual birds sang only the commoner of the two songs sung by his father. He must have heard both songs, but for some unknown reason sang only one of them. These two examples show that sons do not always sing only what they hear from their fathers. The first example is particularly instructive in showing that they can produce other songs apparently acquired from other birds.

Even more striking evidence, though still circumstantial, for the last point is the occasional occurrence of birds singing heterospecific song. On Daphne Major, individuals of the two resident species, *G. fortis* and *G.*

scandens, have sung each other's song and no other. On Pinta the same has been noted for individuals of *G. difficilis* and *G. fuliginosa*. On Genovesa a male *G. conirostris*, or possibly a hybrid, has been recorded singing only *G. magnirostris* song. Other examples of heterospecific singing have been recorded by Bowman (1983). In none of these cases is it known how the birds acquired the heterospecific song, but a reasonable suggestion is that they lost contact, as fledglings, with their fathers, were fed by a male of another species, and became misimprinted on his song. They may even have acquired their strange song, initially, in the nest. Ratcliffe (1981) observed a *G. fuliginosa* male on Española regularly feeding the nestlings of a warbler finch for several days, in between bouts of singing, courting, and nest-building with a female *G. fuliginosa* a few meters away. A third possibility is that they were reared in the nest of another species. Egg laying in nests other than the mother's is not known in Darwin's Finches but could possibly occur at a very low frequency. Finally, in some but not all cases, they may have been F_1 hybrids who resembled their mothers most but who sang their fathers' song.

Apparently even more rarely a male will sing both conspecific and heterospecific song. A male tree finch, presumed, from its measurements, to be a hybrid offspring of *Certhidea olivacea* and *Camarhynchus parvulus* parents (Bowman 1983), was tape-recorded singing derived songs of both hypothetical parental species (Fig. 71). His two songs can be explained by supposing that he learned the morphological features of both parents, and one song from his father and the other from one or more males that resembled his mother. It is consistent with the hypothesis that offspring imprint on both sets of cues.

Misimprinted birds are interesting because they broadcast not one identity but two: a false one at long distance and a true one at short distance. Their behavior, and the behavior of other birds towards them, further illuminates the significance of the two sets of cues. On Pinta a *G. fuliginosa* individual sang a *G. difficilis* song in an area lacking *G. difficilis*; it chased all *G. fuliginosa* intruders. A *G. difficilis* individual with a *G. fuliginosa* song countersang with neighboring *G. fuliginosa*, but only chased *G. difficilis* intruders (D. Schluter, pers. comm.). Much the same behavior has been observed on Daphne Major, both in natural encounters and in responses to experimental playback of song. A *G. fortis* individual that sang a *G. scandens* song responded more strongly to playback of *G. scandens* song than to playback of conspecific song, in other words more strongly to the type of song he himself sang. He was observed to chase all *G. fortis* intruders in his territory, but only two out of five *G. scandens* intruders. Possibly, therefore, he perceived *G. fortis* individuals as conspecifics, and certainly he was perceived by them as such because he was chased by several of them, as well as by *G. scandens; G. fortis* rarely chase *G. scandens*. The

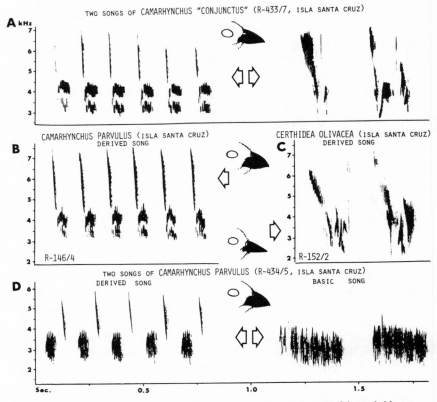

FIG. 71. The song of a presumed hybrid (A) compared with typical songs of the probable parents, *Camarhynchus parvulus* (B) and *Certhidea olivacea* (C). Some *C. parvulus* sing two songs (D), but they are not the same as the song of the presumed hybrid (A). All song recordings were made on the south side of Santa Cruz. From Bowman (1983).

misimprinted *G. scandens* individual differed from the *G. fortis* individual in its response to song. It responded more strongly to playback of *G. scandens* song than to playback of *G. fortis* song, and chased a significantly higher proportion of intruding *G. scandens* than would a normal *G. fortis*. In other words, although he sang a *G. fortis* song he recognized other *G. scandens* as conspecifics, and he was recognized as a *G. scandens*, because he was never chased by a *G. fortis*. He may have learned to associate true *G. scandens* song with *G. scandens* morphology, and to disregard songs that resembled his own (Ratcliffe and Grant 1985).

The crucial question is how do females respond to misimprinted males? If they use only one cue in recognition of conspecifics, consistency in the pairings of misimprinted males is to be expected. Either the males pair with heterospecifics if the false cue (song) is used, or they pair with conspecifics if the true cue (morphology) is used. On the other hand, if both cues are used, a mixture of pairing patterns is to be expected; and in fact this is observed.

On Daphne Major, three misimprinted males of each species (*G. fortis* and *G. scandens*) have been recorded during a ten-year study. One of the *G. fortis* males and two of the *G. scandens* males paired and bred with the other species. The remainder paired and bred with conspecifics, although some difficulty in acquiring a mate is indicated by the fact that the *G. scandens* individual which was tested with playback remained without a mate for at least 7 of the 9 years. The three interspecific pairs constitute 50% of all *fortis* × *scandens* pairs observed during the study on that island. On Pinta, the misimprinted *G. fuliginosa* and *G. difficilis* individuals had conspecific mates, whereas on Genovesa the only bird known to sing a heterospecific song—a *G. conirostris* male, but possibly a hybrid—bred with a *G. magnirostris* female, treated *G. magnirostris* males as intruders, and ignored *G. conirostris* individuals of both sexes.

BEYOND SPECIES-RECOGNITION: MATE CHOICE

Although the subject of this chapter is species-recognition, it is appropriate to conclude by considering how mates are chosen from members of the same species. Exclusion of heterospecifics as potential mates is not the only mating decision a finch takes, even though it may be the first. Selectivity in mating is expected when members of the selected sex vary in quality due to genetic factors, experience, or where they live (e.g. male territories), and when members of the selecting sex have the appropriate powers of discrimination. The rules governing mate choice have engendered much interest and debate in proportion to the difficulties of deciphering them (e.g. see Bateson 1983). Here I will concentrate on what has been learned from observing Darwin's Finches in the field.

With females imprinting on their fathers' song, and having a choice of two song types when subsequently breeding, one might expect them to pair mainly with males that sing the same song as their fathers', thereby giving rise to a partially subdivided structure of the population. Assortative mating by song type was raised as a possibility by Grant and Grant (1979), then asserted as a fact by Bowman (1983), but unfortunately it is no more than an interesting idea that has turned out to be wrong; or, more correctly, the available evidence from the studies of *G. fortis* and *G. conirostris* discussed above shows that females pair randomly with respect to father's song type when they breed for the first time (Grant and Grant 1983, B. R. Grant 1984, Millington and Price 1985).

Assortative mating by bill characteristics might also be expected for the same reasons. The first examination of field data provided some support for this possibility. In 1976 on Daphne Major *G. fortis* were paired nonrandomly. We found positive correlations between mates in some bill and body dimensions, but they were weak and barely significant (Boag and Grant 1978). Measurements were not correlated between mates in subsequent years, nor were they correlated in the well-studied *G. conirostris* population on Genovesa (Grant and Grant 1980a, Grant 1981a, 1983a).

A nonrandom pattern of pairings, such as the one seen on Daphne Major in 1976, can arise from selective mating or from random mating within groups that differ. The latter explanation is probably applicable to the Daphne observations, since pairing occurs primarily within cohorts, and cohorts differ in the mean size of their dimensions owing to different growth or selection among years (Boag 1983, Price and Grant 1984). In other words, it was probably an artifact.

A proper test of the biological hypothesis of assortative (selective) mating requires a knowledge of the number and types of potential mates available when the pairing process occurs. This information is known for part of the study of the *G. conirostris* population on Genovesa. The two types of song in the population were classified into eight subtypes on the basis of note structure (B. R. Grant 1984). The song subtypes of the 59 unmated males in the study area were known during the period when 14 daughters, whose fathers' song subtypes were also known, paired with 14 of the males. None of the females paired with males that sang the same song subtype as their fathers', whereas exactly 2.38 would be expected if pairing was random. The difference, as assessed by a binominal test, is almost statistically significant ($P = 0.07$, B. R. Grant 1984). Possibly, therefore, females use song subtypes as a means of recognizing kin and avoiding mating with them.

The strongest opportunity for selective mating occurs when the sex ratio, which is normally 1:1, is distorted, for then members of the minority sex have the broadest range of potential mates and the greatest scope for choice. A severe distortion took place on Daphne Major in 1977 when survival of males during a drought was as much as six times higher than survival of females. It has already been mentioned in the previous chapter that *G. fortis* females paired nonrandomly in subsequent years. This happened despite the apparent readiness of all males to accept a mate, to judge from their singing and other territorial behavior. Under these circumstances the pairs that were formed were probably largely the result of female choice (Price 1984a). The objects of their choice were large males, with fully or almost fully black plumage, and with large territories. Since these three properties of the males covaried it was not possible to identify the one, if there was only one, to which the females responded (Chapter 8). The females might have used one, two, or all three factors in choosing their mates, or some other factor such as general physical vigor with which all three were correlated.

From these observations it appears that the two sets of cues used in species recognition, song and morphology, are also used when finches choose a particular conspecific mate. The recognition and exclusion of heterospecifics as potential mates is therefore one element of a general process of selecting a mate. In addition to song and morphology, other factors contribute to the choice of a mate: territory size, and plumage state which is an indicator of age. We may speculate that the reason for these choices being made is that both sexes are attempting to maximize their reproductive fitness through the avoidance of breeding with kin, with the possible attendant risks of inbreeding depression, and through the acquisition of a mate who will provide good parental care (B. R. Grant 1984, Price 1984a). According to this view, however, it is better for a bird to breed with kin, or with a member of another species (Chapter 8; see also Wilson and Hedrick 1982), rather than not breed at all, providing that two conditions are met: that some reproductive success can be expected, and that future reproduction is not jeopardized thereby.

This brings us back to the initial observation, introduced at the beginning of the chapter, that focused attention on the question of species-recognition; individuals of two sympatric species may be more similar to each other morphologically than either is to average members of their own populations. Despite the similarity they rarely interbreed. The rarity can be explained by the arguments of this chapter. Early experience, especially in association with parents, enables young birds to learn the morphological and song characteristics of members of their population, and later, when they are adults, they use these characteristics when choosing mates. The morphological characteristics are graded, while the song characteristics are more discrete, and in combination they provide an unambiguous description of the identity of the population and the basis of a specific-mate recognition system (Paterson 1980). When interbreeding does occur it is not between the morphologically most similar members of two species (Chapter 8), nor is mate choice of members of the same species influenced by morphological similarity (above). Thus mate choice, at least initially, appears to be based on a dichotomous classification of birds of the opposite sex into acceptable and not acceptable categories. It is the major reason why species persist as discrete species.

SUMMARY

Hybridization of Darwin's Finch species is very rare, and so too is the occurrence of courtship behavior that is misdirected towards other species.

The question addressed in this chapter is how finches discriminate between members of their own species and members of other species.

Neither plumage nor courtship behavior is specific enough. Possible cues to species-identity are song and morphological features of the bill and body, although even these are very similar among certain sympatric species. Experiments with either stuffed specimens or playback of tape-recorded song have demonstrated that males and females of the *Geospiza* species can make fine discriminations between conspecifics and heterospecifics on the basis of these two sets of cues. It is argued that both sets of cues are used by finches when choosing mates.

The sexual responses of males to female specimens give the clearest indication of discrimination in a reproductive context. Males also discriminate between conspecific and heterospecific advertising song, but the responses are aggressive because only males sing. Females did not respond to song playback in the experiments, but since males can discriminate between songs, females can probably do so as well. Females discriminated between conspecific and heterospecific male specimens in experiments, but their responses, like those of males, were aggressive and not sexual. Again, arguing by analogy, I suggest that if males discriminate in a reproductive context by using visual cues alone, females, which can make the same discrimination, probably do so in that context as well.

I also suggest that both males and females acquire their discriminating abilities by imprinting on fathers' song and on the morphological features of both parents when they are being fed by them. Individual males of the ground finch species typically sing only one song, and sons copy their father's song precisely. Two main types of song are present in many populations of most species. Sons rarely misimprint on the song of another male of the same or of a different species. Misimprinted birds sometimes pair with conspecifics and sometimes with a member of another species whose song they sing, which is further indirect evidence that both acoustical and visual cues are used in species-recognition and mate-choice.

CHAPTER TEN

Evolution and Speciation

EVOLUTION

I shall now use the patterns of variation in morphology and the ecological processes governing the contemporary lives of finches to address the question raised in the Introduction: how did the species evolve?

Actually there are more questions concerning the history of the species to be answered than just this one. To begin with, where did Darwin's Finch species come from, and when? How many species were formed, when, where, and by what means? Why did they evolve in the ways that they did and not in other ways? How many species arose and then became extinct, what were their properties, and why did they become extinct? If we could answer all these questions we could say we understand the evolution of Darwin's Finches. In fact we can answer only some of the questions, and the answers range from the confident to the speculative.

I shall first consider the origins of the finches and their pattern of speciation before discussing how speciation took place.

ORIGINS

The fact that all Darwin's Finch species are more similar to each other than any one is to a continental species makes it likely that they were all descended from a single ancestral species of emberizine finch. We can be confident that the ancestral species colonized the Galápagos or Cocos Island by overwater flight from South or Central America some time in the last five million years (Chapter 1), and probably fairly recently (see later). But who was the ancestor?

In the absence of fossils that might give us the answer we can only identify the continental lineage descended from the ancestor. This is worth attempting because, if we assume that evolutionary change in the continental lineage has been modest in comparison with the large amount of change postulated to have occurred on the Galápagos, the closest modern relative on the continent will give us the best picture of the original colonist. It could also help us to determine which traits are primitive in Darwin's Finches and which traits have evolved recently.

The question is thus rephrased: which species now living on the continent is the closest relative to the ancestral species? The species can be identified by systematic studies that reveal the closest relative to modern Darwin's Finches among modern continental finch species. Various candidates have been suggested. In recent times, three species have been given special attention: *Melanospiza richardsonii*, now restricted to the island of St. Lucia in the West Indies but possibly more widely distributed in earlier times (Bond 1948, Bowman 1961); *Tiaris* species, currently distributed widely in the West Indies and Central America; and *Volatinia jacarina* whose distribution extends through the Pacific lowlands of South and Central America (Steadman 1982). A fourth species, *Coereba flaveola*, suggested by Harris (1972), is not an emberizine (Tordoff 1954, Paynter 1970) and for this reason it is not likely to be related to Darwin's Finches. Males of the other three species have black plumage and the females are dull brown; hence they resemble the ground finches on Galápagos and the Cocos Island Finch. To varying degrees their internal anatomy has features shared by Darwin's Finches. For example, the syrinx (voice box) and its musculature in *Melanospiza* (Cutler 1970) and certain skeletal features of *Volatinia* (Steadman 1982) resemble the corresponding features of Darwin's Finches. Also the song produced by *Melanospiza* males has the characteristics of basic (type *B*) song of Darwin's Finches (Bowman 1983).

There is uncertainty about the origins of Darwin's Finches because not one of these three candidate species is clearly more closely related to them than are the others. Steadman (1982) has discussed the problems associated with accepting either of the first two species. One problem is that the features they share with Darwin's Finches are also possessed by some other continental species. Another problem is convergence; phylogenetically distantly related species may share features not because of common ancestry but because they have independently evolved under similar selection regimes in similar environments. The song of *Melanospiza* and its resemblance to Darwin's Finch song is a possible case in point. Bowman (1983) has amply documented the phenomenon of convergence in song by showing the close resemblance between certain songs of Darwin's Finches and songs of North American species in different families. The fine structure of song could be a guide to phylogenetic affinities, but this needs to be explored more fully.

Volatinia jacarina is an interesting possible relative because some aspects of its molting resemble the molting of Darwin's Finches. Males of the ground finches acquire black plumage in successive molts in a systematic way, from the head posteriorly (Fig. 10). The Cocos Island finch is unique among Darwin's Finches in acquiring black plumage in a mosaic fashion over the body (Lack 1945), apparently gradually and over a long period of time (T. W. Sherry and T. K. Werner, pers. comm.). Populations of *Vola-*

tinia show the same variation, with those in Peru resembling the Galápagos finches and those in Central America resembling the Cocos Island finch. Furthermore, the Cocos Island finch and the Central American *Volatinia* have a shiny black plumage, whereas Galápagos finches and the Peruvian *Volatinia* have a duller black plumage. Steadman (1982) drew attention to these features and suggested that Cocos Island and the Galápagos were colonized separately by representatives of the different *Volatinia* populations.

This hypothesis explains the variation in molting pattern in Darwin's Finches, whereas the alternatives involving *Melanospiza* and *Tiaris* do not. On the other hand, *Volatinia jacarina* differs markedly from all Darwin's Finches in four ways, which make me doubt the suggested close relationship between *Volatinia* and Darwin's Finches, and doubt even more the independent colonization of the Galápagos islands and Cocos Island by separate *Volatinia* stocks. First, *Volatinia* males, unlike Darwin's Finch males, have a white axillary patch under the wing, which is conspicuous in epigamic display. Second, males molt into a female-like plumage after breeding (Dickey and van Rossem 1938), unlike Darwin's Finches and the other two suggested relatives. Before the next breeding season they molt into blue-black plumage (Darwin's Finches are black). Females also molt twice each year. Third, males have a characteristic, vertical, initially jumping, display flight that differs conspicuously from the more horizontal display flight of all Darwin's Finches. Finally, nests of *Volatinia* are cup-shaped, whereas nests of all Darwin's Finches (and *Tiaris* and *Melanospiza*) are dome-shaped with entrances on the side. It would be remarkable if all these features which set Darwin's Finches apart from *Volatinia*, especially the last two features, evolved in parallel on Cocos Island and the Galápagos.

The climatic and vegetational aspects of the environments of *Volatinia* in the arid lowlands of South America and Darwin's Finches on the Galápagos are similar, so it is not clear why such marked behavioral differences evolved, if *Volatinia* really is close to the ancestral stock. In contrast to this, similar variation in molting pattern among Darwin's Finches and *Volatinia* populations may be accounted for by the difference in climate between Cocos Island and the lowlands of Central America on the one hand, and Galápagos and lowland Peru on the other. There should be a much weaker seasonal entrainment of timing and sequence of feather replacement in the northern populations, because their environments are relatively aseasonal, than in the southern populations. In addition, the more glossy plumage of finches on Cocos Island than on the Galápagos may be related to the greater need for waterproofing, since the climate is much wetter. The Cocos Island finch has an unusually conspicuous uropygial gland (Nicola Grant, pers. comm.).

This is a typical problem in systematic studies. Species *B* resembles species *A* in trait 1 but not in trait 2, whereas species *C* resembles species *A* in trait 2 but not in trait 1. How is the relative importance of the two resemblances to be gauged, particularly in those cases, as here, when they involve different kinds of traits such as molting pattern on the one hand and nest structure on the other? There are different schools of thought on how to proceed, and different procedures to be followed, but so far no attempt has been made to apply them to the Darwin's Finch problem in a comprehensive and quantitative manner.

Since studies of phenotypic resemblance have not led to the unambiguous identification of the closest relative, greater success may be achieved through the use of biochemical techniques to estimate genetic relatedness. Such genetic studies have not yet been undertaken with tissues from Darwin's Finches and from continental and West Indian species, but one indication of what may lie ahead is provided by the results of "hybridizing" the DNA from different continental species to measure their genetic similarity.

DNA hybridization experiments with material from numerous species of passerine birds in South America have yielded a surprising result. Many species currently classified as emberizine finches are more closely related to members of the tanagers (Thraupidae) than to emberizines (Sibley and Ahlquist 1984). Presumably they look like finches because they have convergently evolved finch-like traits in habitats typically occupied by finches elsewhere. If Darwin's Finches are found by the same method to be closely allied to one or more continental or West Indian species of tanagers traditionally classified as emberizines, they will need to be renamed Darwin's Finch-Tanagers (Grant 1984c).

THE NUMBER OF SPECIES

It is very unlikely that the fourteen extant species of finches are the only ones ever to have evolved. Others probably arose, then became extinct. We could only know this by discovering their fossils. It always seemed to me that the volcanic terrain of the Galápagos was the worst possible environment for finding fossils, yet the improbable has been realized. David Steadman (1981, 1985) has assembled and identified a remarkable collection of bones from cracked lava tubes on Floreana and Santa Cruz. Many of the fossils are believed to be derived from the disintegrated pellets of barn owls. Although modern studies on Galápagos show that barn owls feed almost exclusively on rodents (Abs et al. 1965), finches may have been the principal prey on the rodent-free island of Floreana prior to the arrival of the black rat. The dated fossils are no older than 2,400 years, but some undated ones are probably much older because they were recovered from lower levels of the substrate. Extinct forms of *G. magnirostris* (Floreana) and *G. difficilis* (both islands) are well represented in the fossil collections as are most of the extant species. While many of the bones have not been studied yet in sufficient detail to be identified with certainty, no new species has been found so far. At present we can only say that the diversification of Darwin's Finches involved at least the current number of recognized species. But the potential for new discoveries is a sufficient stimulus for continued searches. New fossil material may resolve some of the distributional problems raised in Chapter 3, and may also document recent evolutionary changes in currently existing species. Especially desirable for the latter purpose are fossils from before 3,000 years ago when the climate was drier than it is now (Chapter 2).

THE PATTERN OF SPECIATION

The ancestor-descendant relationships among the fourteen species, together with the time at which speciation events took place, constitute the pattern of speciation. I will first describe the phylogenetic pattern, and then place it in a time framework.

Lack (1947) used the similarities and differences among species in plumage, size, and shape (especially of the bill) to construct a tentative phylogenetic tree (Fig. 4). The main features of the tree are two episodes of major differentiation: (a) an early differentiation of warbler-like finches, and (b) subsequent divergence of the tree finch stock from the ground finch stock. Each branch of the tree then differentiated into twigs (congeneric species).

Although not developed in any formal or quantitative way, Lack's phylogenetic tree is essentially a model whose construction rests on the assumption that a constant relationship exists between the morphological difference between species and the time elapsed since they diverged from a common ancestor. Thus the warbler finch, which is morphologically the most distinctive, was considered to have arisen the earliest.

There are other assumptions. One is that an original generalist species gave rise to successively more specialized species. Another is that plumage similarities are a better guide to affinities than are morphometric similarities (Snodgrass and Heller 1904, Swarth 1934, Lack 1945). Although reasonable, this could be wrong. An alternative model could be constructed largely on the basis of morphometric similarity. The results of a multivariate analysis of size and shape depicted in Figure 21 indicate how species might be grouped in this alternative scheme. For example, the small ground finch might be most closely related to the small tree finch; the medium ground finch might be most closely related to the medium tree finch; and the large ground and tree finches might be most closely related to each other. Accordingly, it would be assumed that plumage differences between members of a pair evolved secondarily, and repeatedly. Since small genetic differences can produce large differences in the plumage coloration of chickens and of budgerigars, this is not an extreme assumption to make.

We need an independent measure of affinities to distinguish between these models. The desired measure has been provided by an analysis of protein polymorphisms, whose connection with plumage and morphometric characters is not known but is likely to be tenuous at best, but whose connection with simple genetic factors is highly probable considering the results of parallel work with other organisms (e.g. Milkman 1981).

Ford et al. (1974), Yang and Patton (1981), and Polans (1983) have used the technique of electrophoresis to assess variation in protein structure in Darwin's Finches. The results confirm the basic features of Lack's model, and provide no support for my suggested alternative. The warbler finch is the most distinctive species biochemically (Fig. 72), and hence diverged from the rest at the earliest time. The six ground finch species cluster together and are distinct from the group of four tree finch species assayed; material was lacking from the other two tree finch species, and from the Cocos Island Finch. There is no support from biochemical data for a much earlier view that *Cactospiza pallida* is the oldest species on the grounds that its buffy unstreaked plumage is likely to be primitive (Snodgrass and Heller 1904).

Within each of the two major groups, the ground finch group and the tree finch group, species are so similar to each other that relationships cannot be assessed reliably, although certain pairs of species are more likely to be closely related than others. For example, the small and the large ground finches are probably more closely related to the medium ground finch, by virtue of their phenotypic similarities with it, than they are to each other. But whether the evolutionary sequence was small–medium–large, large–medium–small, medium–large and medium–small, or an alternative involving a fourth species, cannot be determined. Perhaps investigation of mitochondrial DNA, which is known to evolve more rapidly than nuclear DNA (e.g. Brown 1983), and of sequences within the DNA molecule, might help to resolve the issue of the temporal pattern of speciation events within genera. When continental relatives are known, these investigations could also clear up the problem of whether the Cocos Island finch gave rise to the Galápagos finches, was derived from one of them such as *G. difficilis*, or had an independent origin.

Lack (1947) used plumage features to hazard a guess as to the characteristics of the original colonists of the Galápagos. Some sharp-beaked ground finches (*G. difficilis*) have buff colored undertail coverts and wing coverts. In this they resemble the Cocos Island finch and some individuals of the unusual *G. magnirostris* population on Darwin (Chapter 4), but no others.

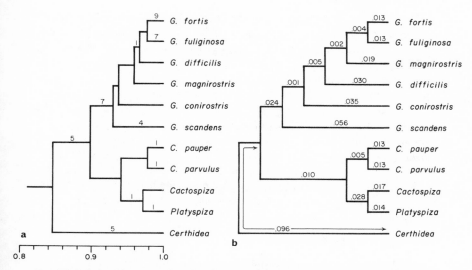

FIG. 72. Relationships among Darwin's Finch species based on an analysis of protein polymorphisms and constructed with Rogers' similarity measures (*a*) and Rogers' distance measures (*b*). Numbers along branches in *a* indicate the number of presumptive alleles unique to two or more of the terminal taxa connected by that branch. The clusterings are very similar but not identical. Both place all ground finch species together, and all tree finch species together (compare with Fig. 4). From Yang and Patton (1981).

Other species have white undertail coverts (males), and their wing coverts are a duller brown (females) or grey-black (males). These traits vary among individuals within a population of *G. difficilis*, and the frequency of individuals possessing the buff color differs greatly between populations. Lack (1947) interpreted this pattern to signify a gradual loss of buff color in the two coverts in the absence of natural selection, and believed that *G. difficilis* was phylogenetically old because only old species would have such non-adaptive traits. Their widely distributed and well differentiated populations (Fig. 11; also Yang and Patton 1981) can also be interpreted as evidence of long establishment in the archipelago.

This interpretation of plumage variation was contested by Bowman (1961). He argued that variable traits like the color of wing and tail coverts could as easily signify gradual acquisition as gradual loss of buff color, and that there may be an adaptive significance to the buff color that has so far escaped detection. Neither functional significance, in a context of communication for example, or adaptive significance has been suggested. But if variation in the traits is neutral with respect to selection pressures, the buff color phenotype is as likely to be increasing as decreasing in the populations.

The results of electrophoresis do not confirm Lack's assignment of greatest age to *G. difficilis; G. scandens* would appear to be closest to the tree finch group (Fig. 72). Nevertheless the ground finches as a group are the most similar in appearance and behavior to typical finches on the continent, and all three continental species thought to be close relatives have black (males) plumage like the ground finches. Therefore it is not the oldest species, the warbler finch, that is the best indicator of the ancestral finch. The warbler finch is the most highly derived, unfinch-like, member of the whole subfamily. Rather, the ancestor may have resembled one of the ground finch species living today, and *G. difficilis* and *G. scandens* are plausible candidates for different reasons. *G. fuliginosa* was also suggested as a candidate by Lack (1945) because of its unspecialized feeding habits, but this hypothesis receives the least support from the biochemical study (Fig. 72).

THE TIME FRAMEWORK

In addition to throwing light on phylogenetic affinities, the biochemical differences between species established by electrophoretic analysis of proteins can be used to estimate the times when speciation events occurred. The method of dating assumes a long-term average constancy in the rate of substitution of selectively neutral alleles that code for the proteins. It is calibrated against other information, including datable fossils. Yang and Patton (1981) have applied Nei's method of dating to their own results of electrophoretic analysis of Darwin's Finch material, carefully considering the uncertainties of the assumptions upon which the method rests (see also Thorpe 1982).

According to their analysis the warbler finch split off from the ancestral stock about 570,000 years ago. It is not known how long the Galápagos were tenanted by the original colonists before this initial speciation event took place. If speciation occurred fairly soon after the arrival of the colonists, on the order of a few thousnd years perhaps, then to a rough approximation it may be said that the diversification of the finches took place in the most recent 10–15% of the history of the Galápagos.

The first differentiation might have occurred earlier, as early as 15–20 million years ago, if more conservative assumptions (Sarich 1977) about the equivalence of electrophoretic distance and time are correct (Yang and Patton 1981). But then the age of the first differentiation would be way out of line with the age of the Galápagos islands themselves. When these more conservative assumptions are adopted they give rise to problems of interpreting speciation in other groups of Galápagos organisms as well. Among the reptiles, the divergence times of taxa in the lava lizards (*Tropidurus*), geckos (*Phyllodactylus*), and the iguanas (land and marine) would be as much as four times greater than the age of the Galápagos (Wright 1983, Wyles and Sarich 1983). The discrepancies cannot be simply rationalized by invoking speciation on ancient islands that have since disappeared (submerged), and dispersal from them to the Galápagos, because geological evidence for the existence of such islands is lacking (Cox 1983). On the other hand, divergence times of these taxa calculated by Nei's method fall within the last five million years. Because they are more compatible with the geological evidence for the age of the Galápagos they can be accepted with more confidence.

Whatever the absolute time scale, the temporal pattern of speciation can be established with the numerical values for biochemical differences between species (Grant 1984c). The pattern is shown in Figure 73 (upper), the time scale being based on Nei's method of dating. The figure shows an increase, with the passage of time, not only in the number of species but also in the rate of speciation up to about 50,000 years ago, and no speciation events thereafter.

Many major groups of organisms show a burst of speciation in their early evolutionary history, followed by a period of approximate taxonomic constancy, and then a decline. Darwin's Finches differ from this pattern in exhibiting a slow initial rate of speciation. This ought not to be for lack of ecological opportunity, because all major islands had been formed by the time finch speciation began (although it must be admitted that the vegetation and arthropods on the islands at this time are not known and may have been depauperate). Rather, the slow initial rate of speciation may be more apparent than real. Many species could have arisen early in the history of the group, then become extinct, and we would not know about them because we lack their fossils. We can only work with extant species. Figure 73 (lower) provides a scheme that corrects the temporal pattern for hypothetical missing data. Since the oldest species are the most likely to have become extinct, and at least some of the oldest are the most divergent (e.g. warbler finch), the full adaptive radiation may have encompassed a distinctly broader array of morphological and ecological modes than is currently known to us. This is especially likely if Galápagos habitats underwent major changes.

It is curious that no species has arisen, or is known to have arisen, in the last 50,000 years. Alternative dating methods would lengthen, not shorten, this interval. Correction for one obvious bias would achieve the same result—that is, the bias that arises from gene exchange between species through hybridization. It has the effect of making species biochemically more similar than they would otherwise be, thereby yielding an estimate of their divergence times more recent than they really are. For example, at two localities, one on Santa Cruz and one on Santiago, *G. fortis* and *G. fuliginosa* are more similar to each other biochemically than either is to conspe-

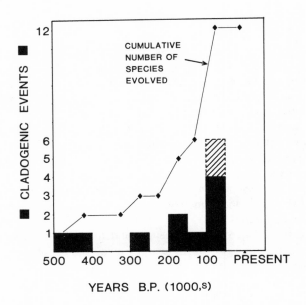

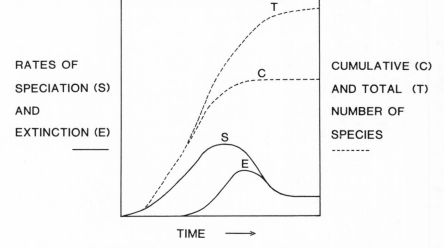

FIG. 73. The temporal pattern of speciation of Darwin's Finches. Upper: The number of speciation (cladogenic) events occurring in each 50,000-year period, from data in Yang and Patton
(1981). The hatched part of the histogram refers to two species, *Camarhynchus psittacula* and
Cactospiza heliobates, not studied by Yang and Patton (1981). The time of their formation has
been estimated on the basis of their morphological similarity with congeners of known times of
origin; it is supported by biochemical data from one specimen of *C. psittacula* (Polans 1983).
Lower: Hypothetical curves of speciation (S) and extinction (E) and the accumulation of species
in total (T) and at any one time (C). It is assumed that there is a fixed maximum number of species sustainable in the archipelago, and that this maximum is approached in dampened fashion
(C) as a rise and fall in the speciation rate (S) and, with a lag, a rise and fall in the extinction rate
(E). The difference between T and C is solely attributable to E. The C curve has approximately
the same form as the cumulative species curve in diagram A; the difference is that the C curve is
the product of speciation and extinction whereas the equivalent curve in diagram A is estimated
from extant forms only. From Grant (1984c).

cifics on other islands (Yang and Patton 1981). This may be the result of hybridization between them; selectively neutral alleles that specify structural proteins may become incorporated into the gene pool of each species relatively easily, whereas alleles governing metric traits would not (Chapter 8). To estimate divergence times more accurately we need to weight the biochemical differences between species by the frequency of gene exchange in some way, but we lack the information on hybridization frequency to do this.

Thus the absence of recent speciation appears to be real. Fossil evidence may one day show this to be wrong. For example, in the recent fluctuations of climate and the extent of arid zone and mesic habitat (Chapter 2), it is not difficult to visualize both speciation and extinction taking place. Nevertheless, the available evidence suggests that under current conditions the Galápagos are saturated with species of finches; ecological information (Chapter 12) is needed to explore this possibility more fully. Saturation, however, does not mean absolute constancy. As represented in Figure 73 (lower), very slow rates of speciation and natural extinction may balance to give an equilibrial number of species in time, just as immigration and extinction are postulated to do in both time and space in the standard theory of island biogeography (MacArthur and Wilson 1967, Williamson 1981).

ALLOPATRIC SPECIATION

The derivation of thirteen or more species on the Galápagos from a single ancestral species was the product of repeated splittings of single lineages into two or more noninterbreeding lines of descent. How did these events take place?

The first modern attempt to answer this question was made by Stresemann (1936). The basic idea he promoted is one of separate populations of a single species on different islands evolving in different directions. Dispersal between islands and the interbreeding of residents and immigrants tended to retard the process of divergence, and to increase the variation in each population. As divergence proceeded further, population variation increased, but then it decreased as interbreeding itself decreased until eventually the difference between residents and new immigrants was so great they did not interbreed. At this point they had become separate species. Since the differentiation of populations of a single species occurred in allopatry, the process is known as allopatric speciation.

Lack (1945, 1947) elaborated this basic idea but without paying explicit attention to changes in the level of population variation during a cycle of speciation events. I quoted a summary of his views at the beginning of this book, and now I will express them in my own language, more fully and more formally, using a five-step model (Grant 1981c) to incorporate the essential features.

The model. In step 1 birds colonized one of the Galápagos islands, following an overwater journey from either Cocos Island or the continent (Fig. 74). They reproduced, the population increased, and at some point, perhaps as the carrying capacity was approached or reached, a few individuals dispersed and colonized another island. This gave rise to step 2, the establishment of allopatric populations of a species, and it may have happened several times. A small amount of genetic change took place in each of the new populations.

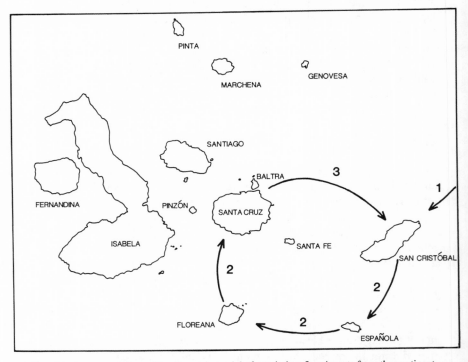

FIG. 74. A representation of the allopatric model of speciation. Immigrants from the continent colonized an island, chosen arbitrarily here to be San Cristóbal, in step 1. Then dispersal to other islands and establishment of new populations took place, probably repeatedly, in step 2. In step 3 members of the original and one of the derived populations encountered each other; here it is shown to occur through the dispersal of members of a derived population to the originally colonized island. Little interbreeding occurred; the two sympatric populations constituted two species, formed from one. The cycle of events was repeated many times, each involving an allopatric phase (step 4) and a secondary contact phase (step 5), and resulting in the formation of thirteen species, possibly more (see Fig. 73). Slightly modified from Grant (1981c, 1984c).

Step 3 is the establishment of contact between previously allopatric populations, through the dispersal of some members of one population to an island supporting another population. In Figure 74 this is illustrated with the dispersal of some members of a derived population to the island supporting the original population of colonists. The flow of individuals could just as easily have gone in the opposite direction; choice of islands to illustrate the process is arbitrary. What is important is the degree of difference between original and derived populations at the time of contact. If the differences were small, interbreeding probably occurred freely and the groups fused into one. If the differences were moderate to large, individuals of the two groups might have shown only a slight tendency to interbreed, and those that did interbreed may have produced offspring less fit than those that bred within their respective groups. Thus natural selection would have favored those individuals that bred "true," resulting in the enhancement of differences between the groups in the signals and responses which were involved in pair formation and mating.

Now, it is important to emphasize that speciation could occur at either step 2 or step 3. Evolutionary differentiation in allopatry could be so strong that by the time secondary contact was made there was no reproductive "confusion." Breeding occurred entirely within groups and therefore reproductive isolation between them was complete; they were two species. Alternatively, speciation might have occurred in step 3; the differences evolved in allopatry were not sufficient to isolate the groups reproductively, but the differences were then reinforced in sympatry under a regime of divergent selection. Stresemann (1936), who did not visit the islands, laid stress on the former process, speciation entirely in allopatry. Lack (1940a, 1940b, 1945) initially did the same. Subsequently, while acknowledging that some populations may have evolved intersterility in allopatry, he dwelt largely on the reinforcement of reproductive isolation in sympatry, and hence favored the view that the speciation process began in allopatry and was completed in sympatry (Lack 1947).

Steps 4 and 5 are a repetition of steps 2 and 3, with the newly formed species themselves going through cycles of allopatric differentiation (step 4) and enhancement of differences in subsequent sympatry (step 5). The distinction between the cycles is small but important. In steps 4 and 5 a species encounters other species formed in an earlier cycle, whereas in steps 2 and 3 it does not. This distinction will be useful in later discussion.

In Lack's version of the process, the evolutionary shifts in sympatry involved ecological specialization and a consequent reduction in competition between the established and newly arrived populations. An implicit assumption is made here that the original species was an ecological generalist and all derived species were, in comparison, specialists restricted to certain aspects of the very broad niche of the original species. This assumption was

adopted in the construction of a phylogenetic tree (p. 10). It is an oversim-
plification. That the original species combined, for example, the blood-
drinking habits of the sharp-beaked ground finch, the tool-using habits of
the woodpecker finch, and the leaf-eating habits of the vegetarian finch is
very doubtful. A better conception is one involving directional change in
morphological features at each speciation event, accompanied (and perhaps
preceded) by a directional change in feeding ecology, to judge from the
strong associations between morphology and ecology observable today
(Chapter 6). Each ecological change facilitated further, directional, ecological change, whether it resulted in specialist feeding habits, as in the cactus
finch (*G. scandens*), or generalist feeding habits, as in the medium ground
finch (*G. fortis*). In this way successive speciation events gave rise to the
full adaptive radiation, with different lineages radiating in very different directions from the common ancestral stock.

An example. To illustrate the processes, Lack pointed to the pattern of morphological differentiation and geographical distribution of the tree finches,
Camarhynchus psittacula and *C. pauper* (Fig. 75). *C. pauper*, the medium
tree finch, is restricted to Floreana. The large tree finch, *C. psittacula*, occurs on Floreana too, as well as on the central, western, and some northern
islands. Birds on the central islands are the same size as Floreana ones;
northern birds have longer beaks and western birds are smaller.

Lack took the morphological distinctiveness of *C. pauper* and the lack of
distinctiveness of the sympatric *C. psittacula* to mean that Floreana was colonized initially by *C. pauper* or the species which gave rise to it, and later
by *C. psittacula*. He then argued: "Formerly, *pauper* and *psittacula* (sens.
strict.) were geographical races of the same species, but by the time that
they met on Charles [Floreana] they had become so different that they did
not interbreed, and so they have become separate species" (Lack 1947, p.
128). Interpreted literally, as was presumably intended, this is an example
of speciation fully in allopatry, with no interaction at secondary contact.
Lack did not offer an example of speciation that involved divergence in
sympatry, despite laying emphasis on the process. I believe the reason for
this lies in the difficulty of identifying colonization routes in an archipelago
of islands arrayed like the Galápagos. Note, for example, the uncertainties
in Figure 75 reflected in the double-headed arrows. For a linear chain of islands like the Hawaiian islands, which moreover were formed in sequence,
the task is much simpler (e.g. Bock 1970, Sibley and Ahlquist 1982, Carson
1983).

There are enough uncertainties to cloud the interpretation of the tree finch
pattern and leave room for alternative explanations. It is easy to envisage
the opposite sequence of colonization of Floreana. *C. pauper* may have

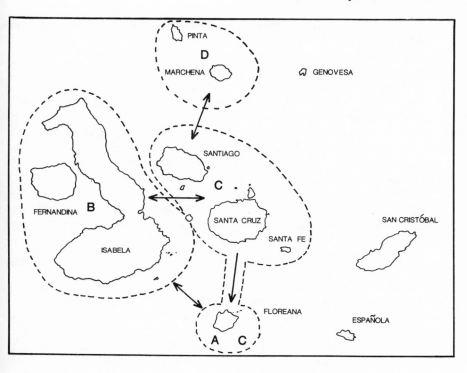

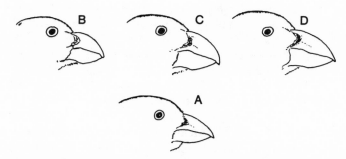

FIG. 75. A possible example of allopatric speciation provided by two species of *Camarhynchus*. *C. pauper* on Floreana (A) may have given rise to the *affinis* form of *C. psittacula* on Isabela (B), from which were derived the *psittacula* form of *C. psittacula* on Santa Cruz (C) and the *habeli* form on the northern islands (D). Recent invasion of Floreana by *psittacula* from Santa Cruz has given rise to coexistence of the original (*C. pauper*) and the derived (*C. psittacula*) species. Uncertainty concerning directions of movement are reflected in the double-headed arrows. Redrawn from Lack (1947).

evolved recently, originating from the small *affinis* form of *C. psittacula* on Isabela and colonizing Floreana after the large form of *C. psittacula* had done so from Santa Cruz. Lack (1969) himself raised the possibility of a recent origin of *C. pauper* to account for its absence on other islands. If it was the second colonist and not the first, its subsequent morphological change can be explained as the result of natural selection minimizing competitive interaction with *C. psittacula*, and possibly facilitated by the absence of *Cactospiza pallida*; in other words, an explanation strictly in terms of Lack's own model, of interaction at secondary contact. We have no means at present of distinguishing between this reconstruction and Lack's. For example, the comparative rarity of *C. psittacula*, to judge from the small number of specimens in museum collections (Lack 1945, Fig. 5 in Grant 1984c), the observations of Felipe Cruz (pers. comm.), and the absence of fossils (Steadman 1985), can be used to support Lack's hypothesis or my alternative.

Even the starting supposition of Lack's scheme may be wrong; *C. pauper* may have been derived from or given rise to the smaller *C. parvulus*, despite resembling the larger *C. psittacula* more in its proportions (Plate 28 and Fig. 21). Biochemically *C. pauper* and *C. parvulus* show a close affinity with one another (Fig. 72). From the biochemical analysis of a single specimen of *C. psittacula* (Polans 1983) we can conclude only that *C. pauper* and *C. parvulus* are not known to differ from each other any more than either differs from *C. psittacula*. Affinities within the genus remain to be established.

This example, with its several interpretations, shows it is much easier to discuss the principles of allopatric speciation with hypothetical examples than it is to find real examples that illustrate successive stages unambiguously. Perhaps with the refinement of biochemical techniques in the future it will be possible to identify which population gave rise to which (Grant 1980), and thereby determine more reliably the sequence of evolutionary events. And with the aid of a good fossil record it will be possible to establish which species were present when those events took place.

The importance of isolation. Archipelagos are suitable environments for speciation because they provide conditions for the establishment of many populations of a species, and the opportunity for them to evolve independently. This may be called geographical opportunity. Mockingbirds provide a clear illustration of one response to this opportunity. Four distinctive forms (Plate 62), apparently different species (Bowman and Carter 1971, Grant 1984b, 1984c), occur entirely allopatrically (Fig. 76), having been derived like Darwin's Finches from a common ancestor. They are *Nesomimus melanotis* on San Cristóbal, *N. macdonaldi* on Española, *N. trifasciatus* on Floreana (now extinct), Champion, and nearby Gardner, and *N. ga-*

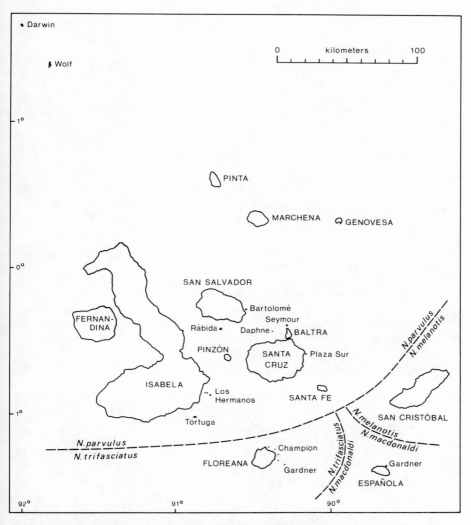

FIG. 76. Allopatric distribution of the four species of mockingbirds (*Nesominus*) on the Galápagos islands. San Salvador is an alternative name for Santiago (Appendix). From Grant (1984c).

PLATE 62. Differentiation of the mockingbirds, genus *Nesomimus*. Upper: *N. parvulus*, Genovesa. Lower: *N. melanotis*, San Cristóbal (*photos by R. L. Curry*).

Upper: *N. trifasciatis*, Champion (*photo by H. Hoeck*). Lower: *N. macdonaldi*, Española.

lapagoensis on all remaining major islands. They differ in plumage, eye color, size, and proportions (Abbott and Abbott 1978), but these are minor variations upon a basic mockingbird plan, and are quite insignificant in comparison with the diversity of the finches.

In addition to geographical opportunity, Darwin's Finches have had ecological opportunity on the Galápagos in the form of diverse environments (e.g. arid, transitional, and humid forests), lacking, or nearly lacking, other bird species. For example, there are no woodpeckers, wrens, orioles, or representatives of many other families of tropical birds on the Galápagos, and, assuming that they never have been there, their absence must have been an important factor facilitating the adaptive radiation (Sushkin 1929, Gulick 1932, Swarth 1934, Stresemann 1936, Lack 1945, 1947). Since Darwin's Finches have radiated much more than any other species on the Galápagos, it is tempting to suppose they had the maximum possible ecological opportunity by arriving first. Alternatively, they may have arrived no earlier than the mockingbirds, flycatchers, and others, but evolved faster (Salvin 1876). These alternatives could be explored in the future with biochemical techniques. Of particular interest is the possibility that the warbler finch evolved in a warbler-free environment. It is supported by fossil evidence suggesting that the yellow warbler (*Dendroica petechia*) is a recent arrival (Steadman 1985).

Within the archipelago there is modern evidence among the finches for the importance of geographical isolation in the speciation process. The incidence of endemism (subspecies and species combined) is highest on the most isolated islands (Rothschild and Hartert 1899, Swarth 1934, Stresemann 1936, Lack 1945, 1947). Through the use of multiple regression analysis, Hamilton and Rubinoff (1963, 1964, 1967) have been able to explain statistically 67–98.5% of the variation in endemism among islands by the single factor of distance to the nearest island. They argued from this that speciation has occurred mainly on the most isolated islands, followed by dispersal to the larger central islands which generally support larger numbers of species. In contrast, the inner islands of the archipelago would be unsuitable for the major episodes of differentiation because of increased dispersal between islands, interbreeding, and obliteration of the initial steps of differentiation. This is a modern statement of Stresemann's (1936) hypothesis that speciation is retarded by gene exchange between populations.

There are two shortcomings with this purely geographical theory. The first is that occasional interbreeding with immigrants from a partly differentiated population has relatively little effect on the mean size of metric traits of the resident population (Chapter 8). The second is that it ignores ecological factors. As the events described in Chapter 8 show, the effects of immigration can be overwhelmed by natural selection. Local ecological circumstances have the potential of determining the characteristics of local finch populations much more powerfully, under the forces of natural selection, than gene flow.

Some correlations further hint at that potential. Two islands have similar bird faunas in proportion to the similarity of their floras (Power 1975). What determines the similarity of the floras? In large part it is a product of the similarity in the number of plant species on each island (Connor and Simberloff 1978), and plant species number is in turn most strongly correlated with island area (Hamilton et al. 1963, Johnson and Raven 1973, Connor and Simberloff 1978). Since islands of similar size tend to be near each other (Hamilton et al. 1963), especially the larger ones, near islands have similar floras and bird faunas, and well isolated islands tend to have dissimilar floras and bird faunas (Rothschild and Hartert 1899, Power 1975, Abbott et al. 1977). There is some uncertainty in these patterns of correlation because, in addition to the limitations of multiple regression analysis (Connor and Simberloff 1978), vegetated areas of some islands (i.e. effective island area) have not been precisely measured (Chapter 2), and plant lists for islands have changed since these analyses were done. Therefore I will draw only a tentative conclusion. It is that endemism is highest on the most isolated islands in part for reasons of geographical isolation, but also in part because of the distinctive ecological conditions on those islands.

I will return to the idea of ecological distinctiveness in the next chapter.

Debatable elements of the model. The most controversial aspects of the allopatric model are the postulated reproductive and ecological interactions occurring at the secondary contact phase of each cycle (steps 3 and 5). The first debatable matter is whether any interactions took place at all, and the second is whether, if they did, they were reproductive, ecological, or both. These are issues of broad significance. Both sides of the argument on reproductive interactions were laid out nearly fifty years ago by two geneticists, Dobzhansky (1940) and Muller (1940), and the argument continues (Paterson 1980). The argument about ecological interactions has had a parallel career (Andrewartha and Birch 1954, Brown and Wilson 1956, Grant 1972b, Abbott 1980, Arthur 1982, Grant and Schluter 1984, Simberloff 1983a, 1983b, 1984).

I will have to postpone answering these questions until Chapter 13. The issues and the evidence are complex, and will be discussed at length in the next three chapters. For the remainder of this chapter I will consider alternatives to the allopatric speciation model as a whole.

ALTERNATIVE MODELS OF SPECIATION

It is possible for speciation to occur without an allopatric stage. In theory, one species may split into two species under two alternative, non-allopatric, sets of conditions (Bush 1975). In the first, populations of a species are separated in space but are geographically contiguous and therefore their members are in contact with each other. The populations are said to be parapatric, and if they become separate species, through selection, drift, or both, the process is described as parapatric speciation. In the second, a species splits into two through adaptation of two segments of the population to different niches in the same habitat. This is sympatric speciation.

PARAPATRIC SPECIATION

There is little scope for this to occur on Galápagos islands because it requires a geographical distribution of populations of a species across a strong gradient of environmental conditions. These conditions are conceivably met on the larger islands, especially along altitudinal gradients, but despite this there is extensive movement of birds along the gradients (Chapter 7), especially in the nonbreeding season when selection pressures are potentially strong, which means that individuals are subjected to more similar conditions over a lifetime than they are within a single season. As a result, geographical trends in morphological variation are weak (Chapter 4). It seems unlikely that speciation in the finches occurred parapatrically.

If there is an exception to this general conclusion it is provided by the two species of *Cactospiza*. They constitute the only example among Darwin's Finches of two congeneric species occupying different habitats. Populations of the woodpecker finch and mangrove finch occur on the same island, Isabela, but in separate habitats: arid zone and mesic forests on the one hand and mangroves on the other. They could be descended from a single species that once comprised contiguous populations along a habitat gradient. But, when the split occurred, presumably many thousands of years ago, they could as easily have been isolated from each other geographically, as is largely the case today. The mangrove stands occupied by the mangrove finch are essentially coastal islands, surrounded by sea on one side and by sparsely covered or bare lava on the other. Therefore, even in this most favorable example, there is doubt about the geographical contiguity of the original populations. Furthermore, even the complete ecological separation is in doubt, as the woodpecker finch has been found breeding in mangroves on the east side of Isabela (Gifford 1919). Neither species has been studied in any ecological detail in modern times.

Lack (1947) considered the possibility of speciation occurring in this way, that is with a single population becoming subdivided in two separate habitats on the same island, but then dismissed it because of "two insuperable objections." The first was the incompleteness of the isolation of the two subdivisions of the population. The second was the absence of any instances of birds known to be in the process of differentiating in adjoining habitats. Neither objection has so much force nowadays. The lack of separation is not necessary, in principle, for speciation to occur (e.g. Endler 1977, Lande 1982). Partly differentiated populations in adjoining habitats can be found elsewhere in the world, although there is always a question of whether the differentiation began when the populations were in contact or not. Nevertheless, in no case is the empirical evidence strong for parapatric speciation on the Galápagos.

Sympatric Speciation

Sympatric speciation can occur under a more stringent set of conditions than pertain to the parapatric origin of species. According to a model developed by Maynard Smith (1966), the conditions are as follows: the presence of genetic polymorphisms (or, more generally, genetic variation), separate regulation of the numbers of each morph in the "niche" it occupies, large selective advantages to the morphs in their respective niches, and the evolution of a mechanism causing reproductive isolation between the morphs. These conditions have been explored theoretically and extended by others (e.g. Dickinson and Antonovics 1973, Felsenstein 1981), but the chances of them being met in nature do not seem to be high because additional factors must be incorporated to explain how ecological and reproductive isolation arise. This is easier to do if the relevant traits vary discretely (Pimm 1979) than if they vary continuously (Felsenstein 1981). The stringent conditions, together with the fact that purported examples of sympatric speciation can be easily attributed alternatively to allopatric speciation, has given risen to general skepticism towards the possibility of sympatric speciation occurring or, at most, ever being important (Futuyma and Mayer 1980). Nevertheless, it has been raised twice in connection with the evolution of Darwin's Finches.

Geospiza fortis on Santa Cruz. The sympatric speciation discussed by Lack (1947) has been considered a special case of the Maynard Smith model (Grant and Grant 1979), but in fact it is closer to parapatric speciation. The first discussion of Darwin's Finch evolution in the context of the model was given by Ford et al. (1973). They found evidence of bimodality in the frequency distribution of beak depths in a sample of male medium ground finches (*G. fortis*) measured at Bahía Academía on Santa Cruz. They suggested that the population might be in the process of splitting into two under a regime of disruptive selection in a heterogeneous environment. Certainly there is a broad range of food types in this floristically diverse area (Abbott et al. 1977) that could support a broad range of finch types.

However, as I discussed in Chapter 8, the bimodality can be explained in two other ways. It could be the result of recent immigration of medium ground finches from one of the southern islands such as San Cristóbal, with a tendency for the original and immigrant components of the population to mate assortatively (Grant and Grant 1979); in other words, this could be the secondary contact phase of speciation, according to the allopatric model. Medium ground finches on San Cristóbal are larger on average than those on Santa Cruz, and the ones on southern Isabela are even larger, the largest in the archipelago (Lack 1945, 1947). Bimodality could also be the result of hybridization with the sympatric large ground finch. This was originally suggested by G. L. Stebbins (in Bowman 1961; see also Snow 1966).

Biochemical studies could resolve the issue. The immigration hypothesis would be supported if the larger members of the *G. fortis* population on Santa Cruz were found to be particularly similar biochemically to the finches on San Cristóbal. The hybridization hypothesis would be supported by a similar correspondence between large members of the medium ground finch population and all members of the large ground finch population. The sympatric divergence hypothesis would gain credence in proportion to the demise of the other two.

Whether or not the process of divergence would ever yield two species depends on the mating pattern. It is not even known if mating is assortative with respect to beak form in this population. This fascinating situation deserves further attention. The population would be fairly easy to study in view of the proximity of the Charles Darwin Research Station.

Geospiza conirostris on Genovesa. I shall first summarize the main facts about this population, and then consider their relevance to sympatric speciation.

Males of the large cactus finch sing either song *A* or song *B* (Chapter 9). In 1978 we made the surprising discovery that the two groups of males differed in average beak length (Fig. 77), but not in other dimensions (Grant and Grant 1979). In the dry season of that year, small samples of the two groups were observed to feed in different ways associated with their beak differences. The longer-billed song *A* males drilled holes in *Opuntia* fruits by hammering and tearing, then removed seeds and consumed the surrounding fleshy arils. The shorter-billed song *B* males, did not feed in this manner, but extracted insect larvae from rotting *Opuntia* pads on the ground after ripping them open with a tearing motion of the bill. These observations on morphological and ecological differences suggested that the population was ecologically subdivided.

The additional possibility of a reproductive subdivision of the population was suggested by an unexpected feature of the distribution of breeding

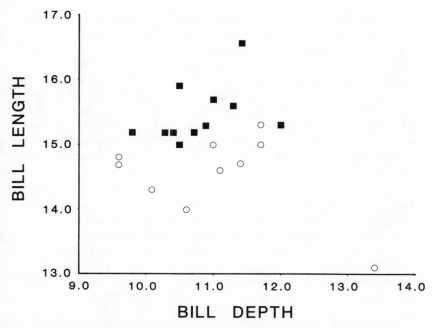

FIG. 77. Beak characteristics of two groups of *G. conirostris* males on Genovesa in 1978, those singing song A (■) and those singing song B (○). The two groups differ significantly in bill length but not in bill depth; measurements of both dimensions are in mm. From Grant (1983a).

birds. No two mated males singing the same song type held adjacent territories. In contrast, unmated males held territories randomly distributed with respect to song neighbors (Fig. 70). Males have fixed territories whereas females are mobile (Chapter 9). Therefore the peculiar difference in the distribution patterns of mated and unmated males can be interpreted as the product of female choice: females may have discriminated between males with homotypic neighbors and males with heterotypic neighbors, and only paired with the latter. Leaving aside the question of why a neighbor's song type would be important to a female (see Chapter 9), the apparent discrimination on the basis of song *could* reflect a tendency for females to pair with males singing the same song type as their fathers'; and if this was the case, the population would be at least partly subdivided into two breeding groups.

Another line of evidence pointed to a subdivided population structure. Nestlings have either pink or yellow beaks (Plate 7). The discrete difference is evident at hatching and persists until melanin is deposited throughout the beak by about two months of age. All species of Darwin's Finches so far examined (Grant et al. 1979), including the Cocos Island finch (T. W.

Sherry and T. K. Werner, pers. comm.), exhibit this polymorphism. Nestlings in a single brood may be all pink morphs, all yellow morphs, or a mixture of the two. The frequency of yellow morphs in nestlings of the song A males (36%) was twice as high as in the nestlings of song B males, and the difference was statistically significant ($P < 0.02$; Grant and Grant 1979). If the color polymorphism is under relatively simple genetic control, as is the case with a similar polymorphism in chickens (Hutt 1964), then a genetic difference in terms of allelic frequencies between the two sets of $G. conirostris$ families is indicated. A two-locus control of the polymorphism with epistasis is suggested by the pattern of inheritance in $G. fortis$ and $G. scandens$ on Daphne Major (Table 15).

TABLE 15. Nestling bill color morphs in relation to parental morphs on Daphne Major in 1983 and 1984. Neither pink × pink or yellow × yellow pairs consistently produced all offspring of one color. This suggests that at least two genetic loci are responsible for the phenotypes. Note the difference between species in their frequencies.

Number of Families	Parents		Offspring		
	♂	♀	Pink	Yellow	% Yellow
1. *G. scandens*					
25	Pink	Pink	60	18	23.1
15	Pink	Yellow	30	25	45.5
2	Yellow	Pink	4	1	20.0
4	Yellow	Yellow	4	21	84.0
46			98	65	39.9
2. *G. fortis*					
104	Pink	Pink	310	33	9.6
17	Pink	Yellow	33	25	43.1
15	Yellow	Pink	45	10	18.2
2	Yellow	Yellow	1	3	75.0
138			389	71	15.4

The study of individually marked birds in this population was extended in order to see if these unusual and unexpected patterns were sustained, and to assess directly the possibility of assortative mating by song. Results were mixed. It was found that individual males sang the same song in successive years, and that sons sang the same song as their fathers' (B. R. Grant 1984). The potential clearly exists for a reproductive subdivision of the population through mating associated with song type. The difference in bill length between the males in 1978 was confirmed; those surviving to 1980 and 1981, recaptured and remeasured, differed significantly again. Furthermore, the difference in frequency of the yellow morph among the nestlings of the two sets of families was sustained in successive years (Grant and Grant 1983).

But, of crucial importance, daughters were not found to pair preferentially with males that sang the same song type as their fathers'. Their first mates were as likely to sing heterotypic as homotypic song, so the initial pairings are random and not assortative. In the population at large also, there was no evidence of assortative mating by bill length. This being so, and given the high heritability of bill length and other traits (Grant 1983a), it was not surprising to find that the bill length difference between the male song groups disappeared in subsequent years as a result of recruitment of new males with intermediate bill lengths to each group. Correspondingly, their diets in the dry season no longer differed. And the alternating territory pattern of mated males broke down, although it did reappear at the beginning of the breeding season in two out of the next three years of study (Grant and Grant 1983, B. R. Grant 1984).

From these results it appears that the structure of the population is in a state of flux. It can be thought of as oscillating between fission and fusion tendencies (Grant and Grant 1983). Fission is induced by disruptive selection under conditions of low, dry-season, food availability, especially in drought years such as the year (1977) preceding this study. There is a tendency for disruptive selection to occur in other dry seasons (B. R. Grant 1985). Fusion is produced by random mating at first breeding, and by the observed tendency (Grant and Grant 1983) for females to re-pair with males of opposite song type to their previous mates' following a breeding failure. The occasional morphological difference between the two groups of males can be explained by large and long-billed females coming into reproductive condition earlier than small ones, and pairing preferentially with males of the rarer group, regardless of their own fathers' song, because such males are likely to be adjacent to males of the opposite song type (see Chapter 9).

Now, to return to the model of sympatric speciation: this population shows evidence of occasional ecological subdivision but no obvious reproductive subdivision. It is clearly a single species exhibiting one weak, and temporary, predisposing condition for splitting into two. In other species of Darwin's Finches, reproductive isolation has arisen as a consequence of divergence in song and appearance (Chapter 9). In this population, the divergence of two groups would presumably have to proceed very much further for reproductive isolation to develop, yet there are forces restraining that divergence. For sympatric speciation to occur, therefore, the most conducive conditions would seem to be extreme reductions in population size, long enough or often enough for chance to play a role in the development of positive assortative mating. The likelihood of this happening seems remote, but is not out of the question. At a more modest level, this mechanism may be the explanation for the origin of the nestling bill-color frequency difference between the two groups of males.

All examples of divergence taking place in sympatry can be easily reconciled with the allopatric speciation model (Futuyma and Mayer 1980). For example, the incipient divergence of two song groups in the population on Genovesa could be explained as the events characterized by step 3 in the model, the recent colonization of Genovesa by one group that has differentiated from the other to a small extent on another island. There is no evidence to support this alternative, although of course fossils of a now extinct population may one day be found on another island and show this alternative to be highly plausible. The nearest population of *G. conirostris* is very far away (>150 km), on Gardner by Española (Fig. 1), and birds on this island and on Española are very different from conspecifics in both song (Bowman 1979, 1983, Ratcliffe 1981) and morphology (Grant and Grant 1982). Elsewhere, the most similar population morphologically is *G. scandens* (cactus finch) on the neighboring island of Marchena. The similarity was strong enough for Lack (1945) to think they might be conspecific. However, in beak structure *G. scandens* on this island is not more similar to one song group of *G. conirostris* on Genovesa than to the other, and in song it is clearly different from both (Fig. 78). In fact, its song and beak proportions more closely resemble those of *G. scandens* on other islands than those of *G. conirostris* (e.g. see Fig. 19). Nevertheless, it is possible that one of the songs of *G. conirostris* originated on another island and was imported to Genovesa by immigrant *G. scandens*. If so, the source was probably Daphne Major (Fig. 78) or Santa Cruz (Fig. 67). Interbreeding probably occurred, for otherwise the new song would not enter into the repertoire of the resident population (Chapter 9). Song, like neutral alleles, may stand a higher chance of becoming incorporated permanently into a new population through interbreeding than the alleles which govern metric traits. Interbreeding and song transfer would explain some striking similarities in the songs of different species (Chapter 9), but they are not sufficient to explain the features of the partially subdivided population of *G. conirostris* on Genovesa.

ALTERNATIVES TO GRADUAL GENETIC CHANGE

Instead of genetic divergence occurring gradually, in small steps, it could have occurred abruptly, in one or more large steps. For example, the doubling of chromosome number in plants at a single step has resulted in two species being formed from a single one. In this case a genetic change has been the cause of speciation. Templeton (1980, 1981) has introduced the

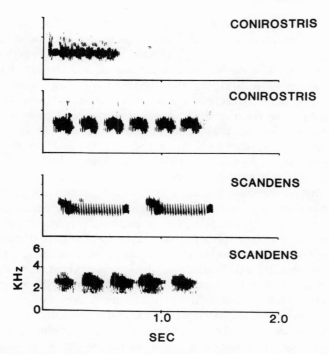

FIG. 78. Songs of *G. conirostris* from Genovesa, and *G. scandens* from Marchena (third from top) and Daphne Major (bottom). The song of *G. scandens* on Marchena is not like either of the *G. conirostris* songs, whereas the main song of *G. scandens* on the more distant island of Daphne Major resembles, in structure and pattern, one of the *G. conirostris* songs (type A). The Marchena sonagram was kindly supplied by R. I. Bowman.

term "genetic transilience" to describe this mode of speciation. It occurs because of genetic instabilities, and despite, rather than as a result of, selection (Templeton 1981). Barrowclough (1983) has suggested that speciation may have occurred among Darwin's Finches by this mechanism, but this seems very unlikely because it is generally expected to be very rare in nature (Templeton 1980), requiring special genetic systems and special population structure (e.g. favoring founder effects) that are not known to exist in the finches. Moreover, uniformity of chromosome characteristics seems to be the rule among Darwin's Finch species (Fig. 58); in both their number and gross morphological structures, chromosomes of the species are not strongly differentiated (Jo 1983).

An older idea is that hybridization has played a role in the development of new species. Lowe (1936) took the extreme position that a small number of original species hybridized to form a "swarm" of similar species. Apparently he misunderstood Swarth (1934) and believed that species were neither ecologically nor reproductively isolated from each other (Stresemann 1936). Lack (1945) initially attributed a hybrid origin to populations of *Geospiza* on Daphne Major, Los Hermanos, and Darwin. Later he changed his mind, realizing that the absence of any known hybridization among the finches is a weakness of even a modified form of Lowe's idea, as is the rarity of the postulated mechanism of speciation in birds in general (Lack 1947). Since there are no exact parallels to Darwin's Finches among other birds anyway, other birds may not be a reliable guide for dismissing this theory. But as Stresemann (1936) pointed out, hybridization would generally prevent, not promote, speciation.

Now that hybridization is known, both within and between genera, the hypothesis needs to be reconsidered. The strongest argument against it is that it is unnecessarily complex. Special circumstances would have to be invoked to explain how species A and C hybridized to form B individuals with roughly intermediate characteristics, and how, from them, a population of interbreeding individuals arose whose members bred rarely, if at all, with members of A and C. A much simpler explanation for the existence of A, B, and C would be directional selection on one of them. For example, all six species of *Geospiza* are so similar to each other that a smooth transition from the smallest to the largest can be made by choosing the appropriate specimens (Fig. 3). Where there are marked morphological discontinuities among three species, the case for the hybrid origin of the intermediate one is stronger, for here it could be argued that it is on an adaptive peak separated from neighboring peaks by deep valleys in the adaptive landscape that cannot be crossed under a regime of directional selection operating on either of the two ancestral species. Given the present array of species, their morphological relationships (Fig. 21), and their times of origin, the only possible example I can identify is the trio, warbler finch–woodpecker finch–large tree finch. But even in this case the intermediate species, the woodpecker finch, is not very different in shape from the warbler finch or in size from the large tree finch (Fig. 21). Especially in the latter case, an adaptive trough between them is not obvious.

Like Stresemann (1936) and Lack (1947), but for different reasons, I doubt if hybridization has resulted in the formation of new species in so simple a fashion. This is not the same as saying it has had no role in speciation. On the contrary, hybridization could have facilitated evolutionary responses to directional selection pressures leading, ultimately, to the formation of new species, in two ways. First, hybridization results in the elevation of genetic variation in populations. Second, it results in novel phenotypes that are determined by unique combinations of genes. Conceivably, enhanced mutation rates (Chapter 8) contribute to the production of novelties. These are potential escapees from the constraints of strong genetic correlations among traits, and as such they could be the starting point of a new evolutionary lineage.

CONCLUSIONS AND SUMMARY

Material in this chapter is organized around the problem of determining how Darwin's Finch species differentiated from a common ancestor into at least fourteen species, and possibly more if some species arose and then became extinct.

Their systematic and geographical origins are obscure. According to geological evidence, the original colonists of the Galápagos must have arrived by overwater flight from the Central or South American continent, or from Cocos Island, some time in the last five million years. They may have arrived as recently as under one million years ago. This is suggested by a time scale of differentiation that is estimated from the electrophoretically detectable biochemical differences among the species. The biochemical evidence further suggests, as does morphological data, that the warbler finch was the earliest lineage to split off from the rest, and that some time after this differentiation in the last half million years the tree finches and ground finches separated from their common ancestor. Relationships among species within each group are much more difficult to establish reliably. Speciation occurred within each group fairly recently, but apparently not in the last 50,000 years. Fossils have been discovered in the last decade, and they may eventually shed light on the question of which species gave rise to which.

Affinities are apparently much stronger among Darwin's Finches than they are between any one of them and species on the continent. Such relationships provide strong justification for believing they were all descended from a single group of colonists, but simultaneously make it difficult to identify the continental lineage from which they were descended.

Gradual change under directional selection was the dominant mode of speciation. There is no evidence for the genetic revolutions that perhaps occur in the founding of populations by a very small number of individuals. Chromosomes are rather uniform among the species, both in number and in gross morphology. Hybridization may have played a role in speciation through the contribution of new alleles to a population undergoing divergent evolution, but not as a means of producing new species by itself.

Lack's model of speciation accounts for the derivation of the thirteen species on the Galápagos from a single ancestor in terms of small differentiation of populations in allopatry, followed by enhancement of differences when two such populations made secondary contact on an island through the dispersal of members of one to the island occupied by the other. This is known as the allopatric model of speciation, since an allopatric phase of differentiation is essential for speciation to occur. It might also be sufficient for speciation to occur, but Lack believed this to be generally unlikely because he thought the islands were not different enough to promote substantial adaptive change. The enhancement of differences in sympatry was driven by natural selection acting against individuals of the two populations who were so similar that they competed with each other for food, and bred with each other with reduced reproductive success. Successive speciation events produced species adapted in very different ways to exploiting the environment for food—different lineages radiating in very different directions from the common ancestral stock.

In theory one species could split into two species without a complete separation of the two derived species. This could happen with populations separated in space but geographically contiguous, so that their members would be in contact with each other to some extent. If the populations experienced very different environmental conditions and selection pressures they could develop into separate species parapatrically. The woodpecker finch and mangrove finch might have formed in this way. They are the only pair of congeneric species that occupy different habitats, but even in this case an allopatric origin is just as plausible.

Speciation could occur sympatrically if a more stringent set of conditions were met; it could occur through adaptation of two segments of a population to different niches in the same habitat. The *G. fortis* population on the southern coast of Santa Cruz and the *G. conirostris* population on Genovesa each show evidence of a small differentiation under disruptive selection. A detailed field study of the latter population has revealed no reproductive isolation between the two groups, however, and I conclude that sympatric separation would have occurred only under a most unusual set of circumstances. The allopatric model is the best explanatory scheme for the speciation of Darwin's Finches.

CHAPTER ELEVEN

Ecological Interactions during Speciation

INTRODUCTION

The allopatric model of the evolution of Darwin's Finches has become a textbook example of what is generally regarded as the principal method of speciation in animals. Its widespread acceptance was not due to its originality; the basic idea of speciation being an exaggerated form of subspeciation, or divergence in separate locations, had been propounded by Darwin (1859) and consolidated by Rensch (1933), Stresemann (1936), Dobzhansky (1937), Huxley (1942), Mayr (1942), and others. The model owed its acceptance to the clarity with which observations were shown by Lack to fit the theory. This was one of his major contributions. Another was to show, by argument, and with some supporting examples, how reproductive and ecological interactions at the secondary contact phase of each speciation cycle could have influenced the radiation (Lack 1947). I shall now examine these interactions in more detail because they are not as clear as they appear to be at first sight. In this chapter and the next one I will discuss ecological interactions, specifically competition for food. In Chapter 13 I will discuss reproductive interactions, and then reexamine the model in the light of evidence for both types of interaction.

ECOLOGICAL ISOLATION

At secondary contact in a cycle of speciation, immigrants and residents are likely to compete for food to a degree dependent upon the initial difference in bill size.

Lack's views on this underwent a dramatic shift between 1939–1940 when he wrote the initial summaries of his findings (Lack 1940a, 1940b) and his monograph (Lack 1945), and 1943–1944 at the time of writing his book (Lack 1947). Initially he thought competition was unimportant. Quoting Gause (1939, p. 255), "if two or more nearly related species live in the field in stable association, these species certainly possess different ecological niches," he went on to argue that *G. magnirostris* and *G. fortis* are exceptions, as are *G. fortis* and *G. fuliginosa* and the tree finches *C. parvulus* and *C. psittacula*. They are exceptions because their diets were thought to be nearly identical (Chapter 6). He was quite emphatic on this point. He responded to Darwin's (1859, p. 401) opinion that "natural selection would probably favour different varieties in the different islands," because they would have to compete with a different set of organisms, with "There is no evidence in favor of Darwin's suggestion. In fact, there is no evidence whatever, in any of the island forms of Geospizinae, that their differences have adaptive significance" (Lack 1945, p. 117; see also Chapter 6).

Lack (1947, p. 62) revised his opinions, apparently influenced by the writings of Huxley (1942) about dietary differences between species of different sizes (Lack 1973), and by conversations with G. C. Varley (Lack 1947, p. 166). "My views have now completely changed, through appreciating the force of Gause's contention that two species with similar ecology cannot live in the same region (Gause, 1934). This is a simple consequence of natural selection." He continued: "In the case of the three species of *Geospiza*, there are similarities, but also established differences, in their diets, and though further evidence is much needed, it is provisionally concluded here that, so far from being unimportant and purely incidental, these food differences are essential to the survival of the three species in the same habitat" (Lack 1947, p. 63). Finally he wrote, with almost as much force as he had earlier given to the opposite conclusion, "To conclude, in Darwin's finches all the main beak differences between the species may be regarded as adaptations to differences in diet" (Lack 1947, p. 72).

The passages have been quoted at length to show how complete was his reversal of opinion on the adaptive significance of beak sizes, and of the differences between sympatric species and the (implied) competition between them. Gause's work was referred to in both instances and no new data were available to Lack when he wrote his book. If the same data can lead to radically different interpretations of past evolutionary processes, those interpretations must be precarious. They have been. Lack encountered strong opposition to his general views on competition when he delivered them to a meeting of the British Ecological Society in 1944 (see Lack 1973). Years later, strong opposition was mounted to the specific role he assigned to competition in the adaptive radiation of the finches (Bowman 1961). Competition has been, and continues to be (Strong et al. 1979, Simberloff 1983a, 1983b, 1984), the most contentious part of the allopatric model.

CAUSES OF INITIAL DIFFERENTIATION

The pivotal role given to competition in step 3 (and steps 4 and 5) of the allopatric model can be appreciated as the resolution of a problem confronted by Lack when trying to explain how and why species diverged.

Darwin (1842, p. 477) wrote: ". . . I must repeat, that neither the nature of the soil, nor height of the land, nor the climate, nor the general character of the associated beings, and therefore their action one on another, can differ much in the different islands.'' Like Darwin, Lack was impressed by the similarities among islands in their physical (volcanic) features and in their habitats at comparable altitudes. ''Environmental conditions in the coastal zone on the Galápagos Islands would seem, so far as they affect the species of *Geospiza* and *Certhidea*, to be similar on the different islands'' (Lack 1945, p. 85). This presented him with a problem when trying to convert patterns of differentiation in space into processes of differentiation in time— when trying to work backwards in time from the present diversity of species, through repeated processes of differentiation, to the initial colonization. How did differentiation begin? If the habitat of a newly colonized island was similar to that of the original island, why did any genetic change take place in a population following the colonization? Adaptation through directional selection could not be the answer if the selection pressures were the same in the two environments. Lack (1945, 1947) believed that most of the current differences in morphological traits between populations of the same species on different islands lacked significance. He labeled the differences, as opposed to the traits themselves, as ''nonadaptive,'' while acknowledging the difficulty of proving anything to be nonadaptive.

One possible answer to the problem is that change occurred through genetic drift (Lack 1945). Lack (1947) later rejected this mechanism on the grounds that contemporary populations comprise several hundreds and probably thousands of individuals, and therefore they are not so small that random processes are likely to predominate. From what we now know about climatic fluctuations (Chapter 7) and the occasionally severe reductions in finch numbers (Chapter 8), genetic drift cannot be dismissed quite so easily, although the argument is broadly correct.

A second possible answer is that it occurred through founder effects (Mayr 1963). By this is meant the founding of a new population by a few individuals that do not possess all of the allelic variation present in the parent population, and whose allelic variation might, by chance, be atypical of the parent population's. Stresemann (1936) was the first to propose a version of this mechanism of initial differentiation, but Lack (1947) argued against it on the grounds that finches move in flocks and were likely to colonize an island in substantial numbers. Again, in view of what we now know of the establishment of new breeding populations (Chapter 8), this argument is overstated. However, the case for the importance of founder effects with polygenic traits has itself been overstated (Lande 1980b, Templeton 1980).

Lack believed the most probable explanation for the differentiation of populations of a species in allopatry to be mutation occurring differently in the separate populations. This is Muller's (1940) theory of differential mutations applied to the finches: different mutations would be favored in the different populations under very similar regimes of selection pressures, resulting in different phenotypes adapted to similar environmental conditions.

By itself this process would not give rise to speciation, except perhaps after an exceedingly long time and presumably through the chance development of some form of intersterility; even this is unlikely in view of the known viability of intrageneric and intergeneric hybrids (Chapter 8). Nor would it be sufficiently pronounced to account for the large differences between species in sympatry. It was for this reason that Lack laid stress on divergent selection in sympatry, impelled in part by competition.

Lack was not entirely consistent on this point, however. He wrote (1947, p.147): "Similar considerations apply wherever two species, originally geographical races of the same species, have become established in the same region. In most cases adaptive or ecological differences probably arose between the two forms during their period of geographical isolation before they met, but their meeting must almost inevitably have resulted in further restrictions of their foods or habitats, and so must have tended to accelerate their adaptive specialization." In this he is arguing for adaptive change in allopatry, in addition to interactions and change in sympatry. Elsewhere he makes it clear that the adaptive change in allopatry comes about through competition with other species (see later); in other words interaction occurs in both stages 4 and 5 of the model when two or more species have already been formed. The key to these views, if I have interpreted them correctly, is that competitive interactions play a major role in adaptive change, and no role in the remaining changes.

AN ALTERNATIVE VIEW

Lack's emphasis on competitive interactions between species, or incipient species, was challenged by Bowman (1961). Bowman pointed out a simple fact which Lack had overlooked: islands differ in their floras. As Beebe (1924, p. 259) put it: "The most astonishing thing about the various islands of the Galápagos is their superficial similarity and their actual diversity." The difference in floras can be appreciated from the early botanical studies of Stewart (1911, 1915), which modern studies have done nothing but confirm (Bowman 1961, Wiggins and Porter 1971, Abbott 1972, Power 1975, Abbott et al. 1977, Connor and Simberloff 1978, Hamann 1981).

This throws a different light on the question of change in allopatry. In stage 2 of the allopatric model, colonizers of new islands would encounter new floristic conditions and be subject to new selective pressures. They would undergo adaptive change. Adaptive change in allopatry removes the necessity of invoking differential mutation as the process giving rise to phenotypic differences between allopatric populations. But more than this, the adaptive change may be so large that when original and derived populations established contact, in stage 3 of the model, they did not interbreed or compete with each other because already they were very different by this time. The same applies to stages 4 and 5.

This is the essence of Bowman's alternative explanation for the adaptive radiation: adaptation to food supplies, without interactions among populations for those food supplies. We thus have two variants of the allopatric model of speciation. In the first (Lack's), differentiation occurs partly in allopatry and partly in sympatry, and in the second, based on but not restricted to Stresemann's ideas and Bowman's more explicitly adaptive arguments, differentiation occurs entirely in allopatry. They have been referred to as the partial allopatric model and the complete allopatric model respectively (P. R. Grant and N. Grant 1983). The distinction between them turns on competition; it plays an important role in the first model and no role in the second. Hence they could alternatively be described as interactive and noninteractive models.

DIFFERENTIATION ENTIRELY IN ALLOPATRY

The simplest application of the complete allopatric model would take the following form. After the initial colonization of one of the Galápagos islands, dispersers successively colonized twelve more islands. Each island had a distinctive flora. Adaptation of the finches to local floras resulted in the formation of thirteen species, which then dispersed and established patterns of sympatry. This is a slightly exaggerated statement of Stresemann's (1931, 1936) view, echoing Rothschild and Hartert (1899), that several species would form in allopatry, each one restricted to one island, before one of them established sympatry with another.

We know this is wrong because the adaptive radiation took place over hundreds of thousands of years, by sequential processes of speciation at an average rate of about one every 50,000 years. A more realistic application of the model would allow for the dispersal of each newly formed species to many if not all of the islands before the next species arose. This means that after the first speciation event there would be an increasing potential for competitive interactions to occur, in both allopatric and sympatric phases of all subsequent speciations, as the differentiation of the group proceeded and the number of species increased. This potential need not have been realized, however, a point of view argued most extensively by Bowman (1961). He could find no reason for believing it is necessary to invoke competition to explain features of the adaptive radiation, but could find plenty of reasons for doubting Lack's arguments and evidence to the contrary. Three quotations summarize, and an additional one illuminates, Bowman's argument:

> . . . on numerous occasions I observed several individuals of at least four species . . . feeding together in the very same shrub of *Scutia spicata* without any apparent interspecific strife. In other words, the birds were not making full utilization of the food resources available to them at this seemingly "critical" time of the year. The food supply was so plentiful . . . that it would seem to have been sufficient to support even a larger population of finches. (p. 274)

The theme of feeding without competing was developed:

> Where we witness two or more species feeding on the same kind of food we must assume that the items taken in common are in plentiful . . . supply. (p. 276)

This leads to:

> The anatomical differences between closely related species of *Geospiza* living in the same locality may be thought of as biological adjustments (adaptations) that prevent these species from competing with each other . . . individuals of one species do not now "compete" for food with certain other individuals of the same species . . . or with individuals of another sympatric species, or at least not in any manner that has evolutionary significance today; and since there is no direct evidence that competition is occurring at the present time, I see no logical reason to assume that it must have occurred in the past. (p. 275)

Notice that neither intraspecific nor interspecific competition for food is believed to occur. Bowman did not explain how and why adaptation to food supplies occurred in the absence of food shortage. Nor did he explain why food supply never limits population sizes, but alluded to it in this final quotation:

> The biological advantages of maintaining the bird population considerably below the maximum level that the food resources of an area can support at any given time, are obvious. Annual variations in the fruiting success of the vegetation necessitates a utilization of the fruits by the birds (and other animals as well) at a level that guarantees a perpetuation of both the vegetation and the animals dependent upon it. (p. 274)

This last statement is an implicit argument that selection acts for the good of the species. Williams (1966) has shown such arguments to rest on false premises; natural selection acts on variation among individuals within a population, blind, as it were, to what is best for the population as a whole (see also Lack 1947, p. 33). But, regardless of this issue, food limitation and competition cannot be assessed by observing whether or not birds feed in the same shrub. Quantitative data are needed on changes in population sizes in relation to changes in food supply. Modern studies that provide such data (Chapter 7) do not bear out the contention that populations remain permanently below carrying capacities set by food supply, and moreover they have produced direct evidence of competition between finch species for food. Therefore it is incorrect to say, "Where we witness two or more species feeding on the same kind of food we *must* assume that the items taken in common are in plentiful . . . supply" (Bowman 1961, p. 276; emphasis added).

The final step in Bowman's argument was, "since there is no direct evidence that competition is occurring at the present time, I see no logical reason to assume that it must have occurred in the past" (Bowman 1961, p. 275). I agree, there is no logical reason to assume that competition *must* have occurred in the past, even though there *is* direct evidence that competition is occurring at the present time; evidence for present competition simply makes past competition more likely. Furthermore, competition in the past is not an assumption, it is a hypothesis that must be treated on its own merits.

Thus there are logical and empirical difficulties with the argument that differentiation always occurred entirely in allopatry. To proceed further we need to know the degree to which variations in food supply alone determine variations in the properties of finches. Then we need to know if additional features of finches are interpretable in terms of competition. Food must be important, competition need not be.

THE FOOD SUPPLY HYPOTHESIS

Certain features of Darwin's Ground Finches are explicable in terms of their food supply, in a manner consistent with the complete allopatric model. Islands differ not only in their floras but in the frequency distributions of seeds and fruits classified into depth-hardness categories (Abbott et al. 1977). An illustration is provided in Figure 79. Genovesa has two large and hard seeds lacking on Daphne Major. The largest and hardest seeds (stones) are those of *Cordia lutea*, and the next category is represented by *Opuntia helleri*, whose seeds are larger and harder than the seeds of *O. echios* on Daphne (Abbott et al. 1977). Therefore, in dispersing from one island to another, the seed-eating ground finches experienced different food supplies. These differences are largely the result of different floral compositions on the islands, and only to a minor extent are they attributable to inter-island varia-

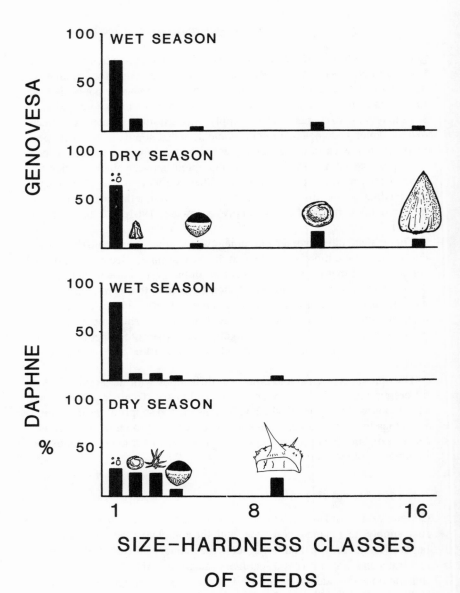

Fig. 79. Frequency distributions of seeds and fruits classified in 16 size-hardness classes (square root of seed depth in mm × hardness in kilograms-force, where 1 kgf is 9.8 newtons). The figure illustrates three points. Islands differ in their frequency distributions. Wet and dry season distributions are similar; but large and hard seeds are relatively more common in the dry season. Use of biomass, instead of numbers, would diminish the relative importance of the small and soft seeds.

tion in the seed size of a given plant species. It is reasonable to assume that the finches would be subject to new selection pressures in the new environment, and to respond evolutionarily, in view of the observed changes known to take place in the *G. fortis* population on Daphne under directional selection when the food supply changed (Chapter 8). Here then is direct evidence for the ecological distinctiveness of islands that I referred to in the previous chapter in the discussion of geographical differentiation of species and the role of isolation.

The importance of food supply is further shown by the fact that the diversity of seeds and fruits consumed by ground finch species, classified by depth-hardness values, is positively correlated ($r = 0.54$, $P < 0.02$) with the diversity of seeds and fruits in their environments (Abbott et al. 1977). In other words, diets are a simple function of food supply as Snodgrass (1902) argued many years ago (see also Chapter 6).

Food supply determines not only diets, but which species occur on an island, and their approximate abundance (Abbot et al. 1977). For example, variation in the *number* of *Geospiza* species occurring on islands is statistically explained, through multiple regression analysis, most strongly by variation in the number of angiosperm plant species occurring on islands, and much less by variation in island area and distance from other islands. Variations in the *diversity* of *Geospiza* species—that is, the number of species on an island, weighted by the relative abundance of each species—was found to be most strongly accounted for in a separate analysis by a similar index of the diversity of seeds classified into depth-hardness categories. In this second analysis, diversity of seeds contributed a significant 64.1% to the total R^2 value, whereas number of plant species at a study site and a measure of each island's isolation contributed a further 3.6% only (Abbott et al. 1977). This second analysis is most meaningful because the food supply of finches is characterized more precisely by seed diversity than by the number of plant species.

Given these correspondences between finches and their food supplies, it might be thought unnecessary to invoke competition in the past: variations in food supply provide a sufficient explanation for variations in the properties of finches. The problem with this attitude is that current finch distributions and diets ought to be determined by food supplies regardless of whether competition has occurred in the past or not. A second, and related, problem is that the food supply hypothesis is close to being unfalsifiable. For example, it explains the absence of a species from an island by the absence of its appropriate food supply, and the presence of a species either by the presence of its appropriate food supply or, in the absence of the appropriate food supply, by adaptation to the local, initially inappropriate, food supply (Bowman 1961). This is dangerously close to being so all-encompassing as to not admit conceivable alternatives (Grant and Grant 1982). To test the hypothesis it would be necessary to know the dividing line between conditions that preclude establishment and conditions that allow persistence long enough for adaptation to take place. Such knowledge is out of reach. Only under unusual circumstances (Schluter and Grant 1982) is it possible to escape from this dilemma. I will discuss later two such circumstances under which the food supply hypothesis can be properly tested, i.e. potentially falsified.

The question that needs to be addressed, therefore, is not can food supply explain properties of finch communities, but does food supply and competition for it provide a better explanation than does food supply alone (see below). Before I leave the food supply hypothesis, one thing should be clear. It is unnecessary to invoke Muller's theory of differential mutation to account for differentiation of populations in allopatry. As Bowman (1961) argued, adaptation to local food supplies is a sufficient explanation. Of course novel mutations are material for selection to act on, through the phenotypes that carry them, but they do not have to be different in allopatric populations for differentiation to occur. To this degree, then, Lack's allopatric model should be modified. In his later writings Lack (1969) acknowledged that islands differed to some extent in food supplies, and that this could account for some of the distributional and dietary features of the finches, which he had previously referred to as non-adaptive. But he never revised his original version of the allopatric model in the light of this concession, and in his last writing on the subject (Lack 1971) he chose to mention only competition.

LACK'S EVIDENCE FOR COMPETITION

Lack (1947) supported his competition thesis with morphological and distributional evidence. It fell into two categories. First, coexisting species of the same genus differ in beak size, with scarcely any overlap in the measurements of two similar species. He interpreted the size differences as reflecting niche differences; this interpretation was supported by his own nonquantitative observations of the feeding of finches and by the analysis of stomach contents performed by Snodgrass (1902). To both Lack (1947) and Bowman (1961), differences between coexisting species represented the avoidance of competition, but Lack alone interpreted the differences as being the evolutionary outcome of natural selection on the species when they competed in the past.

Second, where a species is absent from an island, a congeneric species that is present is morphologically somewhat similar to it. Again Lack interpreted this pattern in terms of feeding niches, supported by rather scanty observations. Assuming that the food supplies are basically the same on the different islands, he argued that the niche of the missing species has been sequestered, at least in part, by the species present.

The major examples of these patterns will now be described. Where congeneric *Geospiza* and *Camarhynchus* species coexist they differ in beak size. Figure 15 illustrates the differences among some *Geospiza* species. It also shows that when one of the species is on its own on an island it is morphologically intermediate between the size it has when coexisting with other species and the size of the most similar congener. The solitary species, be it *G. fortis* (Daphne) or *G. fuliginosa* (Hermanos), has been released from competition and as a result has come to exploit two niches, for which an intermediate beak size is most appropriate (Plate 63).

Other evidence for competitive release in *Geospiza* species is summarized in Table 16. On Genovesa there are three ecological niches filled by three species with very distinctive beak sizes and shapes and distinctive body sizes (Plate 37): *G. magnirostris, G. conirostris* and *G. difficilis*. Two of these three species, *G. magnirostris* and *G. difficilis*, are missing from Española. *G. difficilis* is replaced by *G. fuliginosa*, and the niche of the missing *G. magnirostris* is occupied by *G. conirostris*. The evidence for this is morphological. *G. fuliginosa* on Española and *G. difficilis* on Genovesa have similar beaks. *G. conirostris* on Española has a deeper, blunter and more powerful beak than its relative has on Genovesa, in which respect it more closely resembles the absent *G. magnirostris* (e.g. compare Plates 1, 32 and 33; see also Fig. 80). *G. fortis* and *G. scandens* are also missing from both islands, and their niches are occupied to some extent by *G. conirostris* on both. But the difference between the *G. conirostris* populations resides solely in the use of the large ground finch niche in the absence of *G. magnirostris* (Española) but not in its presence (Genovesa). One can think of this as being the result of competitive release on Española, competitive displacement on Genovesa, or both.

Similarly, according to Lack (1947) the cactus food niche is exploited by *G. difficilis* on Wolf (and Darwin) in the absence of both *G. conirostris* and *G. scandens*, but not on other islands where either of them is present. The bill of *G. difficilis* on Wolf is unusually long and shallow, and sufficiently similar to that of *G. scandens* that this population was once classified as a subspecies of *G. scandens*. *G. difficilis* also fills the small ground finch niche, in the absence of its customary inhabitant, *G. fuliginosa*. On Genovesa, where a cactus-feeder (*G. conirostris*) is present but the small seed-eater (*G. fuliginosa*) is absent, *G. difficilis* has evolved a very *fuliginosa*-like bill and body size, and there occupies the small seed niche.

In these examples *G. difficilis* provides the most profound evidence of morphological and inferred feeding changes, because on the central large

PLATE 63. The classical case of character release. The species are more similar where they occur separately (upper photos) than where they occur together (lower photos). Upper: Medium ground finch, *Geospiza fortis*, on Daphne Major. Lower: Medium ground finch, *Geospiza fortis*, at Bahía Academía, Santa Cruz (*photo by W. Clark*).

Upper: Small ground finch, *Geospiza fuliginosa*, on Hermanos III (*photo by D. Schluter*). Lower: Small ground finch, *Geospiza fuliginosa*, at Bahía Academía, Santa Cruz (*photo by W. Clark*).

TABLE16. Ecological niches of species on outlying Galápagos islands, as suggested by Lack (1947).

Niche	Islands		
	Genovesa	Española	Wolf
Large Ground Finch	G. magnirostris		G. magnirostris
		↑ G. conirostris ↓	
Cactus Finch	G. conirostris		↑ G. difficilis ↓
Small Ground Finch	G. difficilis	G. fuliginosa	
Warbler Finch	C. olivacea	C. olivacea	C. olivacea

islands of the archipelago it occupies humid forest. Relying on information from Gifford (1919) on the feeding and distribution of *G. difficilis* on Pinta, Santiago, and Santa Cruz, Lack (1947) argued that the species was restricted to medium and high altitudes through competition in the arid lowlands with *G. fuliginosa*. The evidence for competitive exclusion is that in the absence of *G. fuliginosa* on the outlying islands listed in Table 16, *G. difficilis* is capable of living entirely in the arid zone; indeed its population densities on those islands are higher than are the population densities of its congeners (e.g. see Grant and Grant 1980a).

Turning now to the tree finches, we find similar patterns among *Camarhynchus* and *Cactospiza* species. San Cristóbal has populations of the small tree finch (*C. parvulus*) and the woodpecker finch (*C. pallida*) but not the large tree finch (*C. psittacula*). *C. parvulus* here has a larger beak than any other island form of this species, and *C. pallida* has a shorter beak than usual. With regard to the shift in beak size of the woodpecker finch, Lack (1947, p. 70), assuming that *C. psittacula* had never been resident on the island (but see Chapter 3), wrote: "This suggests that, in the absence of *C. psittacula*, there has been survival value to *C. pallidus* [pallida] in becoming less specialized in beak, and the Chatham [San Cristóbal] form of *C. pallidus* takes some of the foods which on other islands are taken by *C. psittacula*." A similar interpretation was given to the beak shift in *C. parvulus*. Finally on the northern islands where the woodpecker finch is absent, *C. psittacula* appears to have undergone a small shift towards the *Cactospiza* beak form, being slightly longer and less deep than, for example, on Santiago (Fig. 75).

Like the ground finch examples, these patterns can be explained by com-

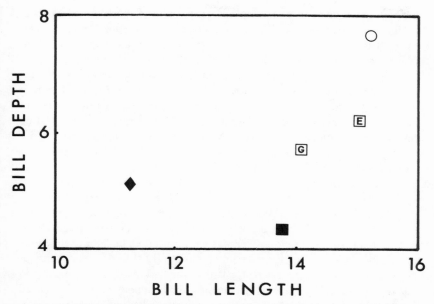

FIG. 80. Mean beak sizes (upper mandibles) in millimeters of four species of ground finches: *G. fortis* (♦) from Daphne Major, *G. scandens* (■) from Daphne Major, *G. magnirostris* (○) from Genovesa, and *G. conirostris* (□) from Genovesa (G) and Española (E). In the absence of all three congeners from Española, *G. conirostris* has an intermediate bill size, closest to that of *G. magnirostris*. In contrast, on Genovesa *G. conirostris* is sympatric with *G. magnirostris*, and it resembles more the missing *G. fortis* and *G. scandens* in bill size. Española and Genovesa, additionally, have a smaller, but different, species, whose mean bill dimensions fall just outside this figure; they are *G. fuliginosa* (3.6, 8.3) on Española and *G. difficilis* (3.3, 9.1) on Genovesa. From Grant and Grant (1982).

petitive displacement in sympatry, competitive release in allopatry, or both. Lack (1947, p. 66) wrote, "The beak modifications of the birds . . . lend further support to the view that such modifications are adapted to the nature of the food." Since the modifications are interpreted solely in terms of the presence or absence of congeners, and not in terms of the presence or absence of particular foods, I believe the adaptive modification in allopatry referred to in his speciation model should be interpreted as driven by competition or facilitated by its absence, as I pointed out earlier.

Lack treated the tree finch and ground finch examples separately because the two groups of species differ largely, but not exclusively, in feeding ecology. They constitute different guilds. Competitive interactions would be expected to occur most strongly within guilds, and the evidence for it is the several patterns of within-guild substitution and replacement I have described. In recent years an example of between-guild substitution has come to light. *G. conirostris* on Genovesa tears bark off the trunk, branches, and twigs of *Bursera* (Plate 60), *Cordia*, and *Croton* trees and shrubs to expose hidden arthropods (Grant and Grant 1980a, 1982). Occasionally it hammers at the bark, with mandibles slightly open. These feeding behaviors are similar to those it uses to open *Opuntia* pads and fruits (Plate 59). Bark-tearing is a major feeding habit of the tree finch, *Camarhynchus psittacula*, and hammering is performed by *Cactospiza pallida*. Neither species of tree finch is present on Genovesa. We have seen bark-stripping behavior (more rarely) on two other islands lacking tree finches: by *G. conirostris* on Española, and by *G. scandens* on Daphne Major.

A question that arises when considering these examples is why certain species are missing from the outlying islands. They may never have reached the islands, or they may have reached them but, being less well adapted than the resident species which ate their foods, they never established themselves. Lack's (1945) first answer was the former, but he later changed his explanation to one of ecological incompatibility (Lack 1947). For example, the absence of *C. pauper* from all islands except Floreana was explained in one of two ways. Either it evolved recently and not enough time has elapsed for it to colonize other islands, or else it evolved earlier and it has been competitively eliminated from other islands by *C. psittacula*, and possibly by *Cactospiza pallida* (Lack 1969). A contemporary example is provided by some ground finches. *G. fortis* stragglers have been recorded on Española and Wolf, and *G. fuliginosa* have also been recorded on Wolf, but, despite suspicions to the contrary (Bowman 1961), they do not have breeding populations on those islands. Such failures to become established might come about because on a small island comparatively specialized feeders, like *G. fortis* according to Lack (1947, 1969), survive less well than species which fill wider ecological niches, like some forms of *G. conirostris* and *G. difficilis*. This would explain why the two cactus finches, the specialized *G. scandens* and the generalized *G. conirostris*, do not coexist on any island.

TESTS OF THE COMPETITION HYPOTHESIS

Lack's aim was to offer a coherent theoretical framework for understanding the adaptive radiation, not to test specific hypotheses. The various manifestations of the competition hypothesis need to be tested, however, because although plausible and widely accepted in the literature they may be wrong. For instance, the patterns of morphological change were described by using species selectively. Inclusion of other species might alter the patterns and so complicate the explanations. To give one example, the intermediate size of *G. fortis* on Daphne can be plausibly interpreted as the result of competitive release in the absence of *G. fuliginosa*, but Lack's treatment of this situation omitted any mention of *G. scandens* which is also present on Daphne, and it did not explain why *G. fortis* had not become larger in the absence of *G. magnirostris*. Bowman (1961) raised several critical objections to the other evidence for competition that was summarized in the previous section.

Testing the competition hypothesis is difficult, for two reasons. First, the hypothesis deals with events in the past. Since we cannot reconstruct those events precisely, we cannot test the hypothesis directly (Grant and Grant 1982); instead it must be tested through its consequences (predictions) or the assumptions upon which it rests. Second, Lack's arguments were not entirely explicit, nor were they presented in the form in which they were originally constructed (Richards 1948, Grant and Grant 1982). To put the arguments into a testable framework we must rephrase them, along the following lines. The observations to be explained are the distribution of species and the inter-island differences in beak size and shape; the hypothesis is that distribution and morphology were causally influenced by interspecific competition for food; the main assumption upon which the hypothesis rests is that the feeding niche of a population is reflected in, and hence adequately indexed by, the average beak characteristics.

Quantitative studies of the diets of the ground finches (Chapter 6) have supported the main assumption, as well as the important corollary that the dietary differences between species parallel the beak differences between them (Fig. 32). I shall now give two examples of an examination of the hypothesis through a test of its predictions.

1. *Competition between G. conirostris and congeners.* We should expect that *G. conirostris* on Española, with mean beak characteristics intermediate between those of the absent *G. magnirostris, G. fortis*, and *G. scandens* (Fig. 80), has an intermediate feeding niche position too. Not only that, it is expected to combine the niches of the three missing species, and consequently its niche should be particularly broad. These are falsifiable predictions because they are not necessarily true. The competition hypothesis would fail if we found that *G. conirostris* did not feed on Española in a manner exhibited by *G. magnirostris, G. fortis*, and *G. scandens* on other islands.

Data to test these predictions were collected in the early and middle dry seasons on Española, Genovesa and Daphne, in 1973–1979 (Grant and Grant 1982). The predictions were supported by the results. A quantitative representation of the feeding niche of *G. conirostris* on Española was found to be more similar to the combined niches of *G. magnirostris, G. fortis*, and *G. scandens* than to the niches of these species considered alone, or to the niche of the conspecific population of *G. conirostris* on Genovesa (Table

TABLE 17. Similarities in foraging for food between *G. conirostris* and selected congeners. Similarity values are Renkonen-Whittaker indices on a scale of 0 (no foraging characteristics in common) to 1 (identical foraging). (From Grant and Grant 1982.)

G. conirostris Compared with Other Populations		Similarity
G. conirostris (Española)	G. conirostris (Genovesa)	.31
	G. magnirostris (Genovesa)	.42
	G. fortis (Daphne)	.38
	G. scandens (Daphne)	.40
	G. magnirostris & conirostris	.53
	G. fortis & G. scandens	.41
	G. magnirostris & G. fortis & G. scandens	.65
G. conirostris (Genovesa)	G. magnirostris (Genovesa)	.14
	G. fortis (Daphne)	.27
	G. scandens (Daphne)	.42
	G. fortis & scandens	.30

17). Moreover, the niche breadth of *G. conirostris* on Española was greater (0.90) than the niche breadth of any other population studied (0.32–0.68).

It is also expected from the competition hypothesis that the niches of *G. conirostris* and *G. magnirostris* on Genovesa should differ substantially: that *G. magnirostris* should take food items exploited by *G. conirostris* on Española but not on Genovesa; and that the niche of *G. magnirostris* should be more similar to the niche of the allopatric *G. conirostris* (Española) than to the niche of the sympatric *G. conirostris* (Genovesa). Quantitative data upheld all these predictions very clearly (see Table 17, and Grant and Grant 1982).

In these several ways feeding niches of the populations were correctly inferred from beak morphology, and the niche differences can be explained by a hypothesis of competition. One additional prediction was supported more ambiguously. It is expected that the niche of *G. conirostris* on Genovesa combines the feeding niches of *G. scandens* and *G. fortis*. In fact, a similarity analysis showed the strongest resemblance to be with the niche of *G. scandens* alone. Despite this, on one matter of detail the prediction is correct; on Genovesa *G. conirostris* crack *Bursera* stones and eat the seeds, which *G. fortis* do elsewhere but *G. scandens*, on Daphne Major at least, do not. Thus *G. conirostris* does combine the niches of *G. fortis* and *G. scandens*, but in its proportional use of elements of the niche, particularly *Opuntia* cactus, it most closely resembles *G. scandens*.

The hypothesis of competition was used by Lack (1947, 1969) to explain not only changes in feeding niches but also the absence of species. He explained the absence of *G. fortis* and *G. scandens* from the two islands, and *G. magnirostris* from Española, in terms of competitive exclusion by the residents. He supposed the residents to be more efficient at exploiting the local foods than were the other species when the latter, as stragglers, occasionally visited the islands. From this general reasoning it is to be expected that *G. conirostris* on Española is more efficient than *G. magnirostris* at dealing with foods on that island. How can that be tested, when the two species do not occur together on Española? The answer is it can be tested by comparing the performance of the two species on a single but probably critical food type, *Cordia lutea* seeds, on different islands: *G. conirostris* on Española and *G. magnirostris* on Genovesa.

In terms of the total time taken to deal with a single seed (stone), the two species were found to have approximately equal efficiency (Grant and Grant 1982). *G. conirostris* has a slight advantage in requiring less food for maintenance, being slightly smaller in body size than *G. magnirostris*. However, *G. magnirostris* has the compensating advantage of being able to crack all stones, regardless of their size, whereas *G. conirostris* feed size-selectively, avoiding the largest ones. Thus the two species have approximately equal rewards per unit effort when feeding on this food type. Nevertheless, the prediction is clearly upheld under one circumstance, the invasion of Española by *G. magnirostris* immatures, because immatures are distinctly less efficient in dealing with *Cordia* stones than are adults. This circumstance is the most likely to arise because almost all stragglers to islands are immatures (Grant et al. 1975).

I conclude that the competition hypothesis has stood up well to these various tests. Inter-island differences in the feeding niches of *G. conirostris* are satisfactorily explained by the inter-island difference in competitive pressure. There is also qualified, and highly indirect, support for the competitive exclusion hypothesis. The evidence is stronger for a feeding efficiency advantage possessed by *G. conirostris* over *G. fortis* and *G. scandens* than over *G. magnirostris* (Grant and Grant 1982). Finally, there is no support for the alternative view that the absence of finch species (from Española and Genovesa) is attributable to the absence of their foods. A list of seed types from each of these islands includes almost all the seeds exploited by *G. fortis*, *G. scandens*, and *G. magnirostris* on other islands (Grant and Grant 1982).

Uncertainty remains concerning whether the few stragglers reaching these islands really constitute potential breeders in all cases, or whether most of them would not stay and attempt to breed but return to the island of their birth regardless of the food conditions and the presence of other species. There is also uncertainty about why *G. magnirostris* has failed to colonize Española but succeeded on Genovesa. If it has been competitively excluded by *G. conirostris* on Española, why has it not been excluded by that species on Genovesa? The answer may lie in the greater richness of *Opuntia* resources on Genovesa than on Española. Assuming that it was the first to colonize these islands, because it appears to be the older species (Fig. 72), *G. conirostris* may have been more of a cactus-feeder on Genovesa than on Española at the time of arrival of *G. magnirostris*, and hence the conditions for coexistence were perhaps more suitable on Genovesa than on Española. Unfortunately this speculation cannot be pursued further because the *Opuntia* population on Española has been substantially reduced by goats that were introduced in the last century (Grant and Grant 1982). This prevents us from knowing whether the current difference between the islands in amount of cactus is solely modern and artificial or a historical condition that has been exaggerated recently.

2. *Competition between G. difficilis and G. fuliginosa.* Twenty low-elevation islands are occupied by *G. fuliginosa* and three more are occupied by *G. difficilis*; none are occupied by both (Tables 2 and 3). Three high-elevation islands (Pinta, Santiago, and Fernandina) have both species (Fig. 81), two more (Santa Cruz and Floreana) did have both before *G. difficilis* became extinct, and yet two more (Isabela and San Cristóbal) possibly did if *G. difficilis* was present, as suspected (Harris 1973), but has since become extinct. The two species on all high-elevation islands were thought by Lack (1945, 1947) to be segregated into lowlands (*G. fuliginosa*) and highlands (*G. difficilis*). Assuming that *G. difficilis* is the phylogenetically older of the two, for which there is modern evidence (Fig. 72), Lack (1947, 1969) supposed that *G. fuliginosa* had competitively excluded *G. difficilis* from the lowlands of high islands and from many of the low islands altogether.

A recent field investigation of these species has revealed that the two species are not segregated altitudinally on the high-elevation islands after all (Schluter 1982a, 1982b, Schluter and Grant 1982, 1984a). On Pinta, for example, they overlap extensively but not entirely in both breeding and nonbreeding seasons (Fig. 82), although *G. fuliginosa* does tend to occur at lower elevations than *G. difficilis* here and on Santiago and Fernandina. These facts were actually known a long time ago (Snodgrass and Heller 1904, Gifford 1919), but Lack, who did not visit these particular islands, misinterpreted Gifford's accurate descriptions.

Even if only some degree of competitive displacement has occurred on the high-elevation islands, instead of the complete displacement argued by Lack, we should still expect *G. difficilis* to feed in a different manner from *G. fuliginosa* on the high-elevation islands but in a manner more similar to *G. fuliginosa* on the low islands where it occurs alone.

Field data uphold these expectations. On Pinta, *G. fuliginosa* feeds pre-

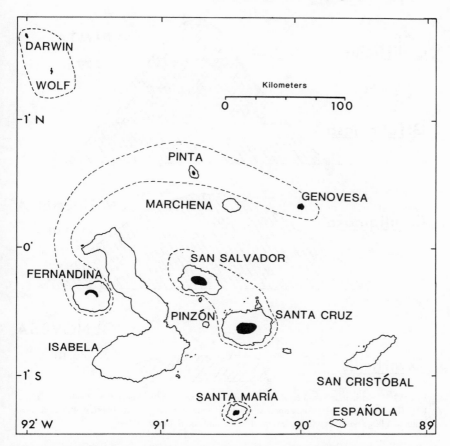

FIG. 81. Distributions of *G. difficilis* (black) and *G. fuliginosa* (white). Dashed lines delimit the ranges of four subspecies of *G. difficilis*; from north to south they are *septentrionalis, debilirostris, difficilis,* and *nebulosa*. Populations of *G. difficilis* have become extinct on Santa Cruz and Floreana, possibly also on Isabela and San Cristóbal (Table 2). Modified from Schluter and Grant (1982).

dominantly on seeds and nectar, whereas *G. difficilis* feeds largely on invertebrates (arthropods and snails) in the leaf litter (Fig. 83). On Genovesa, *G. difficilis* combines the niches of the two populations on Pinta (Fig. 83), but the diet is much more similar to the diet of the absent *G. fuliginosa* than it is to the diet of conspecifics on Pinta (Table 18). Moreover, the abundance of *G. difficilis* on Genovesa is more accurately predicted from a knowledge of the supply of small seeds, a *fuliginosa* food on Pinta, than from a knowledge of the supply of invertebrates, a *difficilis* food on Pinta (Fig. 84).

Instead of explaining these observations in terms of competition, one

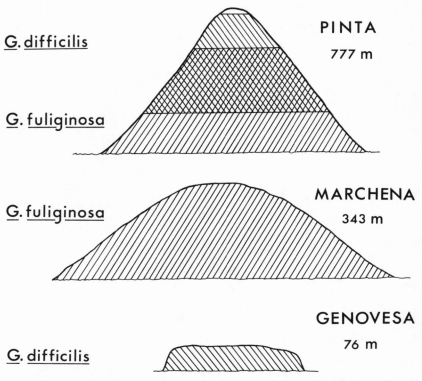

G. difficilis

G. fuliginosa

PINTA
777 m

G. fuliginosa

MARCHENA
343 m

GENOVESA
76 m

G. difficilis

FIG. 82. The distributions of *G. difficilis* and *G. fuliginosa* on three northern islands (see Fig. 81). The islands are drawn only approximately to scale. The two species overlap altitudinally on Pinta to a large but not complete extent. From Grant and Schluter (1984).

could argue that the *fuliginosa*-like diet of *G. difficilis* on Genovesa merely reflects the availability of *fuliginosa* foods and the absence of typical *difficilis* foods. A sampling of the food supply on the islands shows that *fuliginosa* foods are certainly present on Genovesa; the seed density is like that at lowland sites on Pinta (Schluter and Grant 1982). However, Genovesa does not lack typical *difficilis* foods. Although dry-season food supplies are low, and snails especially scarce, the proportional availability of invertebrates is actually higher on Genovesa than on Pinta (Schluter and Grant 1982). Yet despite this, *G. difficilis* consumes nectar on Genovesa, especially from *Waltheria ovata* flowers, just as *G. fuliginosa* does on Pinta, Marchena, and other islands, but which *G. difficilis* does not do on Pinta. *G. difficilis* on Genovesa also shows similar behavior to *G. fuliginosa* when foraging on the ground and beneath rocks for seeds. Therefore its *fuliginosa*-like diet is not explained by the absence of typical *difficilis* foods. It is attributable instead

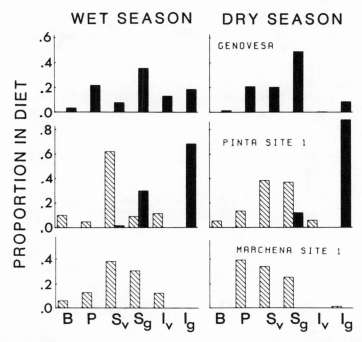

FIG. 83. The diets of *G. difficilis* (solid bars) and *G. fuliginosa* (hatched bars) at lowland sites on three northern islands (see Fig. 81). Symbols: *B*, berries and fruit arils; *P*, pollen and nectar; *Sv*, seeds on the vegetation; *Sg*, seeds on the ground; *Iv*; invertebrates (arthropods and snails) on the vegetation; *Ig*, invertebrates on the ground. The species have similar diets in allopatry, different diets in sympatry. From Grant and Schluter (1984).

to the combination of two factors: a greater profit obtained from the *fuliginosa* foods, and the absence of a population of *G. fuliginosa*.

These results clearly demonstrate a similarity in the diets of the two species in allopatry, in support of the competition hypothesis. They were correctly inferred by Lack (1947) from morphological comparisons, although the evidence available to him was slender. In bill form *G. difficilis* on Genovesa is only slightly more similar to *G. fuliginosa* than is *G. difficilis* on Pinta (Schluter and Grant 1984a). Morphological convergence of *G. difficilis* on Genovesa and *G. fuliginosa* is much more striking in body size, leg length and thickness, and the length of the hind toe and claw (Grant et al. 1985). On Pinta *G. difficilis* scratches in leaf litter for invertebrates, for which strong legs and feet are adaptive (Plate 42). On Genovesa, since nectar is frequently removed from small (*Waltheria*) flowers on thin branches (Plate 42), metabolic efficiency as well as perching ability were probably major selective factors in the evolution of small size. Interestingly, while

TABLE 18. Similarity in the diets of *G. difficilis* and *G. fuliginosa* on Pinta (P), Marchena (M), and Genovesa (G). Values in the table are Renkonen-Whittaker similarity indices on a scale of 0 (no diet items in common) to 1 (identical diets). Those above 0.5 are shown in boldface. (From Schluter and Grant 1982.)

Sites	*G. difficilis* (G) vs. *G. difficilis* (P)	*G. fuliginosa* (P) vs. *G. difficilis* (P)	*G. difficilis* (G) vs. *G. fuliginosa* (P,M)	*G. fuliginosa* (M) vs. *G. fuliginosa* (P)
		WET SEASON		
P1	**.63**	.46	**.72**	**.98**
P2	**.76**	**.69**	**.63**	**.92**
P3	**.72**	.45	**.61**	**.90**
P4	**.75**	**.55**	**.60**	**.89**
P5	**.54**	.32	**.55**	**.86**
P6	.38	—	—	—
M1	—	—	**.75**	—
M2	—	—	**.68**	—
		DRY SEASON		
P1	.21	.18	**.94**	**.92**
P2	**.69**	**.63**	**.91**	**.91**
P3	.27	.25	**.93**	**.90**
P4	.29	.28	**.80**	**.76**
P5	.10	.05	**.72**	.46
P6	.09	—	—	—
M1	—	—	**.82**	—
M2	—	—	**.84**	—

the sharp-beaked *G. difficilis* on Genovesa probes the flowers with the entire beak to obtain nectar, the more blunt-beaked *G. fuliginosa* on Marchena and Pinta uses only the lower mandible (I have seen a vagrant small tree finch, *C. parvulus*, on Genovesa do the same). The evolutionary transition in niches between the *G. difficilis* populations was accomplished by shifts in body size but not in beak size or shape, hence by a change in allometric relations between them.

Explaining the properties of a species is easier than explaining the absence of a species from an island. The absence of *G. fuliginosa* from Genovesa cannot be explained by the absence of an appropriate food supply. We have been unable to identify anything unique about Genovesa that explains the presence of *G. difficilis* but the absence of *G. fuliginosa*. Likewise there is apparently nothing unique about Marchena that explains the presence of *G. fuliginosa* but the absence of a Genovesa form of *G. difficilis* (Schluter and Grant 1982).

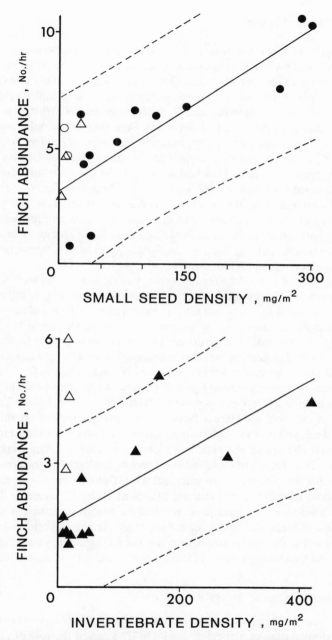

FIG. 84. Abundance of *G. difficilis* (△) in censuses on Genovesa predicted from the relationship between *G. fuliginosa* (●) numbers and the density of small seeds on Pinta (upper). The three values for *G. difficilis* fall within the 95% confidence limits for the estimates of numbers; so do two values for *G. fuliginosa* on Marchena (○). In contrast, *G. difficilis* numbers on Genovesa are not well predicted by the density of invertebrates from a relationship established for *G. difficilis* (▲) on Pinta (lower). From Grant and Schluter (1984).

An alternative explanation for the absence of a species is that it may never have reached the island in question. In the present case the position of Marchena midway between Pinta and Genovesa makes it unlikely that *G. difficilis* has never reached it from one island or the other. Similarly, if a few individuals of *G. fuliginosa* (and *G. fortis*) have reached Wolf, it is almost certain that a few have reached Genovesa. Detecting them would be difficult in view of their similarity in appearance to the resident *G. difficilis*. At least seven *G. fortis* and one *G. scandens* have been recorded on Genovesa during an eight-year study. Thus failure to reach the islands is an unlikely explanation for the absence of *G. difficilis* from Marchena and *G. fuliginosa* from Genovesa. A better explanation is that on these islands, and on Wolf and Darwin, all of which are well isolated, a combination of low number of immigrants and the presence of an ecologically similar species in large numbers, makes it difficult for the immigrants to establish themselves as a breeding population (Schluter and Grant 1984a).

Less is known about the feeding ecology of *G. difficilis* on Wolf and Darwin. Lack (1947) reasoned from the *scandens*-like but smaller bill of these two populations that they had filled the small ground finch and cactus finch niches in the absence of *G. fuliginosa* and *G. scandens* (Table 16). This is partly correct in that *G. difficilis* on these two islands is more similar to *G. fuliginosa* in diet than are the other populations except the one on Genovesa, but the similarity in diet with *G. scandens* is weak (Chapter 6). These two populations certainly exploit *Opuntia* flowers like *G. scandens*, but so does the short-billed Genovesa population (Grant and Grant 1981). The long beaks of the Wolf and Darwin birds may be partly adaptations to this mode of feeding, and partly adaptations to feeding on two other food sources not exploited elsewhere: the eggs and blood of seabirds (Schluter and Grant 1984a). The absence of *G. scandens* cannot be explained by the presence of a superior competitor, and its principal food (*Opuntia* cactus) is present on both these islands and on Pinta and Marchena where *G. scandens* does occur. Therefore it may have never reached the islands in sufficient numbers to colonize them; or, in view of the phylogenetic age and otherwise widespread distribution of *G. scandens* in the archipelago, it may have colonized them and then become extinct for reasons that are not discernible today.

DIFFERENT EXPLANATIONS RECONCILED

I conclude this chapter by returning to the starting point: the need to explain the differentiation of a species. Lack (1947) stressed the role of competition, while Bowman (1961) emphasized the importance of floristic differences among islands. The two explanations are more complementary than antagonistic, and come together in an examination of the question, so far ignored, of why islands differ in their floras.

This was an important problem for Darwin (1842), because in their physical features the various Galápagos islands seemed so similar. I shall now quote his solution from the *Origin of Species* (Darwin 1859, pp. 400–401).

This [dissimilarity of species on similar islands] long appeared to me a great difficulty: but it arises in chief part from the deeply-seated error of considering the physical conditions of a country as the most important for its inhabitants; whereas it cannot, I think, be disputed that the nature of the other inhabitants, with which each has to compete, is at least as important, and generally a far more important element of success. . . . This difference [between island forms] might indeed have been expected on the view of the islands having been stocked by occasional means of transport—a seed, for instance, of one plant having been brought to one island, and that of another plant to another island. Hence, when in former times an immigrant settled on any one or more of the islands, or when it subsequently spread from one island to another, it would undoubtedly be exposed to different conditions of life in the different islands, for it would have to compete with different sets of organisms: a plant, for instance, would find the best-fitted ground more perfectly occupied by distinct plants in one island than in another, and it would be exposed to the attacks of somewhat different enemies. If then it varied, natural selection would probably favour different varieties in the different islands.

In other words, floristic differences among islands originated in chance dispersal events; evolutionary changes, occurring differently in the different islands and arising from various interactions among species, magnified the initial differences. A modern statement of the causes of floristic differences among islands would build upon this foundation. It would stress that physical features of the islands are important. The Galápagos islands vary in area, elevation, and isolation. Factors that govern the number of plant species on an island vary in relation to these three physical features (Hamilton et al. 1963, Johnson and Raven 1973, Connor and Simberloff 1978, Hamann 1984). Islands differ in floras, therefore, because they differ in size, topography, and position.

If floristic differences among islands can be explained by this combination of deterministic and stochastic processes, so too can the differences in arthropods which are dependent on plants. And if plant and arthropod variation can be accounted for, so too can variation in the finches which depend on the lower trophic levels for food. This discussion serves to emphasize that Lack focused too narrowly on competitive interactions and Bowman focused too narrowly on floristic differences among islands. The chapter has provided evidence that floristic differences are real but that competition among finches cannot be ignored. Differentiation of the finches occurred in a manner not fundamentally different from the one sketched out by Darwin.

CONCLUSIONS AND SUMMARY

According to the adaptive radiation model, reproductive and ecological interactions occurred at the secondary contact phase of speciation cycles. This chapter deals with the arguments for the ecological interactions, and with the evidence. The evidence is extensive, complex, and debated.

David Lack assumed that food supplies for finches were approximately the same on different islands. In constructing a model of repeated speciation, he invoked Muller's theory of differential mutation to explain the initial divergence of populations of a species, and laid stress on competitive interactions to explain the enhancement of the initially small divergence. Both mutation and competition hypotheses have been criticized.

In fact, islands differ in the characteristics of their food supplies, at least in seeds and fruits; therefore adaptation to local food supplies is a sufficient explanation for divergence in allopatry. It is unnecessary to invoke Muller's theory of differential mutation, although it may be applicable in some instances.

A parallel argument has been made with regard to competition. The question is whether it is necessary to invoke competition at all—whether adaptations to local food supplies are sufficient to account for the origin and occurrence of all morphological and ecological differences between species.

Lack supported his competition thesis with two sets of evidence. In the first set, coexisting species of the same genus differ in beak size, with scarcely any overlap in the measurements of two similar species. These morphological differences were interpreted as reflecting niche differences, and were believed to be the product of natural selection on the species when they competed in the past. In the second set, when only one member of a related pair of species is present on an island, that member is morphologically intermediate between the size it has when coexisting with the other species and the size of that other species. The interpretation given to this pattern is that the solitary species has been released from competition, and as a result has come to exploit two niches for which an evolved intermediate beak size is most appropriate. The continued absence of the other species is attributable to subsequent competitive exclusion by the solitary species.

Two clear patterns of the sort just described have been subjected to test, using quantitative data on diets, to see if predictions of the competition hypothesis are upheld or not. The first pattern centers on *G. conirostris*. On the island of Genovesa it resembles the absent *G. fortis* and *G. scandens*, while on Española, which lacks these two species and *G. magnirostris* as well, its intermediate bill size has something of the character of the bills of all three missing species. The second pattern centers on *G. difficilis*, principally on the difference in morphology between the population on Pinta, where it is sympatric with *G. fuliginosa*, and the allopatric population on Genovesa.

The ecological data have generally upheld the competition hypothesis by consistent agreement with its predictions. Since the tests are inevitably indirect they cannot be considered conclusive. They are attended by some unresolved uncertainties, such as whether the absence of a species from an island is due to the absence of a sufficient number of potential breeders, or to the presence of a presumed competitor. Nevertheless, they allow rejection of the alternative hypothesis that food supply alone has determined the morphological and distributional features of finches. For example, the resemblance of *G. difficilis* on Genovesa to the missing *G. fuliginosa* cannot be attributed to natural selection arising from the absence of *difficilis* foods and the presence of *fuliginosa* foods; both types of foods are present.

I conclude that inter-island variation in food supply and interspecific competition for food jointly provide a better explanation for patterns in the morphology and distribution of finches than does variation in food supply alone. Interspecific competition for food played a role in the adaptive radiation. There are two possible arenas. One is allopatric: two populations of the same species can be influenced by different sets of competing species on different islands, and evolve under natural selection in different directions as a consequence. The Darwin quotation given on p. 311 refers to this possibility. The other is sympatric: two previously allopatric populations of the same species may compete for food upon establishing sympatry, and evolve in different directions as a consequence. The first occurs in stage 4 of the speciation model, the second occurs in both stages 3 and 5. Distinguishing between them, or assigning relative importance values to each, is clearly impossible in the absence of known colonization routes in the archipelago. Nevertheless, we can learn more about competition by shifting attention from individual pairs of species, such as the *G. difficilis*–*G. fuliginosa* pair, to sets of species in a feeding guild, such as the *Geospiza* species which all feed on seeds and fruits on the ground. This will be the subject of the next chapter.

CHAPTER TWELVE

Competition and Finch Communities

INTRODUCTION

The patterns I discussed in the previous chapter are fairly clear, as are the interpretations involving interspecific competition, despite some inevitable uncertainties about colonization events. Other patterns, such as the ones identified by Lack involving tree finches, have not been investigated. Elsewhere, among other species and in other parts of the archipelago, neither morphological variation nor the presence or absence of species can be explained so simply. There are two reasons for this.

First, as I have repeatedly emphasized, there is the general problem of determining which island population gave rise to which. Second, and more importantly, there may have been multiple interactions among species from which no simple pattern has emerged. At the first step species A may have responded evolutionarily to species B, and a comparison of the two species in sympatry and allopatry at this point would show a simple pattern of enhanced differences in sympatry, in other words character displacement (Brown and Wilson 1956, Grant 1972b). Thereafter the successive immigration of species C, D, and E to some of the islands, and evolutionary changes in these, in A, and in B, would obscure the simple pattern. The result of a series of changes would be the creation of differences among the several sympatric species and differences between populations of the same species. We observe such differences. The question then arises, how could we know that interspecific interactions had contributed to them, *if* they had, when the variation is so complex it might be explained more simply, and possibly correctly, in terms of largely independent evolutionary responses of each population to its local food supply?

In this chapter I shall attempt to answer the question through an examination of properties of groups of species in feeding guilds. The issue is one of detecting the effects of competition within sets of species, and not of establishing the connection between competition and specific stages of the speciation model as I did in the last chapter. I shall ignore doves and mockingbirds, which take some of the foods eaten by finches (Chapter 6), because they are present on almost every island and are therefore a nearly constant element in the finch environments (Grant 1984b, 1985b).

Before proceeding it will be helpful to emphasize the distinction between two types of competitive processes and their results. The first, character displacement, is an evolutionary process: competition between two or more species leads to a divergence by one or more of them in body size or bill size. As a result of the divergence, competition is reduced. The second, differential colonization (Grant 1969), is an ecological process without evolution: species that immigrate to an island are competitively excluded if they are ecologically similar to those present, but establish themselves if they are ecologically different. As a result, a community of finch species on an island is a nonrandom sample of those species arriving on an island. Exclusion takes place before there is sufficient time for competition to be alleviated through character displacement. Case and Sidell (1983) have used the alternative terms of "size adjustment" for character displacement and "size assortment" for differential colonization. The term "size adjustment" is confusing because it includes both the evolutionary process of character displacement and the non-evolutionary process of "invasion and extinction of populations until compatible, stable, species sets are formed" (p. 843). I prefer to use the original terms.

COMBINATIONS OF SPECIES

We can recognize the influence of interspecific competition on community structure by nonrandom patterns in its components, i.e. if the components deviate significantly from some particular mathematical model of randomness, not in any direction but in the direction expected from a hypothesis of competition. With this as a simple starting point, and restricting attention to the ground finches since there are many more populations of them than there are tree finch populations, we can ask if ground finch species are distributed randomly among Galápagos islands. This question can be answered by a probability analysis, providing we can make the assumption that all six species have had sufficient opportunity to colonize all islands. This seems a reasonable assumption to make in view of their age: no one species is restricted to a corner of the archipelago, and the species with the fewest populations, *G. conirostris*, is widely distributed in the archipelago and relatively old (Fig. 72).

An example illustrates how the calculations are made. At most five out of the six species occur together on an island. There are six possible ways of drawing five-species from a total of six species; each of the six species is missing from one of the five-species combinations. More formally, the six possible five-species combinations are derived as follows:

$$\binom{6}{5} = \frac{6!}{5!\,1!} = 6$$

Only one of these is observed in the archipelago: it comprises *magnirostris*, *fortis, fuliginosa, difficilis*, and *scandens*. This group is found on Santiago and Pinta. Until fifty years ago, prior to the extinction of *G. difficilis*, it was also found on Santa Cruz. Thus it occurred on three islands. If we assume that all six possible five-species combinations are equally likely to occur, what is the probability that the same combination (and no other) occurs three times? The answer is:

$$P = \binom{6}{5}\binom{1}{6}^3 = 0.028$$

The chance occurrence of just one five-species combination three times is so low (<0.05) we can conclude it is not simply a random sample of the possible combinations. The probability value drops to 0.00013 if we add to these three islands three more which supported, or almost certainly supported, the same five species last century prior to some extinctions: San Cristóbal, Floreana, and Isabela (Table 2).

The procedure can be generalized and extended to estimate the probability that each of the observed number of *Geospiza* combinations (5, 4, . . . 1) is a random sample of the possible combinations (Abbott et al. 1977, Grant and Schluter 1984). The results are striking (Table 19). The probabilities are extremely low in each case. Future revisions to the list of species on each island might make small changes to the calculated values but would not alter the general conclusion that distributions of *Geospiza* species in the Galápagos are decidedly nonrandom.

Simberloff and Boecklin (1981) performed a modified analysis with the data given by Abbott et al. (1977); Tables 2 and 3 have superseded the original data. Probabilities of combinations of species were assessed in the same computational manner, but each species was weighted by its frequency of

TABLE 19. Probabilities (*P*) that the observed number of each *Geospiza* species combination is a random sample of the possible combinations. (From Grant and Schluter 1984.)

No. Geospiza Species / Island	No. Possible Combinations	No. Islands	No. Observed Combinations	P
5	6	6	1	<.001
4	15	5	2	.004
3	20	4	2	.017
2	15	7	4	.072
1	6	10	1	<.001

occurrence, the argument being that species with few populations are less likely to occur together than are species with many populations. This confuses result with process (Grant and Abbott 1980, Colwell and Winkler 1984, Gilpin and Diamond 1984). Despite this artificiality, significantly fewer combinations of three, of four, and of five species were observed than expected by chance. Using a somewhat similar approach, Alatalo (1982) showed that similarities in the distributions of pairs of *Geospiza* species are significantly lower than expected by chance. These two results confirm the above conclusion that distributions of *Geospiza* species in the Galápagos are decidely nonrandom.

STRUCTURE DETERMINED BY COMPETITION

Certain combinations of ground finch species are highly underrepresented or do not occur at all. Several factors may have contributed to this pattern, including a nonuniform distribution of food types among the islands. If interspecific competition (exclusion) has been partly responsible for this nonrandomness, if, in other words, differential colonization has occurred, members of the combinations of species that do not occur should be more similar ecologically to each other than the members of those that do occur. If competition has not been responsible there should be no such difference.

If we had enough data to characterize the diets of each species precisely we could make a direct test of these two alternatives. However, estimates of dry season diets of most populations studied are subject to errors associated with differences in feeding at different times within a dry season, and differences between dry seasons in different years (e.g. see Schluter and Grant 1984a). Furthermore, many populations of ground finches have not been studied in the dry season. Instead, an indirect test can be performed by using the morphology of species as an index to their diets. This has the advantage that all populations of all species can be morphologically characterized precisely, by using principal components analysis (Fig. 12). The difference between any two species can be represented by the Mahalanobis distance, D^2, in number of standard deviations which separate their means (a measure of relative difference), or by the Euclidean distance as a measure of absolute difference (Grant and Schluter 1984). The use of a morphological distance measure to represent dietary differences rests on the assumption that the two covary positively, for which there is empirical evidence (Fig. 85).

Average D^2 values for nearest-neighbors (most similar species) in combinations of species that do occur and that do not occur are compared in Table 20. In agreement with expectation derived from the competition hypothesis, the mean D^2 for combinations observed is larger than the mean D^2 for combinations not observed, in each of the four sets. In none of the cases

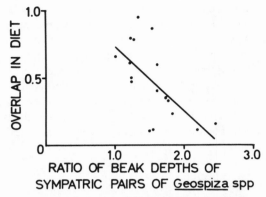

FIG. 85. Overlap in diet as a function of the difference in beak depth between sympatric species of ground finches. When beak depths are almost identical (ratio of ~1.0), two species have almost identical diets (overlap ~1.0); the more two species differ in beak depth, the more they differ in diet (see also Fig. 32). Based on data in the early dry season, from Abbott et al. (1977).

is the difference statistically significant however (Grant and Schluter 1984), so although the trend is correctly predicted the evidence can be considered to provide only weak support for the hypothesis.

Analysis on a finer level can be performed by comparing those pairs of species that occur frequently together with those that occur infrequently together or not at all. There are 15 pairs of *Geospiza* species, and their frequences of co-occurence range from 0 islands (e.g. *G. scandens* and *G. conirostris*) to 15 islands (e.g. *G. scandens* and *G. fortis*). If we assume that all combinations of species are equally probable, the average number of islands jointly occupied by any two particular species is calculated to be 5.7 (Grant and Schluter 1984). This allows a division of the pairs of species into those that occur on more than the expected number, and there are 7 of them,

TABLE 20. Comparison of mean nearest-neighbor morphological distance (D^2) between combinations of species observed and not observed. (From Grant and Schluter 1984).

			Mean Nearest-Neighbor (D^2)	
No. Geospiza Species/Island	No. Possible Combinations	No. Observed Combinations	Combinations Observed	Combinations Not Observed
5	6	1	27.0 >	25.1
4	15	2	36.9 >	29.9
3	20	2	41.2 >	39.8
2	15	4	80.3 >	64.1

and those that occur on fewer than the expected number, and there are 8 of those. The morphological distance measures between members of each pair of species in the two groups are given in Table 21.

It can be seen that D^2 values for the two groups are different: pairs of species occurring together on fewer than 5.7 islands are morphologically more similar, generally, than those occurring together on more than 5.7 islands. The trend is difficult to verify statistically since pairwise distances are not independent, and there are not enough of them to circumvent the problem by drawing a random and exclusive sample for analysis. The two groups are nevertheless different according to a one-tailed Mann-Whitney U-test; the null hypothesis is rejected when D^2 values (Table 21) are compared ($P = 0.014$), and also when Euclidean distances are used instead to avoid the possible complication of unequal variances within groups ($P = 0.037$). These results support the competition hypothesis, as does one additional feature shown in the table: each of the six species occurs together with its morphologically most similar species, and hence presumably its ecologically most similar congener, on fewer islands than expected by chance.

TABLE 21. Morphological distance (D^2) values and the number of islands jointly occupied by pairs of species. Each nearest neighbor, morphologically, is indicated by an arrow and its direction (e.g. the nearest neighbor of *G. magnirostris* is *G. conirostris*). (From Grant and Schluter 1984.)

Pair of Species		No. Islands Jointly Occupied	D^2
A. Occur Together on Six Islands or More			
G. magnirostris ——	*G. fortis*	11	72.4
——	*G. fuliginosa*	11	184.1
——	*G. difficilis*	7	163.4
——	*G. scandens*	10	181.4
G. fortis ——	*G. fuliginosa*	15	26.0
——	*G. scandens*	15	54.6
G. fuliginosa ——	*G. scandens*	13	58.2
B. Occur Together on Five Islands or Fewer			
G. magnirostris ——▶	*G. conirostris*	1	53.3
G. fortis ——	*G. difficilis*	4	22.1
◀——▶	*G. conirostris*	0	21.4
G. fuliginosa ◀——▶	*G. difficilis*	4	10.8
——	*G. conirostris*	2	77.2
G. difficilis ◀——	*G. scandens*	3	19.0
——	*G. conirostris*	1	43.8
G. conirostris ——	*G. scandens*	0	38.4

Case and Sidell (1983) have performed a similar analysis. They used a random sampling procedure to test whether those sets of species that differ largely from each other in bill morphology occur together more frequently than would be expected by chance. They used only bill length data taken from Lack (1947), and found that "size-dissimilar" sets of species were indeed significantly more common than expected, in both the ground finches and, separately analyzed, the tree finches (*Camarhynchus, Cactospiza*, and *Platyspiza* combined). With a different random sampling procedure, Alatalo (1982) found similar evidence for different distributions of *Geospiza* species with similar beak sizes.

Among the various pairs of *Geospiza* species, two pairs do not occur anywhere in the archipelago: *G. conirostris* does not occur with *G. fortis* or *G. scandens*. These exclusive distributions have already been discussed. In the present context, notice in Table 21 that these pairs of species are morphologically among the most similar, in agreement with expectation, but they are not the most similar of all. They would not necessarily be expected to be the most similar, given the imperfect relationship between morphological distance and ecological difference (Figs. 32 and 85), and the fact that the probability of competitive exclusion is not just a simple function of morphological distance but is dependent also on the food supply. In addition, the discussion on pp. 295 and 301 of the ecological relationships between *G. conirostris* and *G. fortis, G. scandens*, and *G. magnirostris* makes it clear that important information may be missed by restricting attention to pairs of species.

In conclusion, assemblages of ground finch species are nonrandom in a direction predicted by a hypothesis of interspecific competition. In the light of this general conclusion, we can now re-examine the pattern of occurrence of *G. fuliginosa* and *G. difficilis* referred to on p. 304 (see also Schluter and Grant 1982). Of the 25 islands lacking a highland zone, 20 are occupied by *G. fuliginosa*, 3 are occupied by *G. difficilis*, 2 are occupied by neither and none are occupied by both. The probability of such a distribution occurring by chance, when 3 populations of *G. difficilis* and 20 populations of *G. fuliginosa* are "placed" randomly and independently on 25 islands, is negligibly small ($P = 0.004$). This probability value is most likely to be an overestimate, for the analysis makes the unlikely assumption that there are only 3 populations of the phylogenetically older species available for colonizing the 25 islands. Since the two species are very similar in allopatry (see also Table 21), the distribution of *G. difficilis* in relation to *G. fuliginosa* is best interpreted as being partly the result of competition, and not just a product of chance.

An objection to this interpretation is that a particular distribution with a low probability is likely to be found, with so many possible combinations of pairs of finch species in the archipelago, even if the distribution is essentially random throughout (D. Simberloff, in Lewin 1983, p. 1412). In fact, the objection has no force in this instance. There are 15 possible pairs of *Geospiza* species. There are another 15 possible pairs of tree finch species. Species in the two groups should not be paired because they are ecologically very different. The maximum number of pairs is thus 30. By chance, and making some simple assumptions, we might expect to find a probability as low as 0.05 once in 20 times, i.e. 5%. In contrast, the probability for the *G. fuliginosa* and *G. difficilis* distributions is 0.004, or one order of magnitude lower, i.e. less than 0.05%. Probabilities for the distributions of the other 29 pairs of species have not been estimated.

MINIMUM DIFFERENCES BETWEEN COEXISTING SPECIES

In the preceding probability analyses I have concentrated on the average differences between species without regard to the particular islands on which they occur. But populations of the same species differ morphologically and ecologically. I shall now examine the differences among groups of sympatric ground finch species, because it is in sympatry that evidence of competition is to be found.

There is little overlap between the frequency distributions of coexisting species along a beak depth axis (Fig. 86). Where three species coexist, the difference between the first two is about the same as the difference between the second two, in other words the frequency distributions are regularly spaced along the axis. *G. fortis* and *G. scandens* on Daphne Major clearly violate the trend towards even spacing, but these two species are quite different in another beak dimension, length, and as a result have discretely different multivariate distributions (Fig. 12). Putting this exception aside, a single axis, that of beak depth, characterizes much of the morphological differences between coexisting species, and much of the ecological differences between them too (Fig. 85).

If competition for food has been partly responsible for the spacing pattern, we should expect a certain minimum difference to exist between pairs of adjacent species on a beak size axis, or on a seed size-hardness axis. This follows from the theory of limiting similarity (MacArthur 1972), which stipulates that if two species are more similar than this critical minimum, one species will exclude the other through competition. It is a simple extension of Gause's principle referred to in the last chapter. Unfortunately, the theory is not precise enough to specify the minimum. Moreover, for various technical reasons (Grant 1981c, Schoener 1984), it has not been possible to develop a satisfactory method of generating minimum differences from entirely random, that is noninteracting, processes, which could then be com-

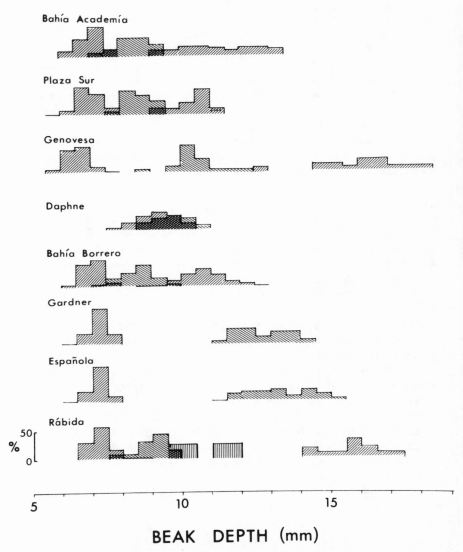

FIG. 86. Frequency distributions of beak depths of ground finches at eight sites in the early dry season. Distributions are relatively evenly spaced, although the spacing varies among islands. Only *G. fortis* and *G. scandens* on Daphne Major violate the trend of regular spacing with little overlap; their beaks differ strongly in length, however (Fig. 16). Most of the overlap between other species is produced by the inclusion of juvenile birds; frequency distributions of adults of one sex rarely overlap (e.g. see Fig. 15). From Abbott et al. (1977).

pared with observed differences to see if the observed ones are significantly greater than the expected ones.

The problem is statistical. Simberloff and Boecklin (1981) assumed a log-uniform distribution of character states as the null (noninteractive) distribution, and drew randomly from it to calculate expected differences between species. Given the widespread occurrence of log-normal size distributions in nature, "a uniform distribution is one of the least sensible models that could be selected for the null model," according to Schoener (1984, p. 275), and cannot be produced by stochastic evolutionary processes (Colwell and Winkler 1984). Opinions differ on this (Simberloff 1983b), but agree on the conservative nature of tests which employ it. In the words of Colwell and Winkler (1984, p. 352): "Since the average (and minimum) distance between random draws from a uniform distribution is always larger than the distance between random draws from a modal distribution with the same range, their tests are biased to an unknown degree against finding any effects of character displacement [this should read interspecific competition]." Yet despite this bias, which Boecklin and NeSmith (1985) estimated to be very small, Simberloff and Boecklin (1981) found that the smallest differences in beak depth between coexisting *Geospiza* species were, statistically, improbably large. Hendrickson (1981) independently obtained the same result: minimum differences in beak depth, as well as in beak length, between coexisting *Geospiza* species almost always exceeded expected values, and to a statistically significant extent, as determined by binomial tests defined on the median. Finally, an analysis of multivariate beak differences gave the same result (Grant and Abbott 1980).

Members of every pair of coexisting ground finch species differ by at least 15% in at least one bill dimension (Fig. 87). This is consistent with the pattern shown around the world by sympatric species of birds on islands derived from an ancestral population by two successive colonizations (Grant 1968, 1972c). It is roughly equivalent to the nonoverlap of two standard deviations either side of the mean of each species, and is consistent with the idea that a limiting similarity exists.

GREATER THAN MINIMUM DIFFERENCES

Beyond this apparent threshold difference of 15% there may or may not be a regular pattern of differences among sympatric populations. For example, the differences between the beak dimensions of *G. fortis* and *G. fuliginosa* are positively and significantly correlated among islands; where the difference in beak length is large, the difference in beak width is also large, and so on (Fig. 88). The same is true for differences in beak length and beak depth. The positive correlation could reflect a limiting similarity that varies

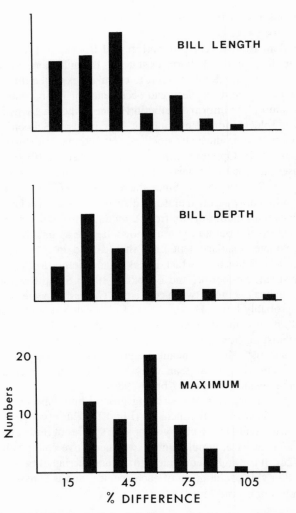

FIG. 87. Differences in mean beak dimensions between all pairs of sympatric populations of ground finch species. The difference between any pair is expressed as a percentage of the smaller mean (grouped into 15% classes). The lower histogram represents the dimension giving the largest difference between pairs of populations, in either bill length or bill depth; all sympatric species differ by at least 15% in one beak dimension. From Grant and Schluter (1984).

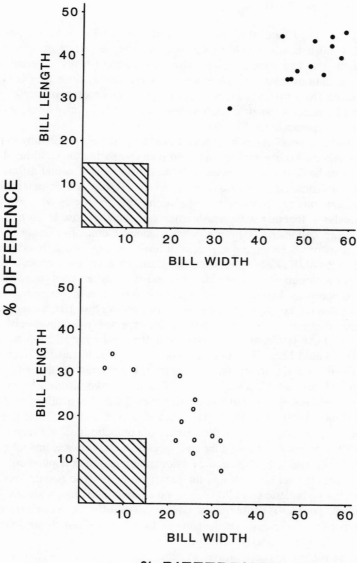

% DIFFERENCE

FIG. 88. Differences in beak dimensions between sympatric populations of *G. fortis* and *G. fuliginosa* (upper) and *G. fortis* and *G. scandens* (lower). Notice the positive correlation among the former and the negative correlation among the latter. The difference between upper and lower figures derives from similar interpopulation allometries of *G. fortis* and *G. fuliginosa*, but different allometries of *G. fortis* and *G. scandens* (Fig. 19). All points lie outside a zone of 15% differences indicated by a box. From Grant (1981c).

from island to island, since we know that food conditions vary among islands, and it is reasonable to suppose that the frequency or severity of droughts that cause food scarcity also vary among islands. Alternatively, the correlations may have no biological significance, since they could be produced from the random pairings of populations of two species with nearly identical allometric relationships between bill dimensions, such as these two species have (Chapter 4).

These two possibilities have been tested by randomly combining in pairs all populations of the two species, and then comparing the resulting distribution of beak-size differences with the distribution of actual differences between sympatric populations of the species (Grant and Schluter 1984). Measurements of 12 sympatric populations of each species were available for analysis, together with measurements of one allopatric *G. fortis* population and 7 allopatric *G. fuliginosa* populations. Cumulative frequency distributions of observed and expected differences (Fig. 89) differed significantly ($P < 0.05$, Kolmogorov-Smirnov test); observed differences between sympatric populations are greater than expected differences, which therefore supports the hypothesis of competitive interaction in sympatry.

The random combination exercise is the only way we can think of to estimate differences between species that are expected in the absence of competition, but it is very artificial and biased. It assumes that all allopatric populations could become sympatric, which is highly unrealistic given their distributions in the archipelago; in contrast, the assumption that all *species* could colonize all islands is realistic. It assumes that recombined populations would not evolve, but ignores why they did evolve in allopatry (Grant and Abbott 1980). Size in allopatry may be a poor and unreliable guide to expected size in sympatry in the absence of competition. Since there are so few allopatric populations of the two species, both sympatric and allopatric populations had to be used in the random combination of populations. This introduces the serious problem that observed (sympatric populations) and expected (sympatric and allopatric populations) distributions are not independent. The bias is conservative, however. The difference between the distributions is minimized, yet despite this the statistical test showed the distributions to be distinctly different.

The problem is much worse for other pairs of species, which is why I have not repeated the analysis for all 13 pairs of coexisting *Geospiza* species. A single example will illustrate the difficulty. Differences between sympatric populations of *G. fortis* and *G. scandens* show a regular pattern, but here the correlation is negative (Fig. 88). The pattern results from different relationships of interpopulation allometry between beak dimensions in the two species (Fig. 20). Sympatric populations are thus moderately different in both dimensions, strongly different in one dimension and weakly different in another, or the reverse. This pattern implies that they differ in

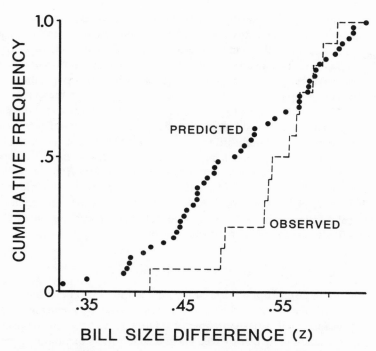

FIG. 89. Bill size differences between sympatric populations of *G. fortis* and *G. fuliginosa* (dashed line) are greater than expected from a random combination of all populations in pairs (dots). The diagram shows the cumulative frequencies of these two sets of differences: the observed and the expected. The bill size difference (Z) is a composite measure of the difference in bill length and in bill depth. From Grant and Schluter (1984).

their diets in different ways on different islands, a possibility which has not been tested so far. The extent to which competition between the species might have contributed to the evolution of these various differences could be examined by randomly combining all the populations. Only 13 populations of each species are available for analysis however, and 12 of each are sympatric, so the problem of lack of independence of observed and expected frequency distributions of differences is especially acute. In fact, the results of this exercise merely show that a reshuffling of already sympatric populations produces no discernible difference in beak length and depth differences (Grant and Schluter 1984). The test is powerless, and, moreover, it takes as given the phenomenon of potentially greatest interest: the difference between the species in their allometric relations, both in slope and in position (intercept). Similar problems have attended other attempts to detect competitive effects by randomly combining all populations of all species (Strong et al. 1979, Grant and Abbott 1980).

An alternative to this pair-by-pair analysis is to examine the possibility that the 13 pairs of coexisting species show a tendency, however weak in each case, to be more different in sympatry than in allopatry (Grant 1981c, 1983c). This is in fact observed. Eleven of the 13 pairs of species are more different in sympatry, on average, than they are when their populations in allopatry and sympatry are combined artificially in all possible combinations (Fig. 90). Such a high proportion (11/13) showing this trend is not expected by chance ($P = 0.02$, sign test). The randomly combined and actually sympatric population differences are not independent, because the random ones contain the actual sympatric ones. The bias is conservative and does not obliterate a real trend, nor does a small departure from normality in the frequency distributions of differences between artifically combined populations (cf. Schoener 1984). The result is strong evidence for enhanced ecological differences in sympatry through enhanced morphological differences—in other words character displacement.

A Digression on Methods of Analysis, and on Bias

The foregoing discussion of tests of the competition hypothesis has made frequent reference to bias in the tests. It arises because we are dealing with populations in a well-isolated archipelago. Differences between coexisting species on islands are to be expected under a wide variety of circumstances, so the problem to be overcome is to find a method that will distinguish between differences attributed to competition and differences attributed to all other causes. The solution to that problem has been to use random models to generate a set of differences expected in the absence of competition. This would be relatively straightforward, although not entirely free from bias, if we could identify a mainland source for the Galápagos island species, for then samples of species from the mainland source could be drawn at random, and their properties compared with those of coexisting species on Galápagos islands. In the absence of an identifiable mainland source, the archipelago has been chosen as the source from which to draw populations at random for comparison with actual, sympatric, populations. This procedure has given rise to problems of detection and interpretation, which I have alluded to in the preceding pages; see also Alatalo (1982), several chapters in the book edited by Strong et al. (1984), and Salt (1984).

Since the artificial sets contain the actual sets, the two are not independent. Regardless of which randomizing protocol is adopted, the lack of independence consistently favors acceptance of the null hypothesis (Grant and Abbott 1980, Colwell and Winkler 1984). For example, Case and Sidell (1983) have shown that the randomization procedure used by Strong et al. (1979) yields expected differences between species that are so similar to ob-

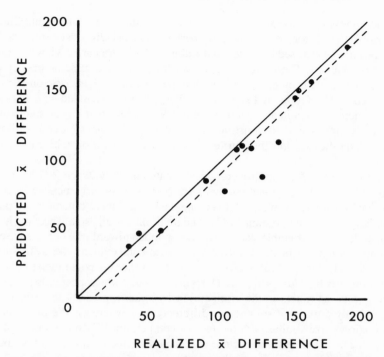

FIG. 90. Prediction of multivariate beak differences, in arbitrary units, between sympatric species of ground finches. Predicted differences are the average differences between species of all island populations, whether they coexist or not. Realized differences are averages of the actual differences between sympatric populations of those species. If predictions were correct in all cases, all points would lie along the solid line. The broken line is a least squares best fit to the points; displaced to the right of the solid line, it shows that realized differences are generally greater than predicted differences. From Grant (1983c).

served differences as to be practically indistinguishable. Setting constraints on the number of populations to be drawn in relation to their frequency of occurrence has a similar effect. If the number of populations of a species is, in part, determined by competitive exclusion, the analysis is insensitive to part of the phenomenon being investigated. Finally, various interpretational problems arise when ecologically irrelevant species are included in the analysis by mixing guilds (Strong et al. 1979); competitive effects have been diluted to the point of disappearance (Abbott and Grant 1980, Colwell and Winkler 1984).

Some recent studies of Darwin's Finch data (as opposed to the finches themselves) have used random models uncritically. They have almost consistently rejected the competition hypothesis, and the authors have invoked randomness itself to explain community properties (e.g. Connor and Simberloff 1978, Strong et al. 1979, Simberloff and Boecklin 1981, Simberloff and Connor 1981; see Hamilton and Rubinoff 1967, for earlier views on the role of chance). These efforts have been criticized and in some cases the same data have been reanalyzed with opposite results (Hendrickson 1981, Alatalo 1982). Gilpin and Diamond (1984, p. 315) concluded an essay on the general subject in a pessimistic vein: ''We believe that it is probably inherently impossible to construct a useful everything-except-competition 'null hypothesis,' and that further efforts in this direction will only sow more confusion.''

At the heart of the controversy is the treatment of evidence for competition. Chitty (1967) pointed out that Lack buttressed his arguments with additional information, instead of seeking data to test them critically and potentially falsify them (Grant 1977). The tradition is well established and has engendered considerable skepticism, not so much towards the occurrence of competition itself, but towards the nonexperimental evidence for it. In Simberloff's (1983b, p. 426) words, ''. . . rarely is the core contention of competition directly addressed. . . . The problem is that when species do not differ in size as much as the competition hypothesis predicts investigators commonly argue that some other difference that *is* large is the difference that allows size similarity in the face of competition.'' In other words, the hypothesis of competition is close to being unfalsifiable. It need not be, but in the way it is sometimes employed, it is.

The recent field studies of Darwin's Finches (Abbott et al. 1977) that preceded, and to some degree precipitated, the controversy were designed with a very explicit awareness of these biases and problems. Both Ian Abbott and I had found supporting evidence for interspecific competitive effects among members of bird communities in some of our previous studies (Grant 1965, 1966, 1968, Abbott 1974a, 1974b) but not in others (Grant 1972b, Abbott 1973, 1974b). In those studies we had set forth alternatives to the competition hypothesis, and in one study dealing with many purported examples of character displacement I found strong evidence for the alternatives and not for character displacement (Grant 1972b). Therefore in designing the field work with Darwin's Finches we attempted to find evidence that would allow clear acceptance or rejection of the competition hypothesis (Abbott et al. 1977).

Despite safeguards, it is always possible for bias to creep in. I have made *ad hoc* adjustments to arguments in the light of knowledge obtained during the course of gathering data to test a hypothesis. To ignore such knowledge is to be blind to reality; but to accept it is to sacrifice some objectivity in the test, and thereby run the risk of giving expression to a bias. These adjustments have been, to my knowledge, rare, and not seriously compromising to any of the tests.

For these various reasons it is important for data to be reanalyzed by others, even at the risk that some data will be misinterpreted. Several sets of Darwin's Finch data have been analyzed by Dan Simberloff and his colleagues. With manifest skepticism towards the evidence for competition, they have constructed statistical tests of a very conservative nature, that is, with a consistently high risk of accepting the null hypothesis when it is false, as has been acknowledged (Simberloff 1984). Most of the tests they have performed fail to provide evidence for competition. A significant minority, however, provide evidence for it. Simberloff (1984, p. 252) concludes, with attendant caveats, "So we appear to agree that Galápagos ground finches are not random subsets, that interspecific competition is likely [to be] at least partly responsible for this, and that the strongest evidence resides in bill features of coexisting races and species."

I believe the evidence from the tests is much stronger than he recognizes, but the important point is the agreement. The tests are too constrained and too artifical to be more exact and quantitative and to allow assignment of numbers to the frequency and importance of competition. Nevertheless, the more limited goal of testing for competitive effects in guilds of finches has been achieved, and the competition hypothesis has been upheld by the results of several of the tests.

PREDICTIVE MODELS

In characteristically forthright language, Haldane (1937) once expressed the predictive value of theories in the following way: "No scientific theory is worth anything unless it enables us to predict something which is actually going on. Until that is done, theories are a mere game with words, and not such a good game as poetry." Whether the prediction covers current processes or past events, the hope is cherished that theory will help us learn more by predicting something previously unknown. The same prediction may also help us to see if the theory is correct through a matching of prediction with observation.

If finch communities are structured by competition, a simple set of rules derived from theory should enable us to predict their properties, e.g. how many species occur on an island, which particular species, their ecological differences, and so forth. The first attempt to use a simple set of rules to predict some of those properties was not very successful. On the assumption that the potential range of beak sizes of finches on the Galápagos is the same as that observed in continental South America, a rule derived from limiting similarity theory was used to predict the maximum number of ground finch species on any one island (Grant 1983c). The number predicted was 7 species, whereas the observed maximum is only 5 species. The discrepancy in number of species can be explained without discarding the theory (Grant

1983c), but this is not a satisfactory test of theory. The theory failed because it was not precise enough; in this application it took no account of food supply. A knowledge of food supply is probably needed to explain why, for example, there is no ground finch smaller than *G. fuliginosa*, even though a "micro-*Geospiza*" is predicted to occur. A comparable species, *Sporophila telasco*, does occur on the mainland (Marchant 1958).

Explicit attention to food supply is needed for another reason. Enhanced differences among coexisting species is certainly evidence in support of the idea that interspecific competition has been at work, but the idea may nevertheless be wrong. Peculiarities of the local food supply by themselves may have led to the same phenomenon, in which case character displacement and/or differential colonization have been mimicked by noncompetitive processes.

I shall now describe some models which predict the beak characteristics of coexisting species from a knowledge of food supply. The models either do or do not incorporate interspecific competitive effects. Thus the actual communities of finches can be compared with predicted communities to see which particular models provide the closest correspondence to actual communities.

Food supplies are known on several islands, and the relationship between beak size and diet is also known. The problem then is to devise a method of analysis that predicts the properties of finch communities from a knowledge of food supplies without merely reassembling the known relationships to predict what is known already. The solution to that problem was developed by Dolph Schluter. It is to estimate and use a third variable: expected population density, as a function of beak depth. The method assumes that the colonization of an island by a species, or the evolution of colonists to a particular size, is dependent on the potential population density attainable by the species with a given mean beak depth (Schluter and Grant 1984b). The analysis is restricted to generalist granivores among the ground finches. Excluded are the cactus specialists *G. scandens* and *G. conirostris* (on Genovesa), and highland populations of *G. difficilis*. These are habitat specialists or take seeds infrequently, so their resource variables (pollen, arthropods, etc.) are impossible to describe on the same axis as the one used for the generalist granivores. Thus I shall be examining the group of generalist seedeaters within the ground finch guild.

Methods of analysis. The procedure is as follows. First, the range of preferred seeds is calculated for a species with any given mean beak depth; additional effects of variance in beak depth upon the acceptable range of seeds (Chapter 6) is assumed to be small and is ignored. The calculation is accomplished by using observations of ground finches feeding in the dry season;

the effect of inter-annual variation in food supply on the calculations is small, as will be explained below. The upper and lower bounds to the range of seeds consumed by a finch species are set by different seed properties; the upper bound is set by hardness, the lower bound is set by size (Fig. 91). The hardness determines what can be cracked and the size determines what can be exploited profitably (Chapter 6). Generally they covary positively. The analysis is restricted to preferred seeds, those whose proportional representation in the diet is at least 50% greater than their proportion in the environment (Abbott et al. 1977). The restriction excludes seeds only casually eaten and unimportant to the finches, and those that are apparently avoided, such as a few species in the Convolvulaceae, which may be toxic to the finches (Grant and Grant 1980b).

The second step is to identify the preferred seeds on a particular island which fall between these limits. Their biomasses are summed. This gives the total seed biomass for a finch species with a given mean beak size. The third step is to convert the seed biomass into potential finch biomass, making use of the relationships depicted in Figure 92. Only three regression lines are drawn, the slopes of which differ significantly from one another. They characterize relationships for the three main granivore species, *G. magnirostris, G. fortis,* and *G. fuliginosa,* with different mean beak depths. Since the magnitudes of the slopes are inversely correlated with average log beak depth of the populations of the three species ($r = -0.99$), estimated intermediate slopes can be used for hypothetical species with intermediate beak depths. The fourth and final step is to convert finch biomass to finch numbers using the known relationship between individual biomass and beak depth given in Figure 93; for a species of given beak depth, expected population density (numbers) is calculated by dividing total biomass by individual finch biomass.

The result of these four steps of calculation is a single expected population density for a finch species with a single mean beak depth. The procedure is repeated in log beak depth increments of 0.5 over the whole feasible range of beak depths on the island in question, and then repeated on all 15 islands for which food data are available; 12 in the mid to late dry season and 3 in late wet season or early dry season.

Expected density profiles. Figure 94 illustrates the expected population density of a solitary granivorous finch species as a function of beak depth on all 15 islands. Expected density is usually not a simple function, but a markedly polymodal function, of beak depth. Moreover, different islands have different profiles as a result of different food supplies. The most important factor responsible for these patterns is the gaps in the frequency distributions of seed characteristics on an island, and the difference among islands

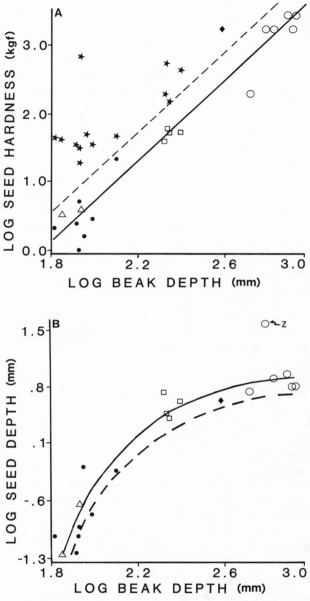

FIG. 91. The hardest seeds and the smallest seeds consumed by granivorous ground finches of different beak sizes. Solid lines are regression lines. Broken lines represent one standard deviation of the residuals added to (A) or subtracted from (B) Y. Symbols: *G. fuliginosa* (●), *G. difficilis* (△), *G. fortis* (□), *G. conirostris* (◆), *G magnirostris* (○). Stars represent seed types not eaten by these populations. Z is an outlier not used in the calculation of the regression relationship in B. From Schluter and Grant (1984b).

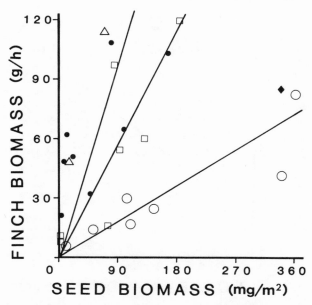

FIG. 92. Biomasses of three species of ground finches, determined from numbers of birds seen per hour of standard census, as functions of seed biomass in the dry season. From steepest to shallowest, the lines are regressions for *G. fuliginosa* (●), *G. fortis* (□), and *G. magnirostris* (○). Points are shown for three other populations: *G. difficilis* (△), and *G. conirostris* (◆) from Española. From Schluter and Grant (1984b).

in the position of these gaps. The effects of these gaps upon expected population density overwhelm the effects of two sources of variation: dry-season variation in relative seed abundance, within or between years, and errors of estimation involved in each of the four basic steps of calculation associated with the use of averages and interpolation (Schluter and Grant 1984b).

Each curve in Figure 94 is analagous to a complex, environmentally determined, adaptive landscape. The peaks represent those mean beak sizes which are estimated to be most efficient at converting available food into numbers; hence natural selection is expected to adjust mean beak size to positions at which maximum numbers are realized. Do species show evidence of having adapted to the landscape? The answer is yes. In almost all cases there is a clear correspondence between the average beak depth of a species on an island and a local maximum population density of a species with that beak size or one very similar to it. Furthermore, only one species is associated with each peak in the landscape. And at the end of the chapter I will give an example of the correspondence between an expected population density curve and a fitness curve for the same population.

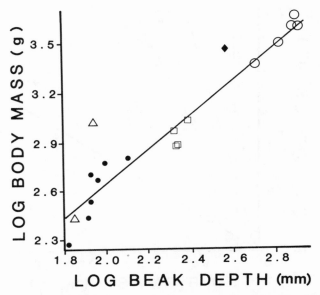

FIG. 93. Allometry between body weight and beak depth (natural logs) of adult male finches. Symbols: *G. fuliginosa* (●), *G. difficilis* (△), *G. fortis* (□), *G. conirostris* (◆), and *G. magnirostris* (○). The line is a least squares best fit to the points for *G. fuliginosa*, *G. fortis*, and *G. magnirostris*. From Schluter and Grant (1984b).

The next question is whether the alignment of species with peaks in the landscape has occurred independently or whether interspecific competition has been involved. There are few pronounced peaks in the frequency distributions of size-hardness classes of seeds and fruits. Therefore the overdispersion of beak sizes suggested the possibility of an interaction among species: they differ from each other more than expected. Peaks in the new function, expected population density, are more conspicuous and necessitate a re-examination of the beak size spacing pattern because the pattern may have been determined solely by these peaks.

To answer these questions I shall use models and start by considering noninteractive processes; then I will introduce competitive interactions. For more detail concerning calculations and assumptions, see Schluter and Grant (1984b).

Models. The development of island assemblages of ground finch species is most simply modeled with the aid of a computer by a process of random colonization, with all phenotypes being equally likely to colonize, within limits set by the food supply. The food supply has no other effect, and com-

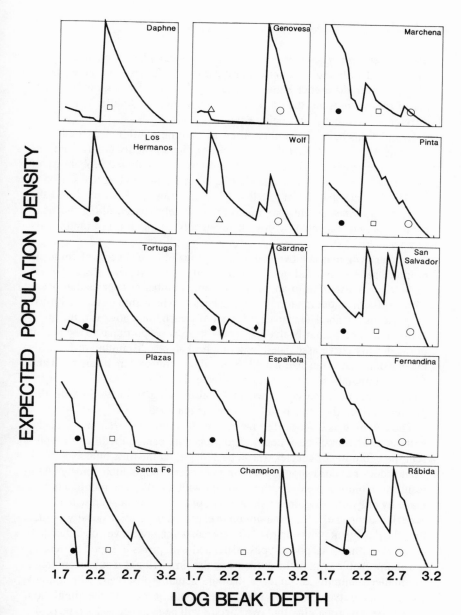

FIG. 94. Expected population density of a solitary granivorous finch species as a function of beak depth on fifteen islands. Absolute heights of curves are scaled to a common maximum: for example, the small local maximum at an intermediate beak depth on Champion is absolutely higher than any of the (unscaled) maxima on Rábida. Mean log beak depths of male ground finches on the islands are indicated by the positions of symbols: *G. fuliginosa* (●), *G. difficilis* (△), *G. fortis* (□), *G. conirostris* (♦), and *G. magnirostris* (○). The beak depth of the extinct *G. magnirostris* from Floreana and San Cristóbal has been placed on the Champion figure (see Fig. 1 for locations of the islands); Champion was first visited by scientists after the large form of *G. magnirostris* had become extinct. *G. magnirostris* is not shown on Santa Fe, but one pair bred there in the 1960's (Chapter 3). From Schluter and Grant (1984b).

petition between species does not occur. The mean phenotype of a species should be distributed as a uniform random variable between these limits. The minimum difference between size-neighbors is a function of the number of species present: the more species there are, the smaller will be this difference, all other things being equal.

Calculations have shown that observed minimum differences tend to be larger than expected, that the probabilities of them individually being as large as they are by chance are low, and that the combined probability of these large values for all 12 islands studied in the mid to late dry season is 0.001 (Schluter and Grant 1984b); other analyses (p. 328) produced the same result. Species are thus much more widely spaced along the beak depth axis than expected by this unrealistically simple form of random assembly.

Greater realism is achieved by making probability of success of colonists proportional to expected density associated with their beak sizes. This is modeled by casting on to an island the same number of species that occur there, 100 times, but randomly varying beak depths within observed limits. A species is assigned a mean depth randomly each time; however, the probability that a particular beak depth is assigned is proportional to the height of the density curve associated with that beak depth. Minimum differences in beak depth between size-neighbors are then compared in the artificial and actual communities. Differences are greater in the actual than in the artificial communities, the combined probability being 0.0002. So with this greater realism, the same result as before is obtained.

The two models discussed so far involve random colonization of islands, without evolution of beak depth. The procedure can be modified to represent the evolution of species whose beak depth, at the time of arrival on an island, does not correspond to a peak in expected population density. As a result of evolution, beak depth is aligned with a peak. The modification is made by only allowing, in the simulation, beak sizes associated with peaks. A peak is defined as a local maximum in expected population density greater than the density at points up to 0.10 log beak depth units away on either side of the maximum. All phenotypes associated with peaks in expected density are equiprobable. Again the outcome of the simulation is a consistent trend of smaller minimum differences than are actually observed, with a combined probability of 0.0005. This comes about because in the simulation there is a high probability that two species will be assigned to a single peak, whereas in actual communities no two species are under the same peak (Fig. 94).

Despite restriction to no more than three *Geospiza* species per island, these three simulations provide the firmest evidence for an overdispersion of beak characteristics of coexisting species, and hence of their niches, compared with those expected from food supply alone.

Exclusive occupancy of a peak is likely to be the product of interspecific competition, since it is not expected by chance, therefore we now need to model the assembly of communities with an explicit allowance for interactions between species. This is done first for colonization with competitive exclusion, but no evolution, in the following simulation. Three species with any feasible beak depths are allowed to colonize a given island. The conditional expected population density of each is then computed as its expected density excluding all foods lying within the preferred range (see Fig. 91) of any of the other species. The process identifies those species (one or more) whose conditional density is zero; the one with the lowest expected density is then eliminated. Conditional densities for the remaining species are recomputed, and serial deletion (extinction) of species continues until all remaining species have positive conditional density. A new colonist with a different beak depth is then added and the entire process is repeated until no additional species can successfully invade. The order of colonization does not matter, for the procedure is entirely deterministic and gives a single result for each island. Coexistence is dependent upon the exclusive portions of the niches of each species, with food abundance having the additional role of determining the species of greatest density when more than one species exploit an array of foods.

Beak sizes of species in artificial and actual communities on 12 islands are now compared: beak sizes in the artificial communities are the predictions from a hypothesis of competitive adjustment of species to their food supply, and these are compared with observed beak sizes. The association between them is very strong ($r = 0.90, N = 23$). A weak association ($r = 0.26$, maximum) is expected by chance, because observed and expected beak depths are assembled in ordered pairs; but when this is allowed for, the statistical significance of the association remains strong (Schluter and Grant 1984b). The hypothesis of competitive exclusion is supported.

An almost identical result ($r = 0.92, N = 32$) was obtained from a similar simulation of a process that allowed for evolution. That combination of beak depths was determined for each island at which the conditional density of each phenotype is maximum, given the beak depth of the other species in the combination. The combination should define the endpoint of evolutionary adjustments of competitor species to a shared food supply, referred to sometimes, and somewhat loosely, as coevolution (Case 1979, Schluter and Grant 1984b; see also Futuyma and Slatkin 1983).

Thus, in conclusion, incorporating competitive effects into the models yields a close correspondence between actual and predicted properties of finch communities. This should not be interpreted to mean that everything is explained by the models. To give one example, the question raised earlier of why there is no "micro-*Geospiza*" is not satisfactorily answered by these analyses. The expected density curves (Fig. 94) show that suitable conditions for such a species appear to exist on some of the islands, although the stability and predictability of those conditions are not known.

The role of competitive exclusion appears to be clear, in that no two species occupy a single peak, and a single peak may be occupied by different species on different islands, such as *G. difficilis* on Genovesa and *G. fuliginosa* on Marchena. Also, the number of granivore species on an island is correlated ($r = 0.79, N = 15, P < 0.01$) with the number predicted by simulation of the competitive exclusion process, but without evolution, as determined by the range of available foods (Schluter and Grant 1984b).

The role of evolutionary adjustments is less clear. Although the competition model that allowed for evolution achieved high success in predicting beak sizes accurately, part of the alignment of species with their peaks may have nothing to do with competition. For example, a species that colonizes an island with a beak depth close to but not at the size associated with a peak in expected population density, will evolve to that size, regardless of the presence or absence of other species associated with other peaks. To assess the role of competition in the evolution of beak sizes of sympatric species, we need to know if the presence of one species has an effect on the closeness of the alignment of another species with its peak. This possibility is examined in the next and final section of this chapter.

THE CLASSICAL CASE OF CHARACTER RELEASE

Geospiza fortis and *G. fuliginosa* are the most suitable species for examining the possibility that character displacement has been responsible for the large differences in beak depth between sympatric species. As I described in the previous chapter, the two species have discrete, nonoverlapping, frequency distributions of beak depths on the several islands they jointly occupy. On Daphne Major the small ground finch, *G. fuliginosa*, does not breed (regularly), and here the medium ground finch, *G. fortis*, has a smaller beak on average than elsewhere (Fig. 15). On Los Hermanos, a group of four small islands off Isabela (Fig. 1, Plates 14 and 15; see also Grant 1975b), *G. fortis* does not breed, and here *G. fuliginosa* has a larger beak on average. Thus the two species are more similar to each other in allopatry than they are in any sympatric location (e.g. see Plate 63). The allopatric populations are not identical, however, which is how we know they belong to their respective species; song characteristics provide additional evidence of their specific identity (Grant 1975b).

Interpretations of this fascinating pattern have varied. Lack (1945) first explained the intermediate sizes of the two species in allopatry as originating from hybridization between *G. fortis* and *G. fuliginosa*. "However," he later wrote (Lack 1947, p. 85) "it would be highly remarkable if two species which occur together on many islands without interbreeding should, just in two places, give rise to a hybrid population.'' He replaced hybridization with a hypothesis of competitive release: "If only one of the two species [on one of the islands] persisted, it might be expected to evolve a beak of intermediate type, since the foods normally taken by both species would be available for it'' (Lack 1947, p. 85). We know now that the species do hybridize. The effect is to increase genetic variance in the *G. fortis* population and to make an evolutionary response to selection more likely, but hybridization by itself is not sufficient to cause a major shift in means (Chapter 8, and Boag and Grant 1984b).

If, as seems likely on geographical grounds, the populations on Daphne and Hermanos were derived from populations on the neighboring large islands of Santa Cruz and Isabela respectively (Fig. 1), their morphological patterns may constitute an example of character release; beak sizes that have changed evolutionarily, as a result of a release from competitive constraints (Grant 1972b). Although Bowman (1961) was critical of several of Lack's examples of competition-mediated bill size changes he did not comment on this particular one. Nevertheless, his arguments can be used to set up the alternative hypothesis that the intermediate bill sizes in allopatry simply reflect adaptation to intermediate seed sizes whereas in sympatry the bill sizes of species differ because their food supplies differ (Boag and Grant 1984b).

The procedures of analysis described in the last section can be used to distinguish between the two alternatives. Since peaks in expected density are determined solely by food, the associated beak sizes provide a standard against which actual beak sizes on different islands can be compared. Additional effects of interactions between particular species may be detected as discrepancies between observed and predicted beak sizes.

Figure 95 shows that the predicted beak sizes of *G. fortis* on Santa Cruz and Daphne Major are very similar, whereas the observed beak sizes are markedly different. Likewise the predicted beak sizes of *G. fuliginosa* on Santa Cruz and Los Hermanos are very similar, but the observed beak sizes are different. These results are not expected from the food supply hypothesis, but they are expected from the character displacement hypothesis.

Interestingly, the larger size of *G. fuliginosa* on Los Hermanos is associated with a shift to a new peak of substantially greater height. This can be interpreted as a shift in beak morphology and feeding niche permitted by the absence of a competitor, *G. fortis*. But the important point is that the observed beak depth is very close to the beak depth predicted by this peak. Thus, but for a small difference in the positions of the peaks on Daphne and Los Hermanos, the populations of *G. fortis* on Daphne and *G. fuliginosa* on Los Hermanos are essentially the same in morphology and ecology. While all this evidence is consistent, a small inconsistency is observed on Tortuga

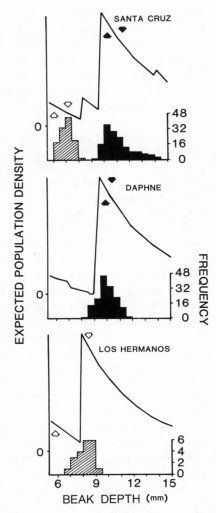

FIG. 95. Beak depths of *G. fuliginosa* (hatched) and *G. fortis* (solid): comparison of observed mean sizes (downward arrows) with mean sizes predicted (upward arrows) by maximum expected population densities which are indicated by peaked curves. Differences between observed and expected means are larger in sympatry (Santa Cruz) than in allopatry (Daphne); on Los Hermanos *G. fuliginosa* has shifted from one peak to another. From Schluter et al. (1985).

(Fig. 94), where the beak size of the allopatric *G. fuliginosa* is somewhat less well predicted by the peak than is the case on Los Hermanos.

Beak sizes are close to those predicted by peaks in expected population density in both allopatric populations, but on Santa Cruz the two sympatric species have displaced beak sizes (Fig. 95). The displacements are general among sympatric populations, as shown in Figure 96. The consistent difference between allopatric and sympatric populations in their proximity to predicted beak sizes is the pattern expected from a hypothesis of competition.

The pattern is explained more fully in terms of the following process. The average beak size in a *G. fortis* population can be thought of as arising from a trade-off between feeding rates on small and soft seeds and on large and hard ones. In sympatry, the trade-off is affected by the presence of *G. fuliginosa* which reduces the dry-season availability of small and soft seeds. Those *G. fortis* individuals too small to crack the large seeds responsible for the *G. fortis* peak compete directly with the more efficient *G. fuliginosa* for the smaller seeds. Disproportionate mortality of these small *G. fortis* individuals leads to a large average size on islands where *G. fuliginosa* is pres-

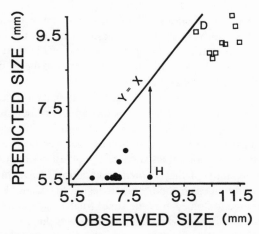

FIG. 96. Observed mean beak sizes of *G. fuliginosa* (●) and *G. fortis* (□) males compared with sizes predicted from peaks in expected population density; an alternative representation of the results in Figure 95. All observed sizes are larger than predicted; points are uniformly displaced to the right of the line with a slope of 1.0. The population of *G. fortis* on Daphne (D) is exceedingly close to the slope; all sympatric populations are strongly displaced. The population of *G. fuliginosa* on Los Hermanos (H), although strongly displaced from the normal peak for this species, is actually very close to a second peak that is typical for *G. fortis*, and this is indicated by the arrowhead. All sympatric populations of *G. fuliginosa* are strongly displaced. From Schluter et al. (1985).

ent; in allopatry, the trade-off is unaffected by *G. fuliginosa* and therefore mortality does not fall disproportionately on the smaller individuals.

We encounter a difficulty when extending this interpretation to *G. fuliginosa*. Like *G. fortis*, it is larger in sympatry than predicted by a peak in expected density, yet there is no smaller species with which it competes. One possible explanation is that its large size in sympatry is the product of selection for efficient exploitation of the small seeds (Schluter et al. 1985). If this factor is sufficient to account for the larger-than-predicted size of sympatric *G. fuliginosa*, it might be sufficient to account for the larger-than-predicted size of *G. fortis* (and *G. magnirostris*; Fig. 94). And if this was the only factor, we would expect to find larger-than-predicted sizes in allopatry also; but we do not (Figs. 95 and 96).

The difficulty disappears if we assume that the discrepancies between observed and predicted beak sizes in sympatry, for all species, arise from the particular way expected density values are estimated. The discrepancies can be removed, for example, by using a procedure that weights available food by a handling-efficiency factor (D. Schluter, pers. comm.). When an arbitrary 5% is subtracted from the beak sizes of all populations, which has the same effect as uniformly adjusting the predicted values, all sympatric species become closely aligned with their predicted sizes, whereas beak sizes of the two allopatric populations become displaced below these revised predicted sizes. This returns us to the main issue, which is the consistent difference between allopatric and sympatric populations in their proximity to predicted beak sizes. The issue of which set is closer to the predicted values, allopatric or sympatric, is secondary to this.

The biological interpretation of the shift in allopatry is that members of the population exploit some of the foods that are consumed in sympatry by another species. They exploit both *fortis* and *fuliginosa* foods; this is Lack's original hypothesis. A testable prediction of this hypothesis is the expectation of disruptive selection on an allopatric population under conditions of food stress. The study of the *G. fortis* population on Daphne during the drought of 1977 provides an opportunity to test it. In Chapter 8 I gave details of directional selection on each of the sexes treated separately. An analysis with the sexes combined showed that not only was there directional selection in favor of large birds, but the expected disruptive selection also occurred (Schluter et al. 1985). Most individuals that survived the drought exploited the large seeds of *Opuntia echios* and *Tribulus cistoides*, which determine the highest peak in expected population density on the island (Fig. 95), but at least six individuals not able to crack these seeds nevertheless survived by exploiting the few remaining small seeds. All six were females, and they formed a large proportion (20%) of the total number of surviving females. Thus the disruptive selection can be attributed to the utilization by members of the Daphne population of both the *G. fortis* and *G. fuliginosa* niches, as defined by the two peaks in the expected density curve (Fig. 95), a view which is supported by the close correspondence between the expected density and individual fitness curves at this time (Fig. 97).

I conclude that character release has occurred in allopatry, and that in sympatry competition with other species tends to restrict a given species to its niche.

The study on Daphne Major also answers the question raised earlier as to why *G. fortis* has not increased in beak size in the absence of *G. magnirostris* (Boag and Grant 1984b). The large, single, peak in expected population density that is determined by food supply (Figs. 94 and 95) not only accounts for the observed beak size of *G. fortis*, it indicates that the smaller *G. fuliginosa* and the larger *G. magnirostris* do not usually have breeding populations on this island because their food supplies are probably not sufficiently large (see Boag and Grant 1984b for details). *G. scandens*, the cactus finch, does breed on the island and it has a competitive influence on *G. fortis* (Chapter 7), but it does not cause a displacement of *G. fortis* from the peak because it only eats one (*Opuntia echios*) of the two seed types which determine the peak (the other is *Tribulus cistoides*).

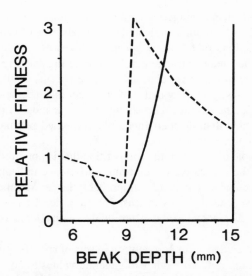

FIG. 97. Relative fitness (solid line) of *G. fortis* individuals (sexes combined) on Daphne Major associated with mortality during a drought in 1977. The trough in the curve shows the effects of disruptive selection. The fitness curve corresponds well with the curve in expected population density (broken line). From Schluter et al. (1985).

CONCLUSIONS AND SUMMARY

Interspecific competition for food has influenced the number of species that occur on an island, the particular species that occur on an island, and their beak sizes and diets. Detecting such competitive effects is fraught with technical difficulties. This chapter explores the evidence for competition through analyses of patterns shown by coexisting sets of ground finch species.

Two types of competitive processes are distinguished: character displacement, which is an evolutionary process of unilateral or mutual divergence in traits such as beak size (and diets) caused by natural selection; and differential (nonrandom) colonization, which is an ecological process of competitive exclusion occurring when a species arrives on an island that already has another species with similar ecological requirements. The result of both processes is an enhanced ecological difference between coexisting species. Some patterns can be attributed to one process but not to the other, whereas other patterns can be attributed to either or both.

Differential colonization is indicated by the occurrence of *Geospiza* species in nonrandom combinations. Some combinations are rare or do not occur on any of the islands in the Galápagos archipelago. Members of these combinations are generally similar to each other in beak sizes and diets. Each one of the six *Geospiza* species occurs with its most similar congener on fewer islands than average, or even none (*G. fortis* and *G. conirostris*).

Character displacement is indicated by comparisons of sympatric and allopatric populations of the same species. Eleven of the thirteen pairs of *Geospiza* species that occur together differ more in sympatry than expected by artificially combining populations of those species in all possible pairwise combinations. This is evidence of character displacement in the aggregate. In one specific case it is also shown; *G. fortis* and *G. fuliginosa* differ more in sympatry than do their randomly combined populations from all parts of the archipelago.

As an example of evidence for competition that is not directly assignable to just one of the processes, the minimum difference in beak size between coexisting species is, statistically, improbably large. Members of all pairs of coexisting species differ by at least 15% in at least one bill dimension. Both character displacement and differential colonization may have contributed to this pattern.

Much of the evidence for competition in sets of species (guilds) has been obtained by using random models to generate values to be expected in the absence of competition, i.e. a null hypothesis. The models have been criticized for their artificiality and for a consistent bias towards failing to detect competitive effects when they are really present. Yet despite these problems the evidence for competitive effects has been strong in several cases.

A more comprehensive and realistic approach to the detection of these effects has been adopted by using models to predict properties of finches from a knowledge of the food supply on each island. The procedure involves estimating expected population density as a function of beak depth for the generalist granivore species. Expected population density is a polymodal function of beak depth on many islands, owing to gaps, or scarcities of certain classes of foods, in the frequency distributions of food types.

Models that incorporate interspecific competition for food are more successful in predicting the number of species on islands and beak sizes than are those that do not. Differential colonization is shown by the fact that there is never more than one species associated with a peak in the expected density function, and that it is a different species on different islands, e.g. *G. difficilis* on Genovesa and *G. fuliginosa* on Marchena. Character displacement is shown by the difference between sympatric and allopatric populations of *G. fortis* and *G. fuliginosa*. Beak sizes of each species are predicted from peaks in the expected density curves to be the same in sympatry and in allopatry, whereas they are not. The two species differ more in sympatry than in allopatry. The role of natural selection in aligning beak sizes with the food-determined optimum was shown in a study of the allopatric *G. fortis* population on Daphne Major during a drought. Disruptive selection occurred, with most of the population aligned with the major (*fortis*) peak but a few aligned with a minor (*fuliginosa*) peak. As a result of selection, fitness curves corresponded closely to the expected density curves.

The overall conclusion from these analyses is that interspecific competition has left its mark on the structure of finch communities in a much more complicated and pervasive fashion than was presented in the previous chapter. This strengthens the view, developed in that chapter, that competition played an influential role in both allopatric and sympatric phases of the cycles of speciation of the ground finches. Presumably it did so in the diversification of the tree finches as well, but the evidence for this is more circumscribed.

CHAPTER THIRTEEN

The Evolution of Reproductive Isolation

INTRODUCTION

In Chapter 10 I identified two debatable issues with the allopatric model of speciation. The first is whether any interaction took place at the secondary contact phase, leading to enhanced differences between original and derived populations. If the answer is yes, the second issue is whether the interactions were ecological, reproductive, or both. The previous two chapters have established a strong case for competitive interactions betwen species, the intensity of which is likely to have varied in proportion to the similarity of the interactants in bill morphology and diet. I concluded that interactions probably did take place at the secondary contact phase of some cycles. The interactions were ecological. The remaining question is whether reproductive interactions took place as well.

The proposition to be considered was well stated by Lack: "Those island forms which met [i.e. steps 3 and 5], but differentiated each other through bill dimensions, have remained segregated (presumably with subsequent intensification of the bill differences), while others which failed to differentiate each other have doubtless merged" (Lack 1945, p. 122). Here are two hypotheses, both involving species recognition by bill characters. One is that the differences evolved initially in allopatry; the other is that the differences became enhanced in sympatry by selection: "Such specific differences have doubtless been intensified by selection in instances where they serve for recognition" (Lack 1945, p. 127).

The claim that birds of different species recognize each other by their bills is well substantiated (Chapter 9). The question I shall consider in this chapter is whether those differences arose in part through a selective intensification, or reinforcement, of initially smaller differences.

EXPERIMENTAL TESTS

A requirement of the reinforcement hypothesis is a potential reproductive confusion between residents and immigrants that are similar in appearance. This condition can be investigated experimentally with stuffed specimens representing immigrants. If the models (specimens) are very different from the local birds they should be ignored, or discriminated against. If they are similar to local birds they should be courted as potential mates. Experiments to test these expectations were conducted with four sets of populations according to the protocols described in Chapter 9 (Ratcliffe and Grant 1983b).

Males of *G. fortis* on Plaza Sur were offered a choice between a local female model and a female model from Daphne where the average size is smaller. This artificially creates the situation of a conspecific but partly different immigrant individual staying to breed. As expected, the responding birds failed to discriminate between the two models (Ratcliffe and Grant 1983a). In contrast, they did discriminate, in a separate set of experiments, between a local female conspecific model and a local female *G. fuliginosa* model (Chapter 9). A similar experiment was conducted with *G. difficilis* on Genovesa, and with similar results. The responding birds failed to discriminate between a local *difficilis* model and a *difficilis* model from Pinta. Since the morphological difference between the resident and immigrant models was small in each experiment, and hence the discrimination task was difficult, all other tests were conducted with pairs of species, albeit morphologically similar ones in most cases.

Males of *allopatric* populations of *G. fuliginosa* on Plaza Sur and *G. difficilis* on Genovesa were offered, on each island, a choice between female models of the two species, thus artificially creating the situation of heterospecific immigrant individuals staying to breed. On neither island did the males discriminate between conspecifics and heterospecifics in these choice experiments. This demonstrates the potential for interbreeding between two morphologically similar species, under admittedly contrived circumstances that allowed no opportunity for females to respond to the males.

A failure to discriminate is unsatisfactory because by itself it may merely reveal the limitations of an experiment and not the workings of a biological process. It is therefore important to place the negative evidence in perspective by contrasting it with the results of parallel experiments that show these birds *can* discriminate, as was done above. On Genovesa *G. difficilis* discriminated against the much larger, sympatric, *G. magnirostris*, and on Plaza Sur *G. fuliginosa* discriminated against the larger, sympatric, *G. fortis* (Ratcliffe and Grant 1983b). On Pinta, where the two species are sympatric, each showed strong discrimination against the heterospecific specimen (Chapter 9). The lack of discrimination in allopatry is therefore a real phenomenon.

I do not wish to imply that there would be no discrimination between conspecific residents and heterospecific immigrants. The experiments do not measure the amount of discrimination. Instead they show (or do not show) a differential response that cannot be attributed to chance within statistically acceptable limits. In this case they show a potential for reproductive confusion between conspecifics and morphologically similar heterospecifics.

They reveal more about differences in discrimination than about absolute levels of discrimination.

A third set of experiments demonstrated the same phenomenon (Ratcliffe and Grant 1983b). When presented with an immigrant *G. fuliginosa* female model from Santa Cruz, *G. fortis* males on Daphne did not discriminate between it and a local *G. fortis* female model. Similarly, *G. fuliginosa* on Española did not discriminate between a conspecific and an immigrant *G. fortis* from Daphne. Once again, it is important to contrast these failures to discriminate with successful discrimination in other contexts. Both Daphne *G. fortis* and Española *G. fuliginosa* did discriminate between *G. fortis* and *G. fuliginosa* models from Santa Cruz, in each case by directing most attention to the conspecific model. Since *G. fortis* is larger on Santa Cruz than on Daphne, the discrimination task was easier in these experiments than in the previous one.

Results of this third set of experiments can be linked more closely with real events than can those of the previous one. *G. fuliginosa* do occasionally immigrate to Daphne from Santa Cruz, a few stay to breed, and some hybridize with *G. fortis* (Chapter 8). The failure of Daphne *G. fortis* to show discrimination in experiments between the morphologically similar *G. fortis* from Daphne and *G. fuliginosa* from Santa Cruz is consistent with the observed hybridization.

In a final set of experiments *G. scandens* on Plaza Sur preferred to court a conspecific specimen rather than immigrants of two other species (in two separate experiments): *G. conirostris* and *G. difficilis* from Genovesa. Conversely, *G. difficilis* on Wolf failed to discriminate between a local conspecific and an immigrant *G. scandens*. The only surprising behavior in all of these experiments was shown by *G. difficilis* on Genovesa. When presented with a conspecific and an immigrant *G. scandens*, these birds mounted and copulated with the immigrant significantly more frequently than with the conspecific; in other words they showed reverse discrimination! Whether this last result is an artifact of the experiment, a chance finding, or a real phenomenon perhaps induced by a "super-normal" stimulation of the tested birds, it is consistent with all other results in demonstrating a potential for reproductive confusion among allopatric birds.

IMPLICATIONS OF THE EXPERIMENTAL RESULTS

These results demonstrate that the condition of potential reproductive confusion is met. If the difference between them is small, residents and immigrants are likely to interbreed; if it is large they are less likely to do so. The likelihood of interbreeding is a little more complicated than this, however, so before discussing the relevance of these results to the speciation model I will consider what more these experiments imply about interbreeding.

The probability of a discrimination being made by a resident is not just a function of the difference in appearance between conspecific residents and immigrants: it is also a function of the difference between conspecifics and other resident species. The probability of discrimination increases in direct proportion to the distinctiveness of the immigrant, but, for a given degree of distinctiveness—i.e. a given difference between a resident species and an immigrant—a resident is more likely to discriminate if it is on an island with several congeners, some of which are similar to itself, than if it is sympatric with only one or two which are substantially different. Interbreeding, therefore, is more likely to occur on islands with a few species than on those with many. Figure 98 illustrates the relationship between the likelihood of interbreeding and the morphological distinctiveness of immigrants on islands with different numbers of species. An alternative three-dimensional representation is given in Ratcliffe and Grant (1983b).

On the species-rich island of Pinta, *G. fuliginosa* and *G. difficilis* each

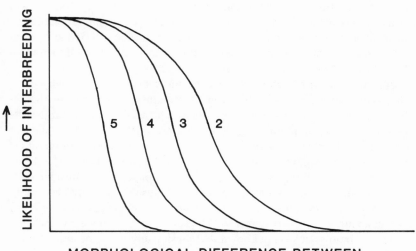

FIG. 98. Likelihood of interbreeding as a function of the morphological difference between a resident species and an immigrant. For a given morphological difference, the likelihood is greater on islands with only 1 or 2 residents species than on islands with 5 species; compare the numbered curves.

recognize the small morphological differences between them, whereas on islands where there are fewer species (Plaza Sur, Genovesa) each species apparently does not recognize the differences. The differences in bill size between the species are approximately constant: there has been virtually no sympatric divergence in bill morphology on Pinta that has facilitated species recognition. The two species differ in body size on Pinta, and the difference may be significant in courtship, but I doubt if it was important in the experiments in view of the similar way in which the models were prepared. It is an open question as to whether the discrimination shown by birds in sympatry but not in allopatry represents genetic differences in perceptual abilities, or whether it arises from experience with similar-looking heterospecifics. Experiments that vary early experience in a systematic manner could distinguish between these possibilities.

Thus interbreeding, following immigration, is more likely to occur on species-poor than on species-rich islands. The three species that were originally considered by Lack (1945) to have originated by hybridization occur on the species-poor islands of Darwin, Daphne, and Los Hermanos. But since the species-poor islands are generally well isolated, the probability of immigration itself is low, so that interbreeding is likely to be rare (Stresemann 1936, Hamilton and Rubinoff 1967), except perhaps on Darwin (Chapter 4) for reasons that are not clear. For instance, immigration of non-resident species to the small but weakly isolated island of Daphne is much more frequent than it is to the more strongly isolated island of Genovesa (Chapters 7 and 8).

An unanswered question arising from these experiments is why there should ever be interbreeding between an immigrant from a species-rich island and a resident on a species-poor island. The resident may court the immigrant, but the immigrant should not respond, if it is a female, and not court the resident if it is a male. Yet hybridization between immigrant *G. fuliginosa* and resident *G. fortis* does occur on Daphne (Chapter 8). Therefore, if there is discrimination exercised by the immigrant, it must be overcome sometimes. It is overcome probably because of the extreme scarcity of mates; rather than not breed at all under these circumstances a bird will accept an unusual mate (Chapter 9). The additional role of song in mate selection of immigrants is not known, but presumably, as in the case of morphological discrimination, song discrimination is occasionally overcome. Song characteristics diverged in allopatry, and there is scarcely any evidence of the enhancement of differences in sympatry (Ratcliffe 1981).

The characteristics by which species can be distinguished may be described as barriers to interbreeding, barriers that are sometimes crossed. It is as if a bird has a range of tolerance to the phenotypes of acceptable mates, and that this range varies from island to island for a given species, and is exceeded only when potential mates within the range are scarce or lacking altogether. Interbreeding between residents and immigrants should be viewed as a probabilistic matter, influenced by many factors including the degree to which residents and immigrants differ in morphology as postulated in the speciation model.

REINFORCEMENT?

One condition for reinforcement exists: there is a potential reproductive confusion between similar immigrants and residents. The mixing of geographically differentiated forms of the same species would lead to interbreeding. Given the reproductive confusion and some interbreeding, for which there is direct evidence (Chapter 8), how is natural selection expected to act? The answer depends on the amount of interbreeding, the difference between the interbreeding forms, and the fitness of the offspring.

For a pre-mating barrier (behavioral mate choice) to develop or be enhanced in sympatry, there has to be a strong post-mating disadvantage. Intersterility (Lack 1945, 1947) is one, but other possibilities are inviability or infertility of the hybrids (Mayr 1963).

Intersterility is not known in Darwin's Finches. Intrageneric hybrids among ground finches are certainly both viable and fertile (Chapter 8), and probably the same is true for intergeneric hybrids between tree finches and warbler finches (see p. 247, and Bowman 1983). Exactly how fit the hybrids are is still not known. There are not enough data to make a quantitative comparison of the survival and reproductive fates of offspring of intraspecific and interspecific matings. The evidence from Daphne Major, so far as it goes, suggests that hybrids may be at a mild disadvantage at most times; possibly they have a slight advantage at times of environment stress. Under conditions of only a mild hybrid disadvantage such as this, the gene exchange between the two interbreeding forms is sufficient to reduce their distinctiveness, and selection in favor of pre-mating isolating mechanisms will become progressively weaker, as will the post-mating disadvantage. If unaffected by other factors this process leads ultimately to the complete fusion of the two forms (e.g. see Futuyma 1979, Templeton 1981).

I therefore conclude that selection for reinforcement of bill differences solely in a reproductive context did not occur in the sympatric phase of speciation cycles.

On the other hand, there is evidence of selection on bill size in an ecological context at this time (Chapters 11 and 12). I am referring to the demonstrated role of natural selection on bill size in the nonbreeding season on Daphne Major, and the various pieces of evidence for ecological character displacement. The two sets of forces (ecological and reproductive) could have acted in concert at the time of secondary contact, with the ecological ones being primary and the reproductive ones being secondary. That is, selection for the minimization of competition for food through divergence of bill sizes would have produced, largely as a correlated effect, an enhancement of reproductive isolation in sympatry (e.g. Moore 1957). A contributing factor to this process could have been the reinforcement of discriminating abilities and mate preferences, for which there is evidence from the field experiments.

The study on Daphne has provided a glimpse of how sexual selection might also be involved. Females of both species paired preferentially with males of large beak size in the years after the 1977 drought (Chapter 8). Such female preferences may be genetic or transmitted through imprinting. In either case the preferences would be expected to become correlated with morphology (e.g. Lande 1981). Since selection in both sexes will result in a correlated response in mate preferences, long and continued evolution under directional selection will lead eventually to reproductive isolation. The reproductive isolation is a byproduct of the selected ecological divergence.

Therefore I conclude that species either do not interbreed, or interbreed rarely, because pre-mating ethological barriers between them arose in allopatry. In some instances the differences became enhanced in sympatry as a correlated effect of ecological character displacement, with sexual selection possibly playing a minor, additional, role.

ABSENCE OF SPECIES FROM ISLANDS

These conclusions have an important consequence. Absences of certain species from islands were interpreted in Chapters 11 and 12 solely in terms of competitive exclusion: a few members of one species arriving on an island occupied by an ecologically similar species would be competitively excluded. We must now add to that interpretation the possibility of interbreeding. If members of the immigrant species bred among themselves, only the ecological process of exclusion would occur. But if the immigrants and residents interbred then the immigrants would become absorbed into the much larger population of residents, with many of the alien genes perhaps being purged at times of occasionally intense directional selection. This is, in essence, one form of the hypothesis of a selection/immigration balance proposed in Chapter 8 to account for the large degree of morphological variation in some populations.

Empirical support for these ideas is provided by observations on Daphne Major. *G. fuliginosa* repeatedly immigrate. They have not become established as a permanent, reproductively isolated, population, partly because they are at a competitive disadvantage with the numerous and relatively small *G. fortis*, and partly because they (or their offspring) are reproductively absorbed by them through interbreeding. This situation is one of those "experiments" Swarth (1934) had in mind when confronting the intergradation among species. It is not unique to Darwin's Finches: a comparable phenomenon occurs with flycatchers on the Swedish islands of Gotland and Öland (Alerstam et al. 1978, Alatalo et al. 1982). An immigrant species, the pied flycatcher (*Ficedula hypoleuca*), has failed to establish large breeding populations on the two islands, apparently as a result of extensive interbreeding with an abundant relative, the collared flycatcher (*F. albicollis*).

The absence of a species can never be explained unequivocally. Beak size appears to be a factor in determining which species coexist on the Galápagos, but even here there is ambiguity in the interpretation because beaks have more than one function. The situation on Daphne Major is exceptional in showing what happens when a species, usually absent, invades and stays to breed. It reveals that both ecological and reproductive interactions occur between residents and immigrants, but whereas the former lead to repulsion of the immigrants the latter lead to their assimilation.

SUMMARY

At the secondary contact phase of each speciation cycle, natural selection may have reinforced the bill size differences between the populations, thereby minimizing subsequent interbreeding. Experiments on five islands with models demonstrate the potential reproductive confusion between residents and immigrants when they are morphologically similar. The tested species showed a lack of discrimination between resident and immigrant models, which stands in contrast to the discrimination they showed between conspecific and resident heterospecific models (Chapter 9).

Reinforcement of the differences in signals (bill size) used in courtship occurs only when there is a disadvantage to interbreeding, either through sterility, or through inviability of the offspring, or their infertility. But interbreeding does occur, and at least some of the offspring are fertile. Intersterility is not known. Under conditions of a relatively weak disadvantage to interbreeding it is unlikely that reinforcement of bill differences would occur solely for reproductive reasons except perhaps to a minor extent. Divergence of bills occurred instead through natural selection which minimized competition (Chapter 12). Therefore reproductive character displacement took place largely as a correlated effect of ecological character displacement.

These results and interpretations cast a slightly different light on some of the arguments for competitive exclusion given in the previous chapter. Competitive exclusion is not the only reason why a species may not have a breeding population on an island. Individuals of this species may immigrate, interbreed with a resident, and become assimilated into the population of residents. On Daphne Major, immigrant *G. fuliginosa* do not form a regularly breeding population, partly because they are at a competitive disadvantage with *G. fortis*, and partly because they are reproductively absorbed by them.

CHAPTER FOURTEEN

Adaptation: Body Size, Plumage Coloration, and Other Traits

INTRODUCTION

In the original colonization of the Galápagos and Cocos Island, the birds that arrived were confronted by a series of problems which they overcame. These problems were associated with finding sufficient food and moisture, avoiding predators if there were any, and obtaining a mate, reproducing, and rearing their offspring to independence. Certain individuals did better than others. Those that were best able to meet their needs by virtue of possessing superior features increased in frequency as a result of natural selection. The populations evolved and became better adapted to their specific environmental circumstances. On the Galápagos the process of adaptive change occurred many times in the many populations that became established on the many islands, whereas on the single and isolated Cocos Island much less adaptive change occurred in the single population.

Not all changes were adaptive. Some changes occurred through random processes, probably most often in the early phase of establishment of a new population whose size was initially small, but also when populations fell to small sizes in times of severe drought. Other changes occurred in some traits as the correlated effects of selection on variation in one or a few other traits.

Observing the large amount of variation that exists today, both among and within species, we would like to know which components are adaptive. To help us find out we need to know the functional significance of a trait, and the significance of variation in that trait in relation to variation in the relevant features of the environment. We would be even better equipped to interpret the variation if we also know its genetic basis, and the identity, direction, and strength of forces of selection acting on the variation. This information is rarely available (e.g. see Endler 1986), although it is available for some of the beak size variation of Darwin's Finches. Several authors have recently commented on the pitfalls for the whole enterprise when carried out with partial information, and have suggested various ways to avoid them (e.g. see Lewontin 1978, Gould and Lewontin 1979, Futuyma 1979, Arnold 1983, Mayr 1983) which I have attempted to follow.

In this chapter I shall discuss the adaptive nature of variation in two major features of the finches that have received relatively little attention since they were described in Chapters 3 and 4: body size and plumage color. The emphasis will be on variation. It is obviously adaptive for a bird to have plumage, for example, but the important question is whether or not differences among populations in the distribution of color in the plumage are the result of different adaptive changes that occurred in the diversification of the finches. The same is true of body size differences. I shall start with a brief description of how ideas have changed about the degree to which morphological variation in Darwin's Finches is adaptive, and then proceed to consider each of the traits separately. Ideas about adaptation in general have been largely expressed in terms of beak size variation in particular, because beak size variation has always been the principal phenomenon to be explained.

HISTORICAL SURVEY

Over the last hundred years or more, our understanding of adaptations of Darwin's Finches can be summarized by the catch phrase, the more you look the more you see. Darwin's writings about adaptation were general and not directed to the explanation of individual traits of the finches. Instead, a good starting point for us is Salvin (1876). He was impressed by the large amount of variation in the beaks, wings, tails, and legs, and degree of black coloration in the plumage of the finches, and seeing no significance in this variation he concluded: "The members of this genus present a field where natural selection has acted with far less rigidity than is usually observable" (Salvin 1876, p. 470).

This view of natural selection being lax persisted for sixty years. For example, Beebe (1924) found no large seeds in the diet of *G. magnirostris* that would help him to understand why the species has such a large beak, and wrote, "Either the mighty beak was developed for coping with some source of food which has now disappeared, or, as seems more probable, the lack of enemies, or an environment inimical to variation, has resulted in a diversity of bill which is not directly adaptive, but reflective of relaxed environmental control" (Beebe 1924, p. 266). Earlier, Snodgrass (1902, p. 381) had declared: "If it be assumed that the various sizes and shapes of bills amongst the *Geospiza* have been developed as adaptations to differences in food habit, then it must be shown that the different species of the genus feed on different species of seeds. This cannot be done." Commenting on this failure to find a correlation between beak sizes and diets, Swarth (1934, p. 229) wrote in a similar vein: "In other words, natural selection was eliminated as a factor in the production of the observed variation, and apparently justly so, for in the amount and sort of differentiation that is seen here, and in the extraordinary amount of intergradation, it is not apparent that there are useful adaptations in the remarkable extremes, nor any lessened fitness in the numerous intermediates.'' These ideas then became amplifieid by Lowe (1936, p. 319): ''There seem to be no differential environmental factors which could be regarded as having a survival value, which could obviously count one way or another; and, therefore, no scope for Natural Selection.'' This sobering conclusion was delivered to the British Association to commemorate the centenary of Darwin's visit to the Galápagos!

Nevertheless, Swarth (1934) made a clear distinction between nonadaptive differences between closely related species on the one hand, and adaptive differences between ground finches, cactus finches, and tree finches on the other. Lack (1945) initially concurred; the genera showed an adaptive radiation, the species within each genus did not. But, as I pointed out in Chapter 11, Lack (1947) changed his views, and came to the conclusion that all species differ adaptively in their beaks. He went further and suggested that a few of the subspecies differed adaptively too, although most differences between populations of the same species were still interpreted as being nonadaptive. He lacked the ecological evidence to argue this point of view; in fact, much of it pointed in the opposite direction, but he had the right insight and appropriate foresight to call for food analyses in the latter part of the dry season ''when food is likely to be least varied and least abundant, and therefore most likely to limit the population density of the birds'' (Lack 1947, p. 63).

Modern field evidence, reviewed in Chapters 6 and 7, has supported Lack's final position on the issue. It has suggested that he may have underestimated the frequency of adaptive differences between populations. It has shown how and why the analyses of Snodgrass (1902) and the observations of Beebe (1924) misled subsequent writers including Lack (1945) about the relationship between bill sizes and diets. It has shown that, to a limited degree, variation within a single population may be adaptive (Chapter 8). And finally it has shown that natural selection, far from being a rare process whose effects are difficult to discern except at the generic level and over very large stretches of time as Darwin (1859) believed, is frequent and can be detected by studying contemporary populations in detail over periods as short as a few months (see also Endler 1986). Moreover, the potential for evolutionary change in modern finches is strong, so species should no longer be viewed as frozen, morphologically, in states determined solely by historical factors, such as founder effects.

I shall now leave variation in beak size and shape and consider whether variation in body size and plumage is also adaptive.

BODY SIZE

Beak size has evolved partly independently of body size. The corollary is that body size has evolved at least partly independently of beak size. For example, the vegetarian finch, *Platyspiza crassirostris*, has the body size of a large ground finch (*G. magnirostris*) but the beak size of a medium ground finch (*G. fortis*). Populations of *G. difficilis* provide a second example. They vary greatly in body size, from a mean of 12 g (Genovesa) to a mean of 27 g (Santiago), but comparatively little in beak size. Why?

Figure 57 provides a helpful scheme for understanding this variation. It shows some identified forces of stabilizing selection acting on body size variation in a single population (Chapter 8). Different populations are held by stabilizing selection at different mean positions on a body size axis. Those different mean positions arise from different strengths of the selective forces, from different types of selective forces, and from different functional relationships between body size and its parts.

It is the differences between populations in their mean positions that need to be explained. To start with the first example given above, if the vegetarian finch evolved from a tree finch that was like modern members of the genus *Camarhynchus*, it was subjected to selection pressures in opposite directions on different traits: body size to increase, beak size to decrease. Just how the species responded depended upon the genetic correlations between the traits, a subject which I will consider in the next chapter. I interpret the small beak with strong curvature and steep profile to be better than a large one at dealing with buds and removing seeds from fruits. This makes functional sense because the major task in food gathering is to manipulate (mandibulate) food or its substrate at the tip of the bill, rather than to crush hard seeds at the base of the bill as the ground finches do; and its ecological counterparts elsewhere in the world, such as the bullfinch (*Pyrrhula pyrrhula*) in Europe (Newton 1967) and the palila (*Loxioides bailleui*) in Hawaii (van Riper 1980), have a similar bill form and body size. The large body size can be interpreted in physiological terms. Vegetarians are limited in their rate of energy intake not so much by consumption as by digestion. The limitation is less severe in large animals than in small ones. The vegetarian finch, in addition to being the largest finch, has a disproportionately large gizzard, long intestine, and disproportionately small heart (Chapter 4). These facts suggest that food-processing rate has been increased and energy expenditure decreased by internal anatomical adjustments produced by selection.

Variations in body size among populations of the same species are small-scale versions of interspecific variation. *Platyspiza* populations represent one extreme; they show almost no geographical variation. The vegetarian niche may be invariant, and the forces of stabilizing selection may be the same on the six islands occupied by the species. What sets the upper and lower limits to size variation in this species is not known. Whatever the factors are they may operate strongly, for population variation is low (Table 7).

G. difficilis populations represent the opposite extreme of pronounced geographical variation. Such variation may be partly adaptive, as there is some association with geographical variation in feeding habits and food supply. For example, the problem of exploiting the supply of nectar in the many small flowers on thin branches of *Waltheria ovata* on Genovesa has been solved by a reduction in body size and not by major beak adjustments (Chapter 11); the same is apparently true for the population of *G. fuliginosa* on Marchena (Schluter 1986). Small-bodied birds have a conjectured metabolic advantage over large-bodied birds in exploiting this resource, as well as a locomotory advantage, and these together provide an explanation for the directional selection that favored birds of low body weights on this island. On other islands the links between food supply, feeding habits, and body size are different. It is possible that the larger, and unmeasured but perceptibly rounder, body form of *G. difficilis* on other islands represents an adaptation to exploiting arthropod supplies on the ground, for the same two reasons of metabolic and locomotory efficiency.

Lying between the *Platyspiza* and *G. difficilis* extremes are several species whose geographical variation is moderate and allometric: one population is larger or smaller than another, in all traits and to approximately the same extent. Moreover, there are species which appear to be little more than scaled-up or scaled-down versions of each other, if we ignore the slight differences in the scaling constants. These include the trio *G. fuliginosa*, *G. fortis*, and *G. magnirostris*. Where body size and body parts covary as a set, the whole set may be fashioned by natural selection; but this is not necessarily so, and if only one component is adaptive there is a risk of misidentifying it. To reiterate the point made in the previous section, one trait, for example beak size, may vary adaptively, and another trait such as body size may vary simply as a correlated, selectively neutral, effect of selection on beak size and not from directly adaptive causes. I doubt if this is true because forces of directional selection have been found to act on body size and beak size independently in the *G. fortis* population on Daphne Major, and I assume that changes within a population can be extrapolated to changes between populations.

For these relatively small differences in overall size between populations of the same species or even closely related species, an adaptive component to the variation may exist in terms of the forces illustrated in Figure 57. I will first consider the possibility that different strengths of survival selection on different islands are responsible for differences in mean body size.

Energy storage capacity increases linearly with increasing body size, the slope of the relationship on a double logarithmic scale being close to 1.0, i.e. the variation is isometric. The rate at which energy is used, however, increases with body size allometrically, the slope of the relationship being distinctly less than 1.0. As a consequence, large birds survive longer than small birds under conditions of deprivation because they have more stored energy to draw upon (Calder 1974). Therefore, large size should be favored in seasonal environments where food shortages occur occasionally (Boyce 1979). If seasonality has stronger effects on one island than another, and in particular if dry season food shortage is more pronounced or occurs more frequently on one island than on another, then larger size may be selectively favored on the first island.

Downhower (1976) explored this idea with beak length data of *G. fuliginosa* and *Certhidea olivacea* taken from Lack (1945). He suggested that food shortage would occur more often on low arid islands than on higher and more mesic ones, because on the latter birds could avoid the effects of a declining food supply in the arid lowlands by migrating to the highlands in the dry season and exploiting the less fluctuating food resources there. In support of this suggestion he found a negative correlation between beak length, taken to be an index of body size, and island size for each of the two species. I have been unable to confirm this result with measurements of body size (weights) as well as beak size from a larger number of populations of *G. fuliginosa* than were available to Lack and Downhower. Correlations were made with island size and, more appropriately, with island elevation (Chapter 4). The *Certhidea* result has not been confirmed either. Body sizes of *Certhidea* are largest on outlying islands, and are not clearly related to island elevation (Chapter 4). Warbler finches are certainly mobile in the dry season, but it is not known if they migrate into the highlands in the way that *G. fuliginosa* are known to do (Schluter 1984a). The hypothesis then is not supported by these data, and for other species there is no correlation between body size and island area or elevation (Chapter 4).

I believe the hypothesis fails because dry season food stresses are protracted and not episodic, and under these circumstances daily food availability in relation to daily requirements is more important in determining survival than is initially stored energy. An example of this point is provided by the studies of natural selection in *G. fortis* on Daphne Major during dry conditions (Chapter 8). It was the particular characteristics of the food supply that were responsible for the nonrandom survival. But despite selection favoring large size, the mean size of the birds on this island has remained the smallest for the species in the archipelago. This emphasizes that the direct influence of food shortage on body size is but one of several determinants of the average body size in a population.

Another determinant is the ability of females to convert an excess of energy into eggs in times of plenty. At the beginning of the breeding season when food supply rapidly increases, small birds can more quickly reach the physiological state at which reproduction becomes energetically feasible than can large birds: this follows from the scaling of energy-use with body size described above. Moreover, as food abundance declines towards the end of the breeding season, small birds should be able to remain longer in that state. Thus, all other things being equal, small birds should start breeding earlier than large birds, they should stop later, and hence breed for longer. There is support for this idea from comparisons of female great tits (*Parus major*) of different sizes (Jones 1973), and from comparisons of different tit species of different sizes (Dunn 1976) in Europe.

Evidence from Darwin's Finches is mixed. Downhower (1976) used the seasonal pattern of percentage of nests with eggs to argue that the small ground finch (*G. fuliginosa*) began breeding earlier than the large cactus finch (*G. conirostris*) on Española in 1973, and ceased later. The illustration of the data (his Fig. 4) actually shows that *G. conirostris* eggs were found first, and that the percentage of nests with eggs was the same for the two species towards the end of the season when observations ceased. In any event, the index is an unreliable one because it presumably includes display nests which are irrelevant to the rearing of young. *G. fuliginosa* were observed to enter the incubation phase of breeding earlier than *G. conirostris*, but *G. conirostris* entered the nestling phase first. This is puzzling because the incubation period is the same for the two species (Downhower 1978). The observed phenologies of the two species appear to have been influenced as much by desertion, caused by the attraction of predatory mockingbirds to nests through playback of tape-recorded song (Downhower 1978), as by differential responses to food supply.

On Daphne and Genovesa the expectation of early breeding by the smallest species has not been upheld, except in the dry year of 1984 when a few pairs of the warbler finch began laying eggs earlier than the other species on Genovesa. Instead the rule has been that cactus finches (*G. scandens* and *G. conirostris*) breed first. On Daphne *G. scandens* is larger than *G. fortis*, and on Genovesa *G. conirostris* is larger than *G. difficilis*. This shows that the all-other-things-being-equal assumption of the argument cannot always be true. The cactus finches come into breeding condition early because, apparently, they gain an energetic advantage over other species from the nectar and pollen of *Opuntia* flowers which bloom in large numbers at the end of the dry season. On the other hand, the largest species on Genovesa (*G. magnirostris*) is the latest to start breeding, and in some years ceases to breed earliest. Its reproductive output is less than that of the much smaller *G. difficilis*, partly as a result of the pattern of timing of reproduction and not because of a difference in production (clutch size) at each breeding attempt (Grant and Grant 1980a).

Among individuals of the same species, small females in the population of *G. fortis* on Daphne Major bred earliest in one year, while large females in the *G. conirostris* population on Genovesa bred earliest in two years.

Altogether there is relatively little evidence for a reproductive advantage to small size. All these considerations, and others discussed in connection with competition in previous chapters, lead me to conclude that adaptive variation in body size, while important and in need of further investigation, is secondary to the adaptive variation in beak size and shape.

PLUMAGE

As described more fully in Chapter 3, plumage variation among Darwin's Finches can be reduced to four main classes: (a) fully black, (b) partly black and partly grey-brown, (c) brown with variable amounts of ventral streaking, and (d) grey-green. In the last three classes the ventral color is paler than the dorsal color, and the body is therefore countershaded.

Pigments in feathers can serve as a protection against abrasion, and as a screen against the penetration of intense radiation. The distribution of pigments determines a bird's appearance, and the appearance can be thought of as a set of signals conveying information to an observer. The identity of a bird—its species, sex, and age—is broadcast clearly to varying degrees. For example, its size can be broadcast somewhat ambiguously if some or all of the feathers are raised, and ambiguity is at a maximum when the bird is viewed against a matching background. Thus there are physiological, structural, and behavioral functions of plumage pigments. Nevertheless, the question of whether variation in plumage color and pattern is adaptive or not has been discussed by other authors almost entirely in one context: the need for finches to avoid being captured by hawks and owls. I will treat this issue first.

Cryptic coloration. Several Galápagos species are somber in their appearance, just like their background (Rothschild and Hartert 1899). This applies not just to finches but to the lava gull, lava heron, and marine iguana (Swarth 1934). Ground finches are darker than tree finches and live against the darker background of black, brown, and red-brown lava and soil (Lack 1947). In the dry season, when much of the arid zone vegetation is leafless, the correspondence between finch colors and background hues is strongest, at least to the human eye. Bowman (1961) noted the difficulties of detecting birds in black and brown plumage on the ground in the arid zone, grey-colored tree finches on the grey bark and trunks of trees in the arid zone, and green-colored finches such as the warbler finch in the foliage of *Scalesia* at higher elevations.

Cryptic coloration would be adaptive if it lowered the chances of detection and capture by predators, even if other cues, such as movement, are the primary ones used in the detection of finches. The chief predators are hawks (de Vries 1975, 1976) and short-eared owls (Abs et al. 1965, Grant and Grant 1980a, Boag and Grant 1984a). From direct observations and from pellet analysis they are known to prey on all the ground finches. Tree finches, warbler finches, and ground finches all mob them, so all may be vulnerable to these predators, although tree finches and warbler finches may be at lesser risk because they forage principally in trees and shrubs.

The circumstantial evidence is thus strong that cryptic coloration is adaptive. Lack's (1947) arguments to the contrary are not convincing. He doubted whether finches formed a large part of the diet of short-eared owls, which is correct (Abs et al. 1965, Grant et al. 1975), and believed the hawk to be almost certainly harmless to adult finches, which is wrong (de Vries 1975, 1976). Although short-eared owls prey on finch nestlings on some islands (Grant and Grant 1980a), both predators concentrate on young birds in immature plumage at the end of the finch breeding season, and feed mainly on other prey types at other times of the year. The fact that adult finches are relatively rare items in their diets does not negate the argument that their coloration is adaptive. They may be at low risk because they are cryptically colored.

The adaptive crypticity hypothesis could be tested experimentally: for example, by painting tree finches fully black, and seeing if non-cryptic ones are preyed upon more frequently than cryptic ones; or by presenting screened pictures of finches against matching or nonmatching backgrounds to a captive hawk or owl, and observing whether the predator distinguishes between them.

The only experiments that have been conducted with Darwin's Finches in a relevant context addressed the different question of whether finch responses to predators are adaptive. Curio (1965, 1969) found that caged finches responded with fear when presented with a stuffed specimen of a hawk or an owl, but did not respond when the test stimulus was a cat. The discrimination was viewed as adaptive because there are no native mammalian predators on Galápagos (cats have been released on some islands). Responses to the avian predators varied among populations. Strong responses were shown on Santa Cruz by *G. fuliginosa*, which are likely to encounter both predators in their environment; weaker responses were shown by *G. difficilis* on Wolf, where neither predator is resident; and the weakest responses were shown by *G. difficilis* on Genovesa where short-eared owls are not only resident but eat finches! Unfortunately, these results do not lead to firm conclusions regarding the threat of predators and their avoidance by finches because the previous experience of tested birds was not known, except on Wolf where there are no predators. The variation appears to be at least partly nonadaptive.

Bowman (1961) has extended the adaptive argument for cryptic color variation to explain the darker female plumage of *G. scandens* on Pinta and Marchena than on other islands, in terms of a suggested darker background on these islands. Certainly there are extensive areas of recent lava flows on both islands (Fig. 6), but the birds feed in the "islands" of vegetation and rarely out on the flows.

A more striking variation in female plumage is shown by *G. conirostris*. On Genovesa female plumage has various shades of brown, as in other ground finch species, whereas on Española it is exceptionally dark, being a sooty color with brown and black tones (Snodgrass and Heller 1904). The adaptive argument leads one to expect comparable differences in background colors (Bowman 1961). I am not sure they exist. Genovesa has unvegetated, dark brown, lava, and Española has dark rocks that are older, smoother because formed in large part underwater (Hall et al. 1983), and extensively covered by pale crustose lichens. Some measurements of background colors in relation to female plumage colors, along the lines developed by Endler (1984) in a study of moths, would greatly clarify this issue. They could also be used to test a similar postulated relationship between warbler finch colors and their backgrounds. Geographical variation in warbler finch color is thought to be adaptive (Lack 1947, Bowman 1961) because the palest colors occur on low islands covered by leafless, grey-barked, trees in the dry season (Santa Fe, Española, and Genovesa), and the greenest colors occur on the other, higher, islands which have trees in leaf at medium altitudes at that season.

I have ignored other potential predators in this discussion because from all that is known about them they appear to be of relatively minor importance. Mockingbirds eat finch eggs and nestlings, and chase adults and immatures, especially small ones, but are unlikely to be a serious threat to healthy adults except perhaps those of the smallest species. Great egrets, night herons and lava herons occasionally capture finches, but, to judge from observations on Daphne Major, they do so largely because inexperienced fledglings blunder into striking range while engrossed in searching for food on the ground; owls capture fledglings feeding out in the open, away from the cover of cactus bushes, in a similar manner. Snakes (*Dromicus*) are possible predators. I have seen one hunting a black male cactus finch on Champion, which repeatedly hopped out of reach while continuing to search for food on the ground. Alf Kastdalen (in Bowman 1961), a long-time resident on Santa Cruz, found a finch nestling in a snake stomach, but possibly it was a very young fledgling. The especially large species of lava lizard on Española, *Tropidurus delanotis*, may prey on finch nestlings (D. Werner, pers. comm.), but the importance of this species and the snake as finch predators, at or away from the nest, is unknown.

The proportion of males in adult plumage. In museum collections, the proportion of male specimens in adult plumage varies substantially from island to island (Swarth 1934). One possible explanation is that the rate of acquiring the fully adult color varies among populations: "On Tower [Genovesa], the proportion of male *G. magnirostris* and *G. difficilis* in fully black plumage is so high that one can be certain that a large proportion of the males must molt into fully black plumage before they are a year old. . . . On the other hand, on Chatham [San Cristóbal] the proportion of partly black to streaked males in *Camarhynchus parvulus salvini* is so low that one can be certain that a large proportion of the males never acquire the black plumage" (Lack 1945, p. 59).

In support of this idea, we have observed that male cactus finches (*G. scandens*) acquire fully black plumage generally faster than the medium ground finch males (*G. fortis*) on Daphne Major. Orr (1945) found that captive males of the same species of ground finch passed through the stages of partial black plumage to the fully black state at different rates. Our observations of banded ground finches on Daphne Major and Genovesa confirm this variation within populations. There is parallel variation in the rate at which the skull becomes completely ossified (pneumatized) (Bowman 1961); it is, on average, an exceptionally slow process.

But the field studies have also shown that the proportion of black birds in a population can vary from about 20% or less, after extensive breeding in an El Niño year, to 70% or more after a drought in which young birds have died. Thus the difference in proportions of males in adult plumage on one island at different times, and on different islands at the same time, can be solely the result of different population dynamics. Lack's certainty is not justified. In fact, his assertion that fully black plumage is acquired by some finches in less than a year on Genovesa is not supported by any of our observations of banded birds there. Nevertheless, his general rate-variation hypothesis stands as a viable, and untested, complementary explanation to the population-dynamic one.

Neither of these hypotheses invokes adaptation, but an adaptive hypothesis has been suggested by Huxley (1955) and Bowman (1961). Lack (1945, 1950) had observed that black male ground finches started to breed earlier than those in brown plumage, and at somewhat higher elevations. Huxley (1955) interpreted these observations to mean that the segregation of different morphs into different habitats, with the early-breeding black birds occupying the presumed optimal interior areas of an island, allowed the population to exploit the environment more fully than would otherwise be possible. In other words, the occurrence and differential distribution of morphs is possibly adaptive.

Bowman (1961, p. 197) endorsed the controversial concept of continuous morphism proposed by Huxley (1955, p. 309; see also Selander 1962, Curio and Kramer 1965b), but argued from his own observations that there were no breeding differences between males in different plumages in either time or place. Instead, in his view, the ". . . continuous range of color phases . . . adapts the population to the broad spectrum of background colors that are found in the arid and transition zones of Galápagos" (Bowman 1961, p. 197), by concealing the individuals from visual predators. This is an explicitly adaptive hypothesis for variation *within* populations because ". . . there is reason to think that certain morphs are genetically fixed" (Bowman 1961, p. 197). It is an explicitly adaptive hypothesis for variation *among* populations, and presumably on different islands whose spectra of background colors vary, because "Natural selection has regulated the population make-up in such a way that during the most critical time of the year (the end of the dry season) there is for each species and for each sex of the species an optimal proportion of the most advantageous plumage morphs" (Bowman 1961, pp. 197–198).

This is the sort of adaptive argument that draws fire from critics of the "adaptationist program" (e.g. Gould and Lewontin 1979). It is rather far-fetched, although not impossible, to expect the proportions of ground finch males in fully black, partly black, and brown plumage to correspond neatly to the proportional areas of background of different colors. Such a correspondence, if it does exist, could be imposed by predators. The adaptive explanation, however, implies that it is not imposed by predators but rather is the result of production of the appropriate number of morphs, this in turn having been guided by natural selection in the past. But there is really no evidence for Bowman's suggestion of genetically fixed differences among birds of different plumage color. The adaptive explanation also implies that birds of a color seek out the appropriately colored background. We have seen no sign of this in our field studies.

The simplest explanation for the variation among populations of males in adult plumage is the nonadaptive, population-dynamic one I described above. Curio and Kramer (1965b) have argued against it because they believe it implies that several populations dominated by birds in fully adult plumage would contain a preponderance of old birds, whereas detailed studies of bird populations show old individuals to constitute a small minority. This objection assumes that the length of life beyond the age at which fully adult plumage is attained is short. The assumption would be appropriate for many temperate zone populations but not for these tropical ones. If fully adult plumage is acquired at age 3 years, the last member of a cohort dies at the end of the tenth year, and annual mortality after the first year is constant, then males in fully adult plumage will be in the majority except at the end of a successful breeding season.

I favor the view that much of the geographical variation in the proportion of males in fully adult plumage is a simple reflection of asynchronous fluctuations in the compositions of populations on different islands. I would assign a minor but not insignificant role to geographical variation in rates of color acquisition. Future field studies may show this to be wrong. Studies of tree finch species are especially needed. If variation in rates is established, and if it has a genetic basis, it might be the product of nonsystematic forces acting on neutral alleles, and hence nonadaptive. Finally, it may be adaptive, but for different reasons from those considered so far (see below).

Intraspecific communication. Salvin (1876) reasoned that since fully black plumage was probably acquired in the third year of life, the scarcity of birds in this condition in museum collections probably means that younger birds breed in developmentally earlier, "mottled," plumage stages. Field observations confirmed this (Rothschild and Hartert 1899, Lack 1945). It gave rise to the view that black plumage of males has no function in courtship display (Lack 1947), a view that can be traced to the extension of Salvin's reasoning: "It would seem, then, that with these singular birds the sexual selection displayed amongst them is such that it is almost a matter of indifference whether the cock birds are mottled or black . . ." (Salvin 1876, p. 470).

Field observations show this tentatively stated conclusion to be wrong (Price 1984a). In the years following a severe drought on Daphne Major, when there was an excess of *G. fortis* males, females paired predominantly with males in fully black, or almost fully black, plumage (Chapter 8). Similar, but less pronounced, sexual selection occurred in the *G. scandens* population on the island. It is possible that females were responding to some other aspect of the male with which plumage state was strongly correlated, but this is ancillary to the demonstrated mating advantage experienced by those males. The fact that males in brown plumage sometimes do gain mates and breed successfully merely shows that conditions, both environmental and populational, are not constant. It cannot be used to argue against the benefit of fully black plumage to males. Even when the sex ratio is approximately 1:1, the first of the unpaired males to gain a mate are usually those in fully black plumage (and older), and they gain a reproductive advantage as a consequence of breeding early, as shown by the study of *G. conirostris* on Genovesa (Grant and Grant 1983, B. R. Grant 1985).

Thus it is advantageous for a male to pass rapidly through plumage stages to the fully adult state. Such males have a higher chance of gaining a mate, or gaining a high quality mate, than those who acquire black plumage more slowly. This raises the question of why all males do not acquire fully adult plumage at the end of their first year. A general answer is that there is some cost to the acquisition of fully black plumage. This is expected from the theory that females choose as mates those males with high fitness, and base their choice on some phenotypic trait (e.g. black plumage) with which that fitness is associated (Hamilton and Zuk 1982, Price 1984a). The cost could be nutritional or energetic, about which little is known among birds in general. Another form of cost is harassment from other males. It may be more difficult for a male to gain and hold a territory, or a territory of high quality, if it is in black plumage as opposed to the female-like brown plumage. Black males may also suffer more harassment than brown males when foraging in the territories of others in the dry season. These ideas on the value of inconspicuous, subadult, male plumage in breeding and nonbreeding seasons have been well explored and tested with other birds (Rowher et al. 1980, Proctor-Gray and Holmes 1981, Payne 1982), but have not been tested with Darwin's Finches. Yet another possible cost in the broad sense is an age-dependent risk of predation. This brings us back to cryptic coloration and the need to integrate the hypothetical forces of selection acting on plumage color and pattern.

Multiple functions. Against their appropriate backgrounds all finches of whatever plumage type can be concealed from the eyes of predators, but against the wrong background they are highly visible. The black plumage of male ground finches is cryptic on black lava but conspicuous when the bird is perched on top of a bush and viewed against the sky. Selective forces promote conspicuousness in some contexts and inconspicuousness in others. Conspicuousness of males, but not of females, is promoted in a reproductive context. Forces acting on individuals of both sexes while feeding promote inconspicuousness. Since they do not feed in tightly coordinated flocks, there is little intraspecific signaling occurring while feeding; exceptions to this are the encounters of individuals at a contested food source, and the tendency for *G. fuliginosa* and *G. fortis* to form loose flocks in the dry season. The important signaling is interspecific, to predators.

In addition to signaling functions, pigments such as melanin can provide some resistance to abrasion of feathers, protect living tissues from radiant energy, and absorb heat. The need for the first two is greatest in open habitats where foraging is largely on the ground. Therefore black plumage of the ground finches may be adaptive in meeting these two needs. Hamilton (1973), in discussing the frequent occurrence of black color in the plumage of desert birds, has argued against the radiant energy protection function. He points out that energy could just as well be reflected by white, or pale colors, as absorbed by black. According to data in Burtt (1979), black and brown feathers are equally effective in screening tissues from ultraviolet radiation. It is also strange that heat absorption is so well served when overheating is the expected problem. Absorption of heat by artificially blackened plumage has been experimentally shown to reduce the energy costs of other finches at low temperatures (Hamilton and Heppner 1967). But at the higher temperatures experienced by Darwin's Finches, absorption of heat would appear to be a disadvantge, and is perhaps an additional cost to those noted previously. It should necessitate foraging in shade during the middle of the day, and black birds should be constrained in this manner more than brown birds. Against these disadvantages there may be an energetic advantage gained by black birds in the relatively cool early morning and later afternoon hours. These possibilities need to be investigated experimentally. All of this reasoning suggests that physiological and mechanical functions are secondary to signaling ones. Furthermore, black occurs in the plumage of the finch on Cocos Island in a humid forest environment which poses none of these physical problems to any marked degree.

Sexual dichromatism, the greater conspicuousness of males, and the different colors and patterns of birds feeding in different locations can be explained in terms of a balance between the selective forces on the interspecific and intraspecific signaling functions. Sexual monochromatism in the mangrove finch, woodpecker finch, and warbler finch is not explained, except by supposing there is no sexual selection at work, but that in itself would need to be explained, and there are no field studies of these species to throw light on the question.

OTHER FEATURES

I shall conclude this chapter with a few remarks on other traits. Several reproductive characteristics of the finches stand in marked contrast to the morphological traits I have discussed in showing little interspecific variation, and therefore little evidence of adaptive change. These include courtship behavior, nest architecture (except for size), and egg color and pattern. These features are either the same as those possessed by the ancestral population when it first colonized the Galápagos, or the ancestral population underwent adaptive change before the initial speciation and the new features remained fixed thereafter. Clutch size, incubation period, and nestling period vary relatively little among the species (on the Galápagos), and if the variation is adaptive it is subtle (Grant and Grant 1980a, Boag and Grant 1984a, Millington and Grant 1984).

Three traits associated with reproduction vary more strongly and in a possibly adaptive way. (I have already mentioned a fourth one, the initiation of breeding, in connection with body size.) First, egg size varies among species, and although this is a function of female size, the slope of the relationship is unusually shallow (Chapter 5). Small species lay relatively large eggs. This may be adaptive, enabling those species to take maximal advantage of a rapid increase in food supply to produce offspring with a high chance of surviving after they have left the nest. An associated long incubation period and long nestling period would be disadvantageous on the mainland where there are many nest-predators, but it is not a problem on the Galápagos except for the larger species (of ground finches) whose nests receive a disproportionate number of attacks from short-eared owls (Grant and Grant 1980a).

Second, some male ground finches occasionally feed the nestlings of other birds of the same species. Such helping behavior could be adaptive, as it is usually thought to be in those species which regularly exhibit the behavior. It was observed and studied in populations of *G. fortis* and *G. scandens* on Daphne Major. An assessment of the genetic relatedness of helpers and helped, and of the consequences of the helping, yielded no evidence of an adaptive advantage, so we concluded that the behavior was misdirected parental care (Price et al. 1983).

The third trait is song. The distribution of sound energy in songs varies in such a way that it might be transmitted best in the particular configuration of vegetation in which each song is sung (Bowman 1979, 1983). There is evidence to support this view, and it has been used by Bowman to develop and comprehensively illustrate the adaptive hypothesis, but it has been used somewhat selectively for the purpose of illustration and is not entirely consistent.

Turning to feeding behavior, the most outstanding example of adaptive behavior is shown by the two *Cactospiza* species. The problem confronted by them is to extract arthropod larvae from deep within woody tissues. It has been solved by these two species with a tool: a twig, leaf petiole, or cactus spine. Some learning is involved in the use of a tool by young finches (Millikan and Bowman 1967). Trial-and-error learning was presumably also involved in the historical development of this most unusual behavioral trait. I can imagine, as the initial step, a frustrated woodpecker finch failing to drop a piece of bark it had just removed from around the entrance to a crevice in a branch, accidentally pushing it into the crevice and touching the prey, and being rewarded by the prey moving towards the entrance and within reach of the bird's beak. *Camarhynchus psittacula* and *Geospiza conirostris* solve the same general problem by ripping off bark with their beaks, but there are no field data to show they are getting the same prey, or prey in the same location, as the spine-wielding mangrove and cactus finches do; so in fact they may be dealing with a slightly different problem (prey and location). The blood-drinking of *G. difficilis* on Wolf and Darwin can be interpreted as similarly adaptive, having evolved from the less bizarre habit of removing ticks (Bowman and Billeb 1965, MacFarland and Reeder 1974) or feeding on hippoboscid flies.

There are a host of other morphological features whose variation can be considered adaptive. Many of them, including bones and musculature, skull architecture and pneumatization, heart weight, gizzard size, intestinal length, and the tips of tongues have been described by Snodgrass (1903) and in much greater detail by Bowman (1961). While patterns of variation in these traits can plausibly be interpreted in terms of adaptation, as Bowman has done, there is little quantitative evidence that they are actually adaptive. Much of the variation is strictly size-related (Chapter 4), and is only adaptive to the extent that the body size variation itself, and the relationship between part and whole, is adaptive. There is still much to be found out about the ways in which morphological, physiological, and behavioral attributes of finches fit them for the tasks they perform and the lives they lead.

SUMMARY

Organisms are adapted to their environments by virtue of possessing structures and functions that enable them to perform specific tasks efficiently and to solve particular problems. Species of Darwin's Finches have become adapted to different environments by the action of natural selection on variation in many morphological traits. Not all changes were adaptive; some were random, and some were correlated effects of selection acting on other traits. This chapter attempts to identify the principal changes that were adaptive, and the reasons for them, in two major traits: body size and plumage color.

Body size has evolved partly independently of beak size, and some body size variation can be interpreted as being adaptive. In one population, natural selection has been demonstrated to occur on body size variation independent of its action on beak size. The large body size and disproportionately large gizzard and long intestine of *Platyspiza crassirostris* are adaptive features which enable these birds to digest relatively rapidly the vegetable matter that makes up a large part of their diet. *G. difficilis* populations vary greatly in body size and comparatively little in beak size. The especially small size of birds on Genovesa is adaptively related to the exploitation of nectar from the small flowers on thin branches of *Waltheria ovata*. Metabolic and locomotory functions are at the core of the adaptive variation, but are generally poorly understood. Energetic considerations suggest that small size is advantageous in enabling birds to breed early, and for a long period, and large size is advantageous under conditions of food shortage;

but field evidence provides only weak support. While some of the body size variation among the species is adaptive in its own right, a component of unknown magnitude can be attributed to the correlated effects of adaptive variation in bill size.

Sexual differences in plumage color (dichromatism), the greater conspicuousness of males, and the different colors and patterns of species feeding in different locations, are adaptively related to (a) the need for all birds to avoid detection and capture by visually hunting predators, and (b) the need for males to attract mates. Circumstantial evidence exists for a cryptic resemblance of birds of a particular plumage type to the background of their foraging station. Geographical variation in warbler finch colors may possibly be adaptively related to geographical variation in the colors of the dry season vegetation in which they feed. While predators promote inconspicuousness in their prey (finches), mate-attraction promotes conspicuousness. Sexual selection in favor of the conspicuous, fully black, plumage of males has been demonstrated in three populations of ground finches.

Variation in other traits has been interpreted in adaptive terms. The relevant morphological features include bones and musculature, skull architecture, heart weight, gizzard size, intestinal length, the tips of tongues, and egg size. Variation in behavioral traits includes the unique use of twigs and spines by the two *Cactospiza* species to extract arthropods from cavities in branches; pecking at the base of developing feathers of boobies by sharp-beaked ground finches (*G. difficilis*) to draw blood which is then consumed; and the production in songs of a distribution of sound energy which may vary in such a way as to be transmitted best in the particular configuration of vegetation in which each song is sung. The adaptive basis to most of this variation has been plausibly argued but not adequately tested. Conspicuously unvarying among the species are two aspects of reproduction: courtship behavior, and breeding characteristics such as clutch size, incubation period, and nestling period.

CHAPTER FIFTEEN

Reconstruction of Phylogeny

INTRODUCTION

"It may metaphorically be said that natural selection is daily and hourly scrutinising, throughout the world, every variation, even the slightest; rejecting that which is bad, preserving and adding up all that is good; silently and insensibly working. . . . We see nothing of these slow changes in progress, until the hand of time has marked the long lapse of ages. . . ." So wrote Darwin (1859, p. 84) about phyletic evolution, with little hope of ever being able to observe evolutionary change. But evolutionary change can be observed under favorable circumstances. We have been fortunate to witness a rapid, small but measurable, change in the morphological features of *G. fortis* on Daphne (Chapter 8). Although I have no reason to believe the change was an initial step on a longterm phyletic trajectory, it can be extrapolated to show the forces of selection necessary to propel a species along such a trajectory.

My object here is to use modern information to reconstruct the phylogeny of the finches. Lack (1947) attempted to do this by comparing morphological phenotypes of species (Fig. 4). We can do better by using recently acquired information on their genetic characteristics. The chapter shows how historical processes of morphological divergence can be quantitatively estimated. It deals with the issue of whether adult or juvenile traits have been the prime targets of selection. It concludes with a discussion of Darwin's Finch phylogeny, which is reconstructed with information on the morphological differences between species and the inheritance of morphological traits.

RECONSTRUCTING THE PROCESS OF MORPHOLOGICAL DIVERGENCE

Evolutionary responses to directional selection depend on the intensities of selection acting directly on morphological characters, the genetic variances of those characters, and the covariances among them. It is possible to estimate the net forces of selection involved in the formation of two species if we know the phenotypic differences between them and the genetic covariance matrix—providing also we are willing to assume that the covariance matrix is the same for the two species, it has been determined without error, and it has remained constant. I shall show how these estimations can be made after first discussing the genetic parameters.

Genetic variation. Genetic variances, obtained from heritability analyses (Chapter 8), are measures of the potential for evolutionary responses to selection. Genetic covariances are similar measures but represent the degree to which genes governing the expression of one character also influence the determination of others, through pleiotropic effects or linkage disequilibrium. They are very important because they determine the evolutionary responses of correlated traits to selection acting on just one of them. Suppose that two traits are strongly and positively correlated, genetically. Selection on one trait will produce comparable evolutionary responses in both of them, in direction as well as in magnitude. If, on the other hand, two traits are strongly and negatively correlated genetically, selection on just one will produce an evolutionary response in the opposite direction in the other trait. As a third possibility, genetic correlations may be very weak, in which case selection on one trait will have little effect on the other.

It follows that the pattern of genetic codetermination of traits will have an important bearing on the evolutionary plasticity of species. Not all directional changes are equally likely, for genetic (as well as environmental) reasons. Some require more selection than others, in terms of intensity, frequency, or both, to overcome genetic constraints. For example, if large overall size is favored on an island, it will be reached under a regime of directional selection fairly quickly when all traits are positively correlated genetically, especially if the genetic variances for the individual traits are high. It will be reached more slowly if traits are negatively correlated genetically, because then the opposite effects of genes will have to be overcome. Similarly, changes in shape are difficult to effect if all traits covary positively and strongly.

Genetic correlations among bill and body size traits of *G. fortis* on Daphne are uniformly strong and positive (Table 22). Those for *G. conirostris* on Genovesa are a little lower, but are positive (Grant 1983a). Genetic correlations are weakest for *G. scandens* on Daphne, and it is possible that some are negative. They can be considered only approximately estimated because they have been derived from small samples of measurements with lower repeatabilities owing to lower phenotypic variances, and because, unlike the situation with the other two species, not all the heritabilities are statistically significant (Table 8).

Evolution. The general formula for multivariate evolution of the mean phenotype under selection has been given by Lande (1979) as follows:

$$\beta = \mathbf{G}^{-1}\Delta z.$$

TABLE 22. Genetic correlations among six morphological traits for three species. After Boag (1983) and Grant (1983a), while correlations involving only beak traits of *G. scandens* are from Price, Grant, and Boag (1984). Correlations are weakest for *G. scandens*; those calculated from nonsignificant heritabilities are shown in parentheses.

	Weight	Wing	Tarsus	Bill Length	Bill Depth
G. fortis: DAPHNE					
Weight					
Wing	.88				
Tarsus	.89	.68			
Bill length	.95	.95	.71		
Bill depth	.87	.87	.75	.90	
Bill width	.94	.78	.61	.89	.93
G. conirostris: GENOVESA					
Weight					
Wing	.59				
Tarsus	.71	.25			
Bill length	.60	.26	.61		
Bill depth	.58	.55	.51	.20	
Bill width	.56	.57	.80	.45	.89
G. scandens: DAPHNE					
Weight					
Wing	(.57)				
Tarsus	(.99)	(.61)			
Bill length	−(.13)	−(.48)	−(.10)		
Bill depth	(.30)	−(.82)	(.77)	.28	
Bill width	(.13)	−(.33)	(.01)	.53	.55

β is the selection gradient (Chapter 8), a vector whose entries can be considered as the forces of directional selection acting directly on each character and independent of correlated responses to selection on other measured characters. G is the genetic variance-covariance matrix which, in this formula, is inverted. $\Delta \bar{z}$ represents the vector of mean differences between the characters of two populations or species. Thus, where $\bar{z}_{i1} - \bar{z}_{i2}$ stands for the difference between species 1 and species 2 in trait i, call it bill length, the vector is a column of all of the mean differences—bill length, bill depth, bill width, etc. This equation is the multivariate equivalent of the formula for the response to selection ($R = h^2 s$) introduced in Chapter 8, with the terms rearranged to show how the forces of selection (β) are estimated.

The selection gradient, or net forces of selection, has been estimated for several transitions between species in the genus *Geospiza* (Table 23) by taking the genetic variance-covariance for *G. fortis* as common to all species

(Price, Grant, and Boag 1984). It is unlikely that *G. fortis*, as currently constituted, gave rise to another species without itself undergoing change, so ideally for these calculations to be made we would compare the transition from ancestor to modern *G. fortis*, the transition from a different ancestor to *G. magnirostris*, and so forth. Nevertheless, there is value in working with only contemporary species in the absence of their ancestors because we can find the relative magnitude of the minimum selective forces involved in species transitions. This possibility holds regardless of which species gave rise to which, and would only be wrong if some species evolved convergently and others did not.

The major result is that the net forces of selection are calculated to be very large in those species transitions involving *G. scandens*; the absolute values of vector entries sum to a large number in every instance (Table 23). In contrast, the net forces of selection are small in the transitions involving pairs of the trio *G. fuliginosa*, *G. fortis*, and *G. magnirostris*. The differences are explicable in terms of phenotypic proportions and the strong and positive genetic correlations. A transition from *G. fuliginosa* to *G. fortis*, for example, involves an increase in the size of all bill dimensions and to an approximately equal extent, which is relatively easily produced given the positive genetic correlations among them. A transition from either of these to *G. scandens* involves selection in opposite directions—antagonistic selection—on bill length and bill depth, because of the strongly positive genetic correlations between the traits (which themselves are likely to undergo change).

An alternative way to view the difference in the net forces of selection is to consider that transitions among *G. fuliginosa*, *G. fortis*, and *G. magnirostris* occur largely along the first principal component derived from the genetic covariance matrix, a size axis, whereas transitions involving *G. scandens* occur largely along the second principal component, a shape axis (Price, Grant, and Boag 1984; see also Fig. 19). The first principal component typically accounts for 4–5 times more of the genetic variance than does the second (Boag 1983, Grant 1983a); therefore evolutionary changes in size are more easily effected by selection than are changes in shape (Grant 1981a).

These conclusions are not affected by a moderate relaxation of the assumptions of constancy of covariances and their error-free estimation (Price, Grant, and Boag 1984). The covariances for *G. scandens* traits are probably lower than those for *G. fortis*, but so are their genetic variances, and these effects tend to cancel. Constancy of covariances is, strictly, improbable. They are likely to have varied to some extent, owing to the effects of mutation, immigration, selection, and genetic drift. But only if characters became negatively correlated genetically, or, alternatively, if the major

TABLE 23. Selection gradients (divided by 100) for six types of species-transitions, based on the genetic variance-covariance matrix of *G. fortis*. Values show the minimum forces of selection on four morphological traits involved in each of the transitions. The vector length is a measure of the net forces of selection, and is calculated by summing the squared values for the four traits, and then taking the square root (i.e. it is the Euclidean distance between the mean morphologies of two species). Notice that greater forces are required in transitions involving *G. scandens* than in the others. (After Price, Grant, and Boag 1984.)

	fuliginosa/ *fortis*	*fuliginosa/* *magnirostris*	*fortis/* *magnirostris*
$\sqrt[3]{\text{Weight}}$	−2.02	0.01	2.03
Beak length	0.93	−0.54	−1.47
Beak depth	0.32	1.46	1.14
Beak width	0.38	0.56	0.18
Vector length	2.28	1.65	2.76
	fuliginosa/ *scandens*	*fortis/* *scandens*	*magnirostris/* *scandens*
$\sqrt[3]{\text{Weight}}$	−12.61	−10.58	−12.61
Beak length	11.16	10.23	11.70
Beak depth	−3.92	−4.24	−5.38
Beak width	1.87	1.49	1.31
Vector length	17.39	15.39	18.07
Observed Selection on G. fortis in 1977			
$\sqrt[3]{\text{Weight}}$	0.08		
Beak length	−0.03		
Beak depth	0.08		
Beak width	−0.04		
Vector length	0.12		

targets of selection during species transitions were not included in the analysis, would the conclusions be seriously in error. Later in the chapter I will take up the question of whether the real targets of selection were traits during growth, with correlated effects upon adult phenotypes.

COMPARISON WITH CONTEMPORARY SELECTION

The selection on the *G. fortis* population on Daphne during the drought of 1977 (Table 9) can be compared with the forces estimated to be involved in species transitions (Table 23). The documented direction of selection was most similar to the selection gradient in the *G. fortis–G. magnirostris* transition (Price, Grant, and Boag 1984). A measure of this similarity is the strong vector correlation ($r = 0.88$) between the estimated selection gradient and the one observed for the two sexes combined. This fits with the observation that surviving *G. fortis* fed on the seeds (*Tribulus cistoides*) that *G. magnirostris* consumed (Chapter 6). The net effect of the observed selection was a 5% shift along the *G. fortis–G. magnirostris* trajectory.

FURTHER EVOLUTION

Another way of expressing this last result is to say that 23 such observed, intense, selection events would be needed to transform *G. fortis* into *G. magnirostris*. The number might be slightly higher if important dimensions were left out of the analysis. Were the transformation to happen on another island, where *G. fortis* are larger, the number of required events would be on the order of 12 to 15. These may still seem large numbers, but if one event were to occur each century without strong depletion of genetic variance, no more than 1,500 years would be required for the transition, and this is a small amount of time in relation to the time course of the adaptive radiation as a whole (Chapter 10).

There are two ways in which these full transitions could be made. Sustained directional selection could produce substantial morphological modification on a single island. For example, all species of *Geospiza* exploit *Opuntia* flowers as a food source (Grant and Grant 1981), but only *G. scandens*, and *G. conirostris* on Genovesa, probe deeply enough to reach the basal nectar, which they can accomplish because they have long and pointed beaks. The long and pointed beak could be the product of sustained directional selection over a long period of time. If the resource itself were undergoing a coevolutionary response to the exploiter species, with flowers becoming longer or seeds becoming larger and harder, the selection period would be prolonged and the evolutionary response protracted. A possible example of coevolution is provided by the especially large form of *G. magnirostris* on the southern islands and the exceptionally large and hard seeds of *Opuntia megasperma* on these islands; the depth-hardness index for *O. megasperma* is ten times greater than the index for *O. echios* seeds on Santa Cruz (Grant and Grant 1982).

The other way involves several small steps on several islands, with one or more steps occurring on successively colonized islands. Since islands have different food supplies, the adaptive peaks for a given species are likely to differ in position on the different islands to some extent. Small islands may be especially important in this process. On each island the population moves fairly rapidly to its peak under directional selection, then is held there, or just off it (Chapter 12), by stabilizing selection. If peaks are narrowly separated by shallow valleys in an adaptive landscape, genetic drift can counter the forces of stabilizing selection and carry the population across a valley to a neighboring peak (Lande 1976b). Populations on Daphne are occasionally small enough for drift to be a real possibility; in early 1979 there were only 28 breeding pairs of *G. fortis* and 21 pairs of *G. scandens*. But measurements of the intensity of selection suggest that valleys are deep (e.g. Fig. 97), which is to say that selection can be very strong and transitions through drift are effectively impossible. On larger islands it is doubtful if populations are sufficiently reduced for drift to be important, except during the very rare occurrence of successive drought years.

ONTOGENY

The discussion of genetic changes among the various species has been restricted to adults, which differ so conspicuously, but juveniles differ too, as do the growth trajectories leading to adult form. Ontogenetic differences among closely related species can be viewed either in terms of selection on morphology, or in terms of shifts in the timing of developmental events (heterochrony) with sometimes profound consequences for adult form (e.g. Gould 1977). All species of Darwin's Finches hatch, fledge, and reach adult size at approximately the same age, and thus show little evidence of heterochrony. These similarities may reflect strong selection for rapid maturation of most of the skeletal elements in all species (Price and Grant 1984, 1985). The question then arises as to whether all differences between species in growth characteristics can be attributed to correlated responses to selection acting on the adults, or whether selection has acted at multiple points along a growth curve, with consequential effects upon the adults. The answer will influence our interpretation of the phylogeny.

Consider, for example, the three *Geospiza* species, *fuliginosa, fortis*, and *magnirostris*. In Chapters 4 and 6 it was shown that adults of these species differ in beak proportions as well as in size. The larger the species the deeper is the beak, both absolutely and relatively. The three species do not lie on a single line of allometry (Chapter 4). Differences in proportions among the adults can be viewed as adaptive (Chapters 6 and 14)—as anatomical reorganizations produced by selection working against small genetic constraints (Table 23) to overcome small but important morphological constraints on feeding efficiency. Differences among adults could also be a partly nonadaptive byproduct of possibly adaptive changes in growth; in Chapter 5 it was shown that the three species do have slightly, but significantly, different growth trajectories. This latter view has been expressed by

Gould (1984) in an article on snails. Having described the difficulties of understanding how selection might act, and might have acted, on an impressive variation in pattern and form of adult snails on some Caribbean islands, difficulties made poignant by successes elsewhere (e.g. Cain 1983), he concluded: "I find it more reasonable to argue that a basic change [in development], which may be selected, churns out a set of [non-adaptive] consequences through forced correlations in growth."

The question of when in the life of finches selection has acted can be answered by using Lande's (1979) formula (p. 376) for multivariate evolution. If G, the genetic variance-covariance matrix for all adult and juvenile characters, is known, we can directly calculate the selection gradient (β) from the vector of mean differences ($\Delta\bar{z}$) among species at comparable growth stages to assess the magnitude of selection acting just on the juvenile characters. However, reliable estimates of genetic variances and covariances are not available for most of the juvenile characters. In the absence of growth data on the parents for comparison with their offspring, heritabilities can only be estimated by full-sib analysis, and these are known to be inflated in several ways, especially through common-environment effects. They are relatively undistorted in old chicks (Boag 1983, Grant 1983a, Price and Grant 1985), and so full-sib analysis can be used to estimate heritabilities of traits in chicks when last handled in the nest (aged 8–9 days), and to estimate genetic covariances between these and adult traits by using measurements of the same birds when recaptured as adults.

Thus estimates are available for genetic variances and covariances of adult characters, and for genetic covariances between adult and juvenile characters. This is enough to evaluate the null hypothesis of no selection acting directly on juvenile characters, in the following way. $\Delta\bar{z}$ for adults + chicks = G (for adults + chicks) β, where the matrix G has been expanded to include genetic covariances between adults and chicks, and the vector β (see Table 23) has been expanded with zeros signifying no selection on the juveniles. The $\Delta\bar{z}$ vector calculated by this method contains the observed differences in mean characters between adults of two species, and predicted differences in mean character values for juveniles of these species under the hypothesis of no direct selection on juvenile characters. Predicted differences can then be compared with observed differences. This is done in Table 24 (Price and Grant 1985 give fuller details).

There is a reasonable correspondence between observed and predicted values in comparisons between $G.\ fortis$ and the other three species, within rather broad statistical limits of confidence. $G.\ fortis$ was used as the starting species because of its intermediate size, and because the genetic variance-covariance matrix was estimated for this species; the results would not be affected by choosing a different species as the starting point. To restate the

TABLE 24. A comparison of observed and predicted log$_e$ sizes of three species at two growth stages. Genetic characteristics of *G. fortis* were used to calculate the predicted values (p. 382). Numbers in parentheses refer to sample sizes of measurements. (After Price and Grant 1985.)

		G. fuliginosa Mean ± SE	*G. magnirostris* Mean ± SE	*G. scandens* Mean ± SE
BEAK DEPTH				
day 3–4	Observed	1.17 ± 0.04 (12)	1.68 ± 0.01 (38)	1.43 ± 0.01 (22)
	Predicted	1.17 ± 0.29	1.87 ± 0.56	1.40 ± 0.86
day 8	Observed	1.44 ± 0.02	1.99 ± 0.01	1.68 ± 0.01
	Predicted	1.38 ± 0.18	2.28 ± 0.54	1.44 ± 0.79
BEAK LENGTH				
day 8	Observed	1.57 ± 0.01 (11)	2.00 ± 0.01 (20)	1.87 ± 0.01 (55)
	Predicted	1.51 ± 0.27	2.18 ± 0.33	1.96 ± 0.45

results, the observed differences between species in growth characteristics that were described in Chapter 5 can be largely accounted for in terms of correlated responses to selection acting solely on adults. We lack the detail, however, to explain why growth characteristics changed in different ways in different species; to explain why, for example, *G. magnirostris* and no other species undergoes a change in post-fledgling growth from a relatively rapid increase in bill depth to a relatively rapid increase in bill length.

Adult characters are strongly correlated genetically with juvenile measures of the same traits, from an early age onwards, and correlated at a somewhat lower level with nonhomologous traits in the juveniles (Boag 1983, Grant 1983a, Price and Grant 1985). This means that if selection acted upon a single growth stage it would produce consequential effects at all subsequent stages in the target trait and in all correlated traits. Selection may have operated at early growth stages to some extent, but probably not to a large extent if, as seems likely, the resulting adult phenotypes departed markedly from the environmentally determined optimum, for then it would be counteracted by selection on the adults.

PHYLOGENY

The formula for multivariate evolution will now be used with adult traits to assess phylogenetic relationships among species of Darwin's Finches. The length of β is a measure of "selection distance" between two species: it estimates the total net forces of directional selection required to produce their observed morphological distances, taking into account genetic relationships among all of the traits considered. Phylogenetic relationships can then be inferred from the selection distances, assuming that relationships are closest when distances are shortest, and further, that distances are proportional to elapsed time since the split between the species occurred. I will return to these assumptions after first describing the method and results.

Schluter (1984b) inferred phylogenetic relationships among *Geospiza* species by computing minimum-length Wagner trees (Kluge and Farris 1969) from the mean characters of species. He used the vector lengths for the selection gradients (β) in species transitions, discussed above, to obtain measures of distances between species. Genetic values for five of the characters of *G. fortis* have been directly determined by Boag (1983). Genetic parameters for the three remaining characters used in morphometric studies (Chapter 4) were estimated by least squares regression. This was possible because genetic variances and covariances are tightly correlated with phenotypic variances and covariances (Schluter 1984b).

The results fall into a pattern not very different from Lack's phylogeny (Fig. 99), with only the positions of *G. difficilis* and *G. scandens* reversed. This is not surprising because Lack based his phylogeny on morphological comparisons. In the absence of plumage differences among the species his tree is a reflection of the degree to which *Geospiza* species differ in beak and body size, ordered along an axis of presumed age. Since all *G. fortis* genetic variances are high and genetic correlations are strongly positive, there is very little weighting of the phenotypic characters by the genetic parameters. And indeed a repeat of Schluter's (1984b) analysis with the phenotypic variance-covariance matrix (**P**) substituted for **G** in the transformation of character means gave an identical pattern of relationships.

Relationships among the tree finch group are less certainly inferred by this method because genetic parameters are not known for any of the species. The phenotypic variance-covariance matrix for *Camarhynchus psittacula* was used by Schluter (1984b) to transform the character means for all species in the tree finch group, on the assumption that it approximates the unknown genetic variance-covariance matrix of all species. Results are once again similar to Lack's arrangement (Fig. 99). The ground finches and tree finches were not combined into a single analysis because their phenotypic variance-covariance matrices are slightly but consistently different (Grant et al. 1985).

These relationships are acceptable only to the extent that the assumptions of the methods are reasonably met. If genetic variance-covariance matrices of species really differ, the relationships are incorrectly inferred. For at least two species, *G. fortis* and *G. conirostris*, matrices are sufficiently similar (Table 22) to allow confidence in the assumption. If traits not included in

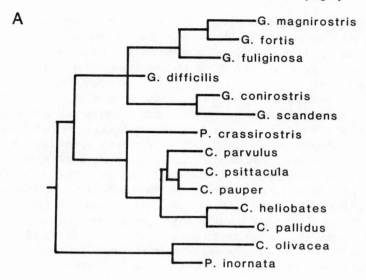

A

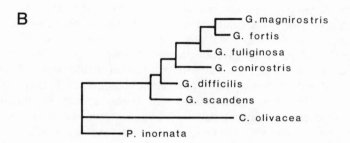

B

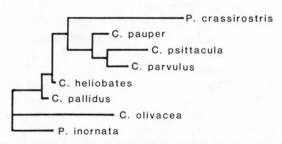

FIG. 99. Relationships among Darwin's Finches represented by minimum-length Wagner trees. (A) Lack's (1947) tree, with branch lengths approximately as he drew them. (B) tree finches and ground finches treated separately, using morphometric information, with *Certhidea olivacea* as the outgroup. From Schluter (1984b).

the analyses were the direct targets of selection during species transitions, then again the inferences will be in error. It is unlikely that juvenile traits were the direct targets, as discussed in the previous section, nor is it likely that other skeletal traits were the targets for reasons considered by Barrow-clough (1983).

The other major assumption is a linear relationship between selection distance and time. This is the most problematical assumption and merits some discussion. We should not expect this relationship to be perfect, because rates of adaptive morphological evolution should not be constant, and directions of evolution may change, resulting in convergence in some cases. There is an additional problem, which can be appreciated by thinking of the time since divergence in two parts. In the first part directional evolution took place. In the second part stabilizing selection prevented further directional change, for once a niche was occupied by the new species there would be scarcely any scope for further change unless the niche itself changed. By the assumption of a linear relationship between selection distance and time, the first part, in which directional evolution occurred, is proportional to the total time since divergence. At the outset this seems unlikely because the length of the stabilizing phase should bear no fixed relationship to the total, yet should be the principal determinant of it: the stabilizing phase should be much longer than the directional phase, given the large amount of time (Chapter 10) and the strong selection pressures known to occur (Chapter 8).

With doubts about its plausibility the assumption should be tested, and can be tested in the following way. If biochemical differences between species reflect the length of time they have been separate species, because the alleles that govern biochemical traits are selectively neutral and mutate at an average constant rate, they can be used as an independent estimate of the divergence time. There should be a positive association between such differences and the selection distances between the same species. Schluter (1984b) tested the assumption and found it to be upheld; there was a strong positive correlation ($r = 0.74$) between Rogers' D values, which are measures of the biochemical differences between species calculated by Yang and Patton (1981; see also Fig. 72), and selection distances. This result supports the linear assumption, and leads to the conclusion that genetic parameters have not only determined the magnitude of evolutionary responses to selection in the diversification of the finches, but also, to some degree, the rates of change and hence the time periods over which those changes occurred (Schluter 1984b). The linear relationship between selection distance and time does not imply constant rates of change. I think the usual pattern of change is likely to have been an initially strong and rapid response to selection for increased or decreased size, followed by a lesser and extremely protracted response in shape owing to the constraining influence of genetic correlations (Grant 1981a, 1983a). The repeated selection of different traits of

G. fortis in opposite directions (Chapter 8) can be interpreted as evidence for the modern persistence of such constraints.

Unfortunately, even here the conclusion regarding the linear assumption must be qualified because hybridization biases the biochemical estimates of divergence times. If all species were equally likely to hybridize with each other there would be no bias, but observations on hybridization show, and results of discrimination experiments indicate, that interbreeding is most likely to occur between phenotypically most similar species (Fig. 98). These may have diverged a long time ago, but their niches would have kept them morphologically similar and gene exchange would have kept them biochemically similar. For instance, the niches of the two smaller species of generalist granivores, *G. fuliginosa* and *G. fortis*, have probably existed on the Galápagos for a very long time, yet the species themselves appear to have evolved only recently. Are they really old species that have maintained genetic continuity with each other, and with other congeners? Or are they modern species that have competitively displaced or replaced the previous occupants of their niches? I see no resolution to this problem without a more accurate and unbiased method of estimating true times since divergence.

In conclusion, we have not one but three models of phylogenetic relationships, based on morphological (Lack 1947), biochemical (Yang and Patton 1981), and morphometric (Schluter 1984b) information. They display largely concordant patterns, so one is not clearly superior to the others. They need to be tested with new data. In particular, the relationships between *Pinaroloxias inornata* and the other species need to be established. As indicated by its plumage, and by skeletal features (Snodgrass 1903, Steadman 1982), it may be more closely related to the ground finches than to the warbler finch, and hence better represented in a phylogenetic tree as an independent offshoot from the ancestral stock. This is the only change I would make to Lack's tree (Fig. 4). And finally, to repeat some comments made in Chapter 10, we need to know the relationships between Darwin's Finches and continental finches. Then the phylogeny will be more securely established.

SUMMARY

Evolutionary responses to selection depend on the intensities of selection acting directly on characters, the genetic variances of those characters, and the genetic covariances among them. In this chapter I have used estimates of genetic parameters and phenotypic differences between species to calculate the net forces of selection involved in the transformation of one species to another. For this purpose it has not been necessary to know which species gave rise to which.

Those morphological characters that have been studied show strong and positive genetic correlations. As a consequence, selection on one of them would produce evolutionary responses in others. This means that a large amount of selective force was involved in the transformation of one species into another which differed substantially in shape, to overcome the constraints imposed by the positive genetic correlations. In contrast, changes in overall size were easily effected by selection, and relatively little selection was involved in the transformation of species that differed only in size, such as *G. fortis* and *G. magnirostris*. The directional selection on the Daphne population of *G. fortis* in 1977 produced a 5% shift along the *G. fortis–G. magnirostris* trajectory. Thus if such events occurred as infrequently as once a century, the full transition from *G. fortis* to *G. magnirostris* would have taken only 2000 years to complete. It would have taken place on a single island or more likely on several islands, with several selection episodes occurring on successively colonized islands. It would have taken place for reasons of ecological forces impinging on populations, with reproductive isolation being a byproduct of the differentiation.

The same approach to evolutionary reconstruction can be applied to the question of whether selection also acted on various juvenile stages of growth. When this is done it is found that selection on adults alone provides a reasonable explanation for differences in the growth characteristics of different species. Adult characters are strongly correlated genetically with juvenile measures of the same traits, from an early age onwards, and correlated at a somewhat lower level with different traits in the juveniles. Thus in the transformation of species, selection on adults produced correlated responses in the juveniles, resulting in differences in their patterns of growth.

Phylogenetic relationships among the species have been inferred from the estimates of selection forces involved in the transformation of one species into another, assuming that relationships are closest when total forces have been smallest. These "selection distances" between species are more informative than phenotypic, or phenetic, distances, because they incorporate the time-dependence of morphological differentiation that arises from the retarding effects of genetic constraints upon the effectiveness of selection. Phylogenetic relationships inferred in this manner agree quite closely with those arrived at through a comparison of morphological phenotypes, and with those arrived at through a comparison of biochemical properties. But it is appropriate to end on a note of caution, for several reasons concerning uncertainties over these inferences, and I can do no better than quote Lack (1945, p. 127): "In the Geospizinae there are probably many extinct intervening forms, hence all reference to relationships are tentative."

CHAPTER SIXTEEN

Recapitulation and Generalization

INTRODUCTION

In previous chapters I have used information about the patterns of diversification of the finches and the processes that govern their lives to suggest how and why they came into being, and how and why they are maintained in their current states. In keeping with this organization I will now summarize the main points, by first dealing with the characteristics of modern finches, and then by discussing the processes in the past which gave rise to them. In this summary I find it convenient to depart from the plan of the chapters in some places in order to rearrange some of the points. I will conclude the chapter by considering the extent to which the lessons learned from Darwin's Finches can be applied to the problem of explaining patterns of evolution in other groups of organisms, many of which are less suitable, or even impossible, to study as living organisms.

PATTERNS AND PROCESSES AMONG MODERN FINCHES

Composition. The fourteen species of finches constitute a monophyletic group derived from an unknown, emberizine, ancestor from Central or South America. Thirteen of the species live on the Galápagos and another occurs on Cocos Island. The ages of these islands, and their volcanic origin far removed from the continental shelf, specify that the ancestral finches reached the islands by overwater flight sometime in the last five million years. Electrophoretic analysis of proteins suggests that their arrival may be as recent as a half to one million years ago.

The finches are currently placed in six genera, and comprise four groups that are distinguished by their appearance as well as by their feeding habits: these are a ground finch group (*Geospiza*) of six species, a tree finch group (*Camarhynchus, Cactospiza,* and *Platyspiza*) of six species, a warbler finch (*Certhidea*), and the Cocos Island finch (*Pinaroloxias*) which has very diverse feeding habits.

Despite several detailed and comprehensive systematic treatments of the finches, the taxonomic status of a few allopatric populations remains uncertain. The sharp-beaked ground finch, *Geospiza difficilis*, is a case in point. It has well-differentiated populations that are grouped together as the same species because they share many features, yet the largest birds, on Santiago (mean weight of 27 g), are more than twice as heavy as the lightest on Genovesa (mean weight of 12 g). It is not known whether birds this different in size would interbreed if the opportunity ever arose. A population of large ground finches on Darwin (*darwini*) has been tentatively assigned to *G. magnirostris*, but it has some of the characteristics of *G. conirostris*. Such uncertainties are typical of monophyletic groups in which differences between some species are little more than magnified versions of differences between populations of the same species.

Distribution and current status. No species is known to have become extinct, but some populations have become extinct as a result of direct and indirect human activity. Clear examples are the especially large forms of *G. magnirostris* on Floreana and San Cristóbal which became extinct some time after Darwin's visit in 1835, and populations of *G. difficilis*, which disappeared from Floreana after 1852 and from Santa Cruz eighty years later. Human influences aside, distributions in the Galápagos archipelago have remained approximately unchanged during this century.

The ground finches occur largely in the seasonally arid coastal zone. Tree finches occur there also, but in addition they occur in transitional and humid forest at higher elevations to a greater extent than do the ground finches. Distributions of tree finches among islands in the Galápagos archipelago are more restricted than are the distributions of the ground finches, since mesic habitats exist on only the large islands. The warbler finch is distributed widely, both altitudinally and geographically.

Geographical distributions of species vary greatly. At one extreme, the medium tree finch, *Camarhynchus pauper*, is restricted to the single island of Floreana. At the other extreme, the small ground finch, *Geospiza fuliginosa*, occurs on more than two dozen islands, including many of the small ones. In general, small and well-isolated islands have few species, large and weakly isolated islands have many species (up to ten). The well-isolated and peripheral islands of the archipelago have a high proportion of distinctive forms, recognized taxonomically as endemic species and subspecies. This pattern has been attributed to the effects on evolution of isolation per se, but is as likely to be due to the different ecological conditions on those islands.

Recent fossil material has confirmed the distributions of some populations before human settlement occurred in 1832. Although not fully analyzed yet, the material has not revealed any extinct, previously unknown, species. It has the potential for documenting recent evolutionary change.

Phenology and population limitation. A dominant influence on the lives of Darwin's Finches on the Galápagos is the food supply, which varies seasonally and unpredictably as a result of wide fluctuations in rainfall and primary production. Rain falls heavily, but erratically, in the months January to April or May. In the remainder of the year, precipitation occurs in the form of light mist (*garúa*), and varies from being generally trivial at low elevations to substantial at high elevations. Breeding usually starts in the first two weeks after the first heavy rainfall, and continues for as long as the supply of caterpillars and pollen remains high. Eggs are laid in clutches of 2–6, and nestlings leave the nest about one month later. In some extreme years so little rain falls that no, or little, breeding occurs. In other years, when the Galápagos are affected by the El Niño phenomenon, so much rain falls, and over such an extended period of time, that breeding continues uninterruptedly for 4 months or more. In the exceptionally wet year of 1982–1983, the breeding season continued for 8 months and individual pairs of ground finches produced up to 8 broods. Two years later (1985) no rain fell during the usual breeding season and scarcely any ground finches bred then.

Population sizes are potentially limited by food supply in the dry season, except in years of heavy rainfall. The evidence for actual limitation is a set of correlations between seed biomass and ground finch biomass (or numbers): among islands in the dry season, among sites within an island (Pinta) during the dry season, and during a drought on another island (Daphne) as both finch numbers and food supply declined. Since diets of different species overlapped at these times, the species were in competition for food. Nevertheless, interspecific competition is reduced by a regular divergence in the diets of species from wet to dry season. Predators, chiefly hawks (*Buteo galapagoensis*) and owls (*Asio galapagoensis*), appear to have a minor effect on population sizes, and the role of parasites and disease is almost unknown.

On Cocos Island conditions are seasonally and annually more stable and wetter. Rain falls throughout the year, and although its seasonality is weak it is sufficient to influence the breeding of finches. El Niño events cause drier conditions than usual, but they are far from being as severe as drought conditions are on the Galápagos.

Morphology. Male ground finches are almost completely black in fully adult plumage; females are brown and streaked. Males of the small, medium, and large tree finch, and of the vegetarian finch, acquire partly black plumage as adults, whereas males of the woodpecker finch and mangrove finch apparently never do. Females of the tree finch group are grey-green or pale brown, and streaked to a varying extent. Males and females of the warbler finch have the grey-green and unstreaked plumage of the woodpecker finch. The Cocos Island finch resembles ground finches in plumage, yet differs from all of them in acquiring black color (males) in a mosaic fashion over the body as opposed to a systematic progression from the head towards the tail.

Species differ much more in size and shape than they do in plumage. In general, species can be distinguished more by their beak sizes and shapes than by body size or by traits like the length of the wing and the leg (tarsus) with which body size is correlated. This applies not only to adults but also to nestlings. Morphological differences arise early in development, for they are already evident at the time of hatching and therefore must originate in the egg. Differences then become magnified with further growth. Relative growth is fastest in those dimensions most pronounced as adults; among the ground finch species, growth in depth and width of the bill varies much more in relation to body size than does growth in other dimensions. Maximal differences among the species are reached when growth ceases after 2–3 months.

Morphology and breeding. The morphological distinctiveness of species is maintained through reproductive isolation: they interbreed very rarely. The cues by which we distinguish the species—their appearance, chiefly of the beak—are used by finches in courtship when choosing a mate. Experiments with stuffed specimens show that male and female ground finches can make fine discriminations between conspecifics and heterospecifics on the basis of appearance alone. While both beak and body characteristics are probably used by birds when making these discriminations, the beak cues are likely to be the more important because they declare species identity less ambiguously than does body size.

Song is also used in courtship and mate choice. There are differences between species in the advertising songs sung by males (only). At some localities the differences are more blurred than are the differences in morphology; nevertheless, experiments show that ground finches are capable of discriminating between conspecific and heterospecific song. Males initiate courtship, and rarely direct their activity to females of another species. Since females do not sing, males must be using morphological cues only, whereas females may be using both sets of cues in the choice of a mate.

I suggest that males and females develop their discriminating abilities by imprinting on the song of their fathers and on the morphological features of both parents when being fed by them. Sons copy the song of their fathers precisely, but occasionally they acquire the song of another species instead, perhaps through misimprinting on an adult male of that species. Misimprinted birds sometimes obtain conspecific mates, sometimes heterospecific mates. The variety of the pairings is a further indication that both visual and acoustical cues are used by females when choosing mates.

Morphology and diet. The large variety of adult beak sizes and shapes reflects a large diversity of feeding activities and diets. This diversity is summarized as follows. As a group, Darwin's Finches rip open rotting cactus
pads, strip the bark off dead branches, kick over stones, probe flowers,
rolled leaves, and cavities in trees, and search for arthropods on the exposed
rocks of the shoreline at low tide. They consume nectar, pollen, leaves,
buds, a host of arthropods, and seeds and fruits of various sizes.

Not all species feed on all things, and the diet of each is determined in
part by its beak size and shape. The beak of a bird is an instrument for reaching, picking up, and dealing with food items in such a way that they can be
swallowed. Different beak sizes and shapes are different designs of this instrument, with different mechanical properties suiting them for different
tasks. Thus, by virtue of their deep beaks, and of the masses and disposi-
tions of the muscles that operate them, ground finches crush seeds at the
base of the bill. In contrast, tree finches apply force at the tips of their bills
to the woody tissues of twigs, branches, and bark, and thereby excavate hidden arthropod prey; here the design of the bill is more appropriately interpreted in the context of getting at the food rather than dealing with it once it
has been secured. The warbler finch, cactus finches, Cocos Island finch,
woodpecker finch, and mangrove finch have relatively long bills which they
use to probe flowers for nectar or holes in woody tissues for arthropods.
Long bills are unsuitable for crushing hard food types. Other subtle features
of the bill, such as curvature, are probably related to subtle features of feeding behavior.

Each species has a unique feeding niche. Differences among genera are
very clear; differences among closely related species are less pronounced
but nevertheless marked in the dry (nonbreeding) season. For example,
large species of ground finches crack open some large and hard seeds (and
consume the kernels) that are beyond the cracking powers of other species
with smaller bills. Species with large bills can also crack some seeds faster
than can species with smaller bills. Large species thus take a broader range
of food sizes than smaller species do. As a result of the relationship between
beak size and diet, the difference in diets between any two sympatric species
of ground finches is a function of the difference in their beak depths. In addition to these quantitative differences in the feeding characteristics of species, there are some outstanding qualitative differences. On the islands of
Wolf and Darwin, sharp-beaked ground finches (*G. difficilis*) have the
unique habit of perching on boobies, pecking at the developing feathers of
the tail and wing, and drinking the blood that flows from the wound. They
also push and kick seabird eggs against rocks, widen the cracks that form in
the shell, and then consume the contents. They do not do this on other is-

lands. Woodpecker finches and mangrove finches have the unique habit of using twigs, cactus spines, or leaf petioles to pry insect larvae and termites out of cavities in the dead branches of trees.

While variation in the size and shape of the bill is pronounced, species also differ in other morphological features that are related to food gathering and processing. Two examples will serve as illustrations. First, the gizzard of the vegetarian finch, when compared with the gizzards of other finches in relation to their body size, is disproportionately large. The intestine of this species is disproportionately long. These features are likely to help meet the need to digest a large bulk of vegetable matter. Second, those ground finches which spend a particularly large amount of time climbing (e.g. *G. scandens*) or scratching (*G. difficilis*) have exceptionally long hind toes and claws.

Intraspecific variation. Populations differ in the degree to which they vary in morphological traits, especially in bill size and body size. Some populations of the three large species of ground finches, *G. magnirostris*, *G. conirostris*, and *G. fortis*, are exceptionally variable.

This broad population variation is ecologically significant because birds with large beaks can crack larger and harder seeds than can birds with smaller beaks. Birds with large beaks are also more efficient than birds with small beaks in dealing with seeds of medium size and hardness. As a result of these phenotypic differences among members of a population, which parallel differences among species, large-billed birds have somewhat different diets from small-billed birds.

The evolutionary significance of broad morphological variation is that it is very responsive to forces of selection. Beak size variation is highly heritable in two populations: *G. fortis* on Daphne and *G. conirostris* on Genovesa. In a third population that has been examined, *G. scandens* on Daphne, phenotypic variation is lower and so too is the underlying genetic variation.

A large population variation is set by the resolution of two opposing processes: introgression of genes, from conspecific or heterospecific sources, which tends to increase genetic variance; and stabilizing or directional selection, which tends to decrease it. Both processes are known to occur in finch populations. Through interbreeding, the *G. fortis* population on Daphne Major receives alien genes from three sources: from conspecific immigrants from nearby Santa Cruz, from *G. fuliginosa* immigrants, and from *G. scandens* residents. Although introgression is rare, that from *G. fuliginosa* alone is sufficient to increase the phenotypic variance by about 6% each breeding episode. Hybrids have bred successfully with *G. fortis*. On Genovesa, *G. conirostris* hybridizes with sympatric *G. magnirostris* and *G.*

difficilis at a slightly lower frequency. The point of balance between introgression and selection varies among populations, and this is why some species vary more than others.

Finch populations experience forces of selection each generation. During a drought in 1977 on the island of Daphne Major, the strongly varying, ecologically generalized, *G. fortis* was subjected to strong directional selection. Large birds, with deep beaks, survived best, because they were best equipped to deal with the large and hard seeds that remained in relatively high numbers after the stock of small seeds had been drastically depleted. Females (the smaller sex) suffered the higher mortality, and the sex ratio became heavily male-biased. In the following breeding season sexual selection reinforced the preceding natural selection: females paired to a disproportionate extent with the largest members of the survivors of the drought, and with those in blackest plumage and hence possibly the oldest. Because heritabilities of all morphological traits are high, the combined effect of natural and sexual selection was an evolutionary response of about 4–5% in the mean size of those traits in the next generation.

While large members of the *G. fortis* population have an advantage during dry conditions, small birds are selectively favored under two different circumstances. Small birds survive best in their first year of life, and small young females come into reproductive condition earlier than large young females. Integrated over the lifetime of a cohort, the selective shifts in opposite directions may roughly balance and be equivalent to a weak form of overall stabilizing selection.

Other possible contributing factors to a broad variation are mutation and genetic drift. Nothing is known about mutation in the finches. Drift is a possibility, since some populations fall occasionally to low numbers, but they do not remain small for very long, and the effects of selection are likely to be stronger than effects of drift. Disruptive selection can contribute to the maintenance of a large variation. Disruptive selection occurred in the total population of *G. fortis*, males and females combined, during the drought referred to above. A few, very small, females survived through exploiting the low supply of very small seeds. Although few in numbers, they constituted a fifth of the breeding females in the following year. Other evidence of disruptive selection has been found in the study of the *G. conirostris* population on Genovesa, where birds with different bill shapes feed to some extent on different food types.

Most populations vary to a small extent because they hybridize very rarely and because they are subject to stabilizing selection (e.g. *G. scandens*). The Cocos Island finch exhibits low morphological variation. Being solitary and extremely isolated it cannot hybridize. Individuals are specialized at different feeding tasks, even though their bill form allows them to perform a variety of tasks. Stabilizing selection may be relatively weak, in which case the low variation is most strongly determined by the absence of introgression.

Competition and the composition of finch communities. The number of species that occur on an island, the particular species that occur on an island, and their beak sizes and diets, have been largely determined by the available food supply, but also by competition for food. Competition is inferred from patterns in the distribution, morphology, and feeding ecology of finches.

Ground finches occur together in nonrandom combinations. Members of combinations that occur rarely or not at all are generally similar to each other in beak size and diet. In contrast, members of all pairs of coexisting species differ by at least 15% in at least one bill dimension, and their frequency distributions scarcely overlap. The large differences between coexisting species are, statistically, improbably large.

These differences are determined in part by the nature of the food supply, which is best understood for the generalist granivore species among the ground finches. Frequency distributions of seed characteristics in the dry season are polymodal, owing to gaps and irregularities in otherwise continuous distributions. Polymodality in the food supply determines adaptive peaks in morphology associated with particular segments of the food supply. An adaptive peak can be recognized by calculating the maximum population density expected for a solitary finch species on a given island with a particular food supply, and by determining the beak size that such a population would have.

Unusually large differences between coexisting species arise through character displacement, an evolutionary process of divergence in traits such as beak size (and diet), or by differential (i.e. nonrandom) colonization, a strictly ecological process. Interspecific competition is involved in both; it contributes to the selective forces producing character displacement, and to the exclusion of ecologically similar species resulting in differential colonization.

A modern sign of the historical occurrence of character displacement is the fact that 11 of the 13 pairs of ground finch species which occur together differ more in beak size where sympatric than expected by chance. In allopatry, some of these species have appropriated the niche of a missing species. Differences in the food supply are not sufficient to account for the dietary and morphological differences between sympatric and allopatric populations. For example, *G. fortis* and *G. fuliginosa* are sympatric on many islands, and are allopatric on Daphne (*G. fortis*) and Los Hermanos (*G. fuliginosa*). If diets were determined solely by food supply, they should be the same in sympatry and allopatry in each case. In fact they are not; each species has intermediate morphology in allopatry, and exploits part of the food supply consumed elsewhere, in sympatry, by the missing species.

Differential colonization of islands is suggested by the fact that no two species are aligned with the same adaptive peak. Although competitive exclusion may have been responsible for this nonrandom pattern, a factor complicating the interpretation is the possibility of interbreeding between a resident species and an ecologically and morphologically similar species that immigrates. The disappearance of the immigrant species, as a species, could be caused partly by a process of "reproductive absorption" into the resident population. The absence of a sustained breeding population of *G. fuliginosa* on Daphne, despite repeated immigration, appears to be due to both competitive exclusion and reproductive absorption.

EVOLUTION

How are the current patterns of Darwin's Finch species to be explained? Why are there 13 species on the Galápagos and not 26, or just one? Why do they vary in body size from 8 g (warbler finch) to 40 g (large ground finch), instead of, say, from 5 g to 100 g? To answer such questions, and others, I will attempt a narrative reconstruction of the history of the group in the following pages, adhering closely to the known facts about the finches and their environments, and to the ideas that logically follow from them.

Some time in the last five million years, perhaps as recently as a half to one million years ago, a group of finches arrived on the Galápagos. We do not know exactly who they were, or whether they came from Cocos Island or the continent, but whatever their geographical point of origin they had completed a very long flight over the sea. The conditions they encountered were much like those we see today: seasonal aridity at low elevations, more mesic conditions higher up. There were probably fewer species of plants then, and even fewer species of other land birds than occur on the islands now. For the right organisms the Galápagos were a land of ecological opportunity. And being arrayed as a series of islands of varying degrees of size, elevation, and isolation, that land was also rich in evolutionary opportunity.

With hindsight we know that the finches were the right organisms to exploit those opportunities. At first the original colonists reproduced and the population built up over the next few years. The propensity to disperse which had brought them to Galápagos led some, or their descendants, to leave the initial island and fly to another. This happened many times, with the result that populations became established on several islands. Each one of the populations was subjected to slightly different selection pressures, because on no two islands were the feeding conditions exactly alike. What types of foods were eaten is unknown. Perhaps this original species was somewhat like a modern ground finch with seed-eating habits, but enough of a generalist to take advantage of other foods, which would have been especially important during occasional food shortages.

Dispersal between islands continued at a low level but regular frequency, and occasionally dispersers stayed to breed with members of the resident population, thereby enhancing the variation in the population in proportion to the difference between residents and immigrants. At this point the species existed as a series of moderately differentiated populations. Then, after an unknown period of time, it split into two. The split began in allopatry, and was completed in sympatry. It happened because immigrants from one island arrived on another island where part of their food supply was not exploited by the residents.

It is easy to visualize this by supposing that the immigrants were larger in size than the residents, being adapted to deal with the larger and harder seeds on their island of origin. At times of food shortage, during droughts, the combined groups (residents and immigrants) experienced the forces of disruptive selection. These tended to eliminate the offspring of mixed parents and of generally intermediate size. The immigrant group evolved to even larger size under directional selection, and became aligned with a separate adaptive peak determined by the food supply. As the two groups diverged in morphology, discrimination between potential mates of the two groups on the basis of morphology became more refined, the incidence of mixed matings declined, and so too did the level of morphologicial variation in each group for reasons of natural selection and diminished gene exchange. The end-point of these processes was a nearly complete reproductive (pre-mating) isolation of the two groups, and coexistence with reduced competition for food: in other words, the formation of two sympatric species.

The important ingredients of speciation were (a) an initial difference between residents and immigrants, evolved in allopatry, (b) an adaptive peak within reach, evolutionarily, of the immigrants, (c) genetic variation among immigrants, allowing an evolutionary response to selection, and (d) forces of selection acting against individuals of the two groups who were so similar that they competed with each other for an occasionally limited supply of food. Thus ecological forces were of primary importance in effecting the split, with reproductive factors such as propensity to interbreed and the fertility of the offspring being secondary.

Over the next half million years or more these processes were repeated, many times: dispersal of individuals of a newly formed species to other islands, formation of allopatric populations, exchange of individuals between them, and eventually the formation of a new species through an enhancement of initial differences between previously separated populations. This was the basic pattern of events, with some variation such as speciation occurring entirely in allopatry; species that disperse between islands relatively infrequently, and encounter different food supplies on different islands, are the best candidates for speciation in this manner. By small steps, different lineages radiated in very different morphological and ecological directions from the common ancestral stock, facilitated by a scarcity of other species and by differences among islands in ecological, principally feeding, conditions. Competitive interactions among species became increasingly complex, and some extinctions occurred through the competitive replacement of old forms by newly evolved species. To give an entirely hypothetical example, the warbler finch and the woodpecker finch, either singly or jointly, may have eliminated a species with intermediate characteristics.

It is likely that in the initial stages of the radiation, the first few species were larger or smaller versions of each other. If the original species was a small seed-eater, for example, then medium and large seed-eating species probably evolved first. A few thousand years would have been enough time for speciation to be accomplished by a series of selection episodes producing small incremental effects on successively colonized islands, culminating in the development of reproductive and ecological isolation in sympatry. Species with markedly different proportions evolved more slowly, in association with the evolutionary development of such novelties as feeding predominantly in trees on arthropods hidden in dead wood. New feeding habits arose by chance and were improved through the trial-and-error learning of young birds that made mistakes but were rewarded—like pecking at a tick on an iguana or tortoise, for example, or failing to drop a twig before reaching into the crevice of a branch, and being rewarded with the larva of a carpenter bee. Adaptive change in beak structure to facilitate such feeding behaviors took a long time, however, because changes in beak shape required prolonged selection to overcome genetic constraints on evolutionary change. The constraints arise from genes governing the expression of two or more traits, such as beak length and beak depth; the two traits do not respond independently to the forces of selection. Nevertheless, invasion of a new adaptive zone such as that occupied by tree finches, through a combination of chance events and selection, freed the invader from competition with other ground finch species, and allowed further speciation and adaptation to different feeding niches within that zone. The need for inconspicuousness, in the face of visually hunting predators (hawks and owls), against the new, arboreal, background of predominantly green and grey colors was met by adaptive change in appearance; a partial or complete loss of black color in the plumage of males, and the acquisition of a greenish cast to the plumage of females.

Some species continued to evolve long after they arose. The warbler finch, for example, started on its evolutionary trajectory early in the history of the group. Its characteristics now are probably very different from those of its earliest ancestors. They are the product of sustained adaptation over many thousands of years to the task of feeding on small arthropods on the leaves and bark of trees. The Cocos Island finch may similarly possess a highly modified set of morphological and behavioral traits acquired as a result of adaptive change over a long period of time.

I raised two difficult questions at the beginning of this section. The first was why there are thirteen finch species on the Galápagos, and not double the number, or just one. The isolation of the islands, their ecological differences, and the passage of time make it easy to understand why the original species evolved into many. It is less easy to say why diversification has not proceeded beyond thirteen. A general answer is that there is limited ecological opportunity for further diversification, and possibly there has not been enough time for the radiation to run its full course as determined by that opportunity. Further diversification is also constrained by the presence of other species, such as the flycatchers and the yellow warbler. If the yellow warbler is of recent origin, and the warbler finch (*Certhidea olivacea*) is ancient, it is not clear why a yellow warbler finch did not evolve from *Certhidea*. Perhaps it did, after all, and became extinct after the yellow warbler arrived in the archipelago.

The second and interrelated question was why the finches do not span a greater range in size. At the lower end of the size range the absence of a micro-*Geospiza* is puzzling because there appears to be a niche for it. At the upper end of the range the absence of a macro-*Geospiza* is explicable in terms of the food supply. I have seen no unexploited hard seeds that a finch of 50–60 g could crack, except for those of *Opuntia megasperma* on Champion and Gardner, and they were probably eaten by the exceptionally large, now extinct, form of *G. magnirostris*. In a sense the Galápagos dove (~100 g) is a macro-finch, but its body plan is so different as to be possibly unattainable by the directional evolution of a finch.

While much is now known about the evolution of Darwin's Finches there is still more to be learned, so I will conclude this section by repeating the major unanswered questions that should guide future research. We do not know the modern relatives of the ancestral stock, and so we can do little more than guess the characteristics of the original colonists. We do not know where the colonists came from, or whether Galápagos and Cocos Island were colonized in parallel or in sequence. The environmental conditions prevailing over much of the group's history are not known in any detail. Extinctions of species probably occurred, perhaps several times, but we do not know of a single case. The pattern of systematic relationships among modern species rests on a fragile foundation. In anatomy, gross morphology, and song all modern species are well characterized, but knowledge of the ecology and behavior of species is unevenly distributed among the group, ground finches being far better known than tree finches. Our largest areas of ignorance lie in the life history characteristics of individuals and the dynamics of populations. Biochemical and paleontological studies can resolve some of the outstanding historical and phylogenetic questions. Field studies of modern tree finches, the warbler finch, and the Cocos Island finch can show the degree to which ground finch populations studied in detail on Daphne Major and Genovesa capture the essential features of Darwin's Finches as a whole.

GENERALIZATIONS

"Science is the activity of finding facts and then arranging them in groups under general concepts" (Bronowski 1977, p. 211), in such a way as to best reveal the relationships among them. Darwin's Finches are unusually suitable for elucidating evolutionary principles because, in the words of David Lack (1947, p. 159), "the evolutionary stage shown by Darwin's finches is both sufficiently advanced to provide a parallel with the more mature evolution of the continents, and sufficiently early for links to remain which reveal the underlying processes." Those early links take the form of extreme similarity of some species in morphology and ecology, the close coupling of ecological and reproductive isolation and, partly as a result of these two features, the large variation in populations of some species. Given these advantages, how far can our understanding of Darwin's Finch evolution stand as a model for the evolution of other groups of animals? I shall offer some answers to this question by first selectively reviewing some of the patterns of evolution displayed by diploid, sexually reproducing, animals, and then by discussing the major theories about their origins.

The radiation of Darwin's Finches is put in perspective by comparing it with patterns of evolution displayed by other birds, and other organisms. Spectacular as it is, it is dwarfed by the radiation of honeycreeper finches in the Hawaiian archipelago (Fig. 100). At least 42 species evolved from a single cardueline finch ancestor (Raikow 1977) in the last six millions years (or conceivably more: Sibley and Ahlquist 1982). There are counterparts among them to some of the Darwin's Finches, and they illustrate the principle of different evolutionary solutions to similar ecological problems. The akiapola'au (Fig. 100), for example, uses its lower mandible (only) in the manner of *Cactospiza pallida* and woodpeckers to remove bark from branches, and its upper mandible to extract exposed insect larvae in tunnels in the wood. In effect the upper mandible of this remarkably unequal bill

does the work of the woodpecker finch's twig or spine. The parrotbill uses its bill more like a large tree finch does in biting and tearing bark and breaking off twigs and even branches, and it too extracts exposed larvae with its upper mandible. The seed-crushing honeycreeper finches are clear counterparts to the ground finches on Galápagos, but, with the exception of the Cocos Island finch, Darwin's Finches do not display any of the rich diversity of curve-billed, nectar- and arthropod-feeding, forms that have evolved in the Hawaiian archipelago. The diversity of all morphological and ecological forms among the honeycreeper finches is far greater than that achieved by Darwin's Finches, a fact which can be explained by the greater ecological richness of the Hawaiian islands and by the birds' longer occupancy there.

This radiation is, in turn, dwarfed by the diversity of fruit flies (*Drosophila*) in the same archipelago. At least 500 species, and perhaps as many as 800, evolved over a much longer period of time (Beverley and Wilson 1985). They diversified into predatory, parasitic, nectarivorous, detritivorous, and herbivorous ways of life. Their vastly greater numbers can be explained in part by their longer history and in part by their smaller individual size. A geographical island is actually an archipelago of vegetation-islands separated by lava flows (Carson 1982a), and whereas birds cross the flows with ease, flies are largely restricted by the barriers and as a result they form many localized populations.

Speciation in *Drosophila* has probably occurred many times on the same island. This is inferred from the observation that all but three of 103 species in the picture-wing group which are endemic to the six high islands are restricted to single islands or island complexes. Therefore they evolved where they are found now. The exact pattern of evolutionary changes has been established by comparing the banding sequences of the five major polytene chromosomes of these species. Variation among the species is due to 213 paracentric inversions. Some of the changes in courtship behavior have also been worked out (Kaneshiro 1976). Using this information and the ages of the sequentially formed islands, Carson (1983) has been able to reconstruct the patterns of dispersal and speciation events in the archipelago, an achievement without parallel in its scope and detail.

FIG. 100. Part of the adaptive radiation of honeycreeper finches in the Hawaiian archipelago. The species are: (1) Mamo, *Drepanis pacifica*; (2) I'iwi, *Vestiaria coccinea*; (3) Crested honeycreeper, *Palmeria dolei*; (4) Ula-ai-hawana, *Ciridops anna*; (5) 'Apapane, *Himatione sanguinea*; (6) 'Akialoa, *Hemignathus procerus*; (7) 'Akiapola'au, *Hemignathus wilsoni*; (8) 'Ākepa, *Loxops coccinea*; (9) 'Amakihi, *Loxops virens*; (10) Parrotbill, *Pseudonestor xanthophrys*; (11) O'u, *Loxioides psittacea*; (12) Creeper, *Paroreomyza maculata*; (13) Grosbeak finch, *Loxioides kona*; (14) Nihoa finch, *Telespiza cantans*; (15) *Melamprosops phaeosoma*; (16) Palila, *Loxioides bailleui*. Redrawn from paintings by H. Douglas Pratt in Raikow (1977).

An outstanding example of multiple speciation on a single island is provided by flightless beetles of the genus *Miocalles* (*Microcryptorhynchus*) on the well-isolated Polynesian island of Rapa. At least 67 species occur in an area of a mere 40 km², most having evolved on the island itself (Zimmerman 1938, Paulay 1985). On the island of Moorea a well-studied parallel is provided by snails in the genus *Partula* (Clarke and Murray 1969, Murray and Clarke 1980), and in the Hawaiian archipelago similar radiations are seen in the achatinellid snails, moths, and cave crickets (Carlquist 1970, Howarth 1980, Montgomery 1982), to name but a few. Allopatric speciation within an island is a plausible explanation for the diversification of these relatively sedentary and habitat-specific organisms.

All of these radiations, and there are countless other instances around the globe, serve to link the early stages of the differentiation of a group of organisms to the large-scale patterns of evolution: the radiation of marsupials in Australia, for example, or the dinosaurs, or fish. Such macroevolutionary patterns are reducible to microevolutionary processes, although there is debate about this (Gould 1980, 1982, Charlesworth et al. 1982, Wright 1982, Carson and Templeton 1984), through the unraveling of ecological and evolutionary forces which impinge on populations and cause new species to be formed. It is in this context that Darwin's Finches have their chief value. They represent an early stage in the diversification of a group, and hence allow us to identify the causes of the origin of an adaptive radiation. By direct study it will never be possible to achieve the same understanding of the marsupial or dinosaur radiation as of the Darwin's Finch radiation, but the principles established for the finches can be applied to the marsupials and dinosaurs.

The other great value of the Darwin's Finch radiation is that the evidence of what happened in the past has not been obscured by human activity. The evolutionary patterns are very clear, and the ecological theater of the evolutionary play, to borrow a phrase from Hutchinson (1965), is still largely intact. Unfortunately, this is not true of the vast majority of islands. The Hawaiian archipelago offers a contrasting example of the more typical condition. Vegetation has been greatly disturbed by human activity and the biota has suffered accordingly. Twenty-seven species of honeycreeper finches survived into historical times, and this number, minus one that was discovered little more than a decade ago (Casey and Jacobi 1974), was thought to represent the sum total of evolutionary diversification (Amadon 1950, Baldwin 1953), until fossils were discovered recently. A minimum of fifteen extinct species must be added to the list (Olson and James 1982), and past attempts to reconstruct the evolutionary pathways (Bock 1970) need to be revised by taking them into account. Since the habitats of the extinct species have been destroyed or radically altered, the exercise can only meet with limited success. The same can be said for many other organisms in other parts of the world, on both islands and continents, on land and in water, where human activity has resulted in major alteration of the habitat.

The key element of evolutionary diversification is speciation (Mayr 1963). Speciation occurs through anagenesis, the gradual but profound transformation from one state to another, but more frequently through cladogenesis, the splitting of a lineage into two, which then evolve independently. Since the development of reproductive isolation between ancestral and derived populations involves the evolution of genetic differences, geneticists have had much to say about the mechanisms of speciation. It is here, in the area of genetics, that studies of other organisms have gone much further than have studies of the finches. Therefore, to complete this survey of evolutionary radiations, I shall now discuss modern ideas on the role and causes of genetic changes in speciation.

Several reviews of the genetic changes accompanying speciation have appeared recently (e.g. White 1978, Templeton 1981, Barigozzi 1982, Carson and Templeton 1984, Barton and Charlesworth 1984), and while geneticists are far from unanimous on various points of view they all stress the diversity of modes and mechanisms of speciation. The important factors for one group are not necessarily applicable to another (White 1978, Templeton 1981, Wright 1982, Mayr 1984). So the manner in which fourteen species of Darwin's Finches came into being—an adaptive differentiation initiated in allopatry—is not universal, although there is broad agreement that it is probably the major one (e.g. Mayr 1963, Templeton 1981).

In other words, there are limits to generalizing from the single example of Darwin's Finches. The limits are set by other population structures, other genetic architectures, and other selection pressures.

The differentiation in allopatry sets Darwin's Finches apart from some organisms. Theories of parapatric (Endler 1977, Lande 1982) and sympatric diversification (Maynard Smith 1966, Bush 1975), while based on restrictive assumptions, allow for the possibility of speciation in the absence of complete geographical isolation of populations. Their results are difficult to distinguish from speciation in allopatry followed by secondary contact, however, and at the present little more can be done than acknowledge their possibility. Predisposing factors for sympatric speciation include close associations between a species and another that it exploits, and enough temporal latitude in breeding seasons to permit reproductive isolation of early and late breeders, as shown by crickets (Alexander 1968), lacewing flies (Tauber and Tauber 1978), and fish (Smith and Todd 1984).

The role of adaptive change sets Darwin's Finches apart from some other organisms. Species with different genetic architectures are likely to follow different evolutionary pathways, and not all of those pathways are adaptive.

For example, chromosomal changes occurring by chance could initiate a speciation event, with little or no adaptive consequences. White (1978) has estimated that the vast majority of all speciation events are accompanied by karyotypic changes, and has argued that structural chromosomal rearrangements occurring in small isolates or local demes of a species (stasipatry) have played a primary role in initiating divergence. Fixation of chromosomal rearrangements could alternatively be an incidental byproduct of small population effects and the unusual selection pressures on such populations (Carson 1978, 1982b). Hence the causal role of chromosomal changes in speciation is not clear (Patton and Sherwood 1983). In any event, the large chromosomal differences among the species of morabine grasshoppers studied by White (1978), species of *Drosophila*, and many mammal species (Patton and Sherwood 1983), are not shared by Darwin's Finch species.

Modern views about nonadaptive changes derive partly from Wright's (1931, 1977, 1980) shifting-balance theory of evolution. It posits multiple selective peaks in a field of genetic variation, and explains how shifts from one peak to another (higher) peak are initiated by the formation of favorable combinations of genes through random drift in a small population with reduced genetic variation when the external environment is constant. The greatest possibility for a shift occurs when the species is represented by small and semi-isolated populations in the same environment; archipelagos might be considered to provide such conditions. After a shift has occurred, natural selection moves the population up the new peak. Dispersal of individuals from this population to others can bring about the eventual transition of all populations to the new peak. In this manner an interplay of selection and drift causes evolutionary change in a species; and if allelic substitutions with major effect occur, it may also cause speciation (Wright 1982).

One reason why I have not given prominence to the shifting-balance mechanism in the evolution of Darwin's Finches, is that polygenic traits depending upon many additive loci of small effect, such as the finches possess, evolve no faster with this population structure of subdivision than under "mass" selection in a single population. Other reasons are the rarity of very small finch populations in which drift is likely to be important, the demonstrated importance of selection, and the ecological differences among islands. Genetic systems with major loci and epistatic modifiers are expected to evolve faster under population subdivision, however (Wright 1977, Templeton 1981). Such coadapted gene complexes determine, for example, color patterns on the wings of butterflies (Turner 1983) and morphological patterns of *Drosophila* flies (Templeton 1980, 1981, Carson 1982a).

The idea that major genetic changes occur through stochastic events has been incorporated into three modern theories of speciation, in addition to

White's. All give importance to the founding of a new population by one (gravid) or a few individuals. In Mayr's (1954, 1984) theory of peripatric speciation, the loss of genetic variation in the founder event itself, followed by further loss in subsequent generations as selection breaks up previously coadapted gene complexes, causes a genetic revolution whose ultimate effect is to allow directional selection on exposed genetic variation to carry the population to a new adaptive peak. Speciation is peripatric, because it takes place in isolated situations (e.g. islands) around the periphery of a broad geographical distribution of a species. Carson's (1968, 1982a, 1982b) founder-flush theory similarly invokes the disruption of ancestral coadapted gene complexes in small founder populations, but under the influence of drift. Novel genetic combinations appear, and as the population approaches carrying capacity and density dependent forces begin to operate, selection, acting on these novel combinations, takes the population to a new adaptive peak. Templeton's (1980, 1981) genetic transilience theory is similar, but places the emphasis on an altered genetic environment and resulting changes in selection pressures in the early stages of a founder event.

All three theories embody the notion of genetic cohesiveness in genomes that disallows certain evolutionary responses to selection pressures. Developmental constraints on evolutionary change are one possible manifestation of genetic cohesiveness (e.g. see Maynard Smith et al. 1985). Cohesiveness is broken in the founder event, and perhaps also in subsequent reductions of population size (so-called bottlenecks), so that, freed from constraints, the population can now respond to certain selection pressures. This is fundamentally different from the adaptive model for Darwin's Finches and many other organisms in that genetic (internal), rather than environmental (external), factors initiate the divergence that leads to speciation. For polygenic traits governed by genes with small additive effects, in contrast, the constraints on evolution are set by genetic variance and covariance for different characters (Charlesworth et al. 1982, Cheverud 1984). These do not protect the genome in any absolute way; instead they determine the rate and direction of response, which could nevertheless be very slow.

A recurring question in the study of speciation is the importance of such random events. Carson (1982b) asserts that disorganization of the gene pool of small populations is the crucial step in all speciation events, yet all models of speciation through founder events have been criticized on theoretical grounds (Barton and Charlesworth 1984). Mayr's theory of genetic revolution has been criticized as overemphasizing loss of genetic variation (Lande 1980b, Wright 1980), and in not having sufficient additive variation when directional selection is required (Carson and Templeton 1984). These objections tend to cancel, and leave unanswered the question of whether founder events are instrumental or irrelevant to the phenotypic changes that cause speciation in isolated populations. For example, isolated populations of birds differ more from relatives on the periphery of a broad geographical distribution than peripheral and central populations differ from each other. This is the geographical pattern that stimulated the development of the theory of peripatric speciation (Mayr 1942, 1954). The environments appear to be similar, so the argument goes, therefore the cause of the differentiation must lie in the genetic systems of populations. Lack's belief in the similarity of the Galápagos islands, contrasted with my own experience of their subtle but important differences, suggests to me that the starting assumption of ecologicial similarity needs to be reappraised, now that the postulated genetic changes have been reappraised (Lande 1980b, Carson and Templeton 1984, Barton and Charlesworth 1984). Passerine birds on islands show strong trends towards the evolution of large size, particularly large beak size (Grant 1965, 1968). These evolutionary trends can be interpreted adaptively in terms of the peculiar environmental conditions on the islands: relatively few species of food types, few competitors (Grant 1965, 1966, 1968), and possibly unusual characteristics of the food types (Abbott 1977; see also Allan et al. 1973, Janzen 1973). For this evolutionary pattern there is no need to invoke founder events.

Founder events may occur frequently as new populations are established, but they rarely result in strong evolutionary changes (Templeton 1981, Mayr 1982). Similarly, the conditions for Wright's shifting-balance model to apply may rarely be met by organisms in nature (Maynard Smith 1984). Where, then, are the conditions suitable for speciation to occur by fairly rapid and major reorganization of gene complexes? According to Templeton (1981), species groups characterized by complex mate-recognition systems subject to intense sexual selection are particularly liable to pre-mating transiliences triggered by founder events.

The Hawaiian *Drosophila* are good candidates. Despite a few host plant shifts, the basic ecological adaptations of members of species-rich groups are very similar, and the complex mating systems have changed much more, as shown by behavioral and morphological factors that pertain to courtship (Carson and Templeton 1984).

Another suitable group are the haplochromine cichlid fishes of the African great lakes (Fryer and Iles 1972, Greenwood 1981a). For example, about 200 species have evolved in Lake Victoria in the last 750,000 years, at maximum (Greenwood 1981a, 1981b, 1984, Dominey 1984, Mayr 1984), and even more have evolved in Lake Tanganyika and Lake Malawi over a longer period of time (Greenwood 1984). Body form has been conservative, but trophic apparatus has become modified much more (Fig. 101), and specialized diets range from zooplankton, benthos, molluscs, and fish to algae and even to fish scales! Despite this broad diversity which is

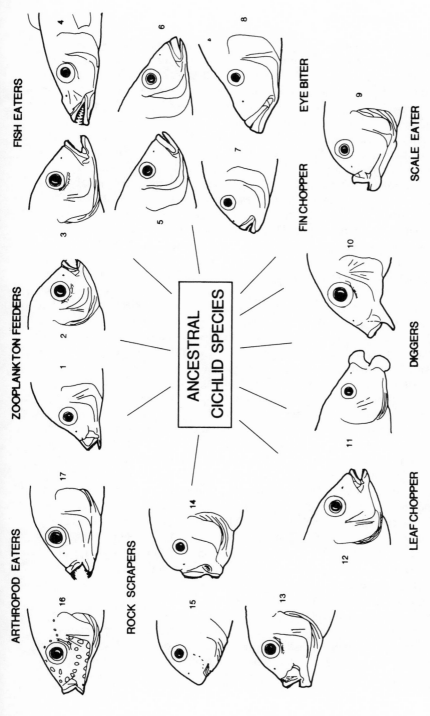

FISH EATERS

ZOOPLANKTON FEEDERS

ARTHROPOD EATERS

ANCESTRAL CICHLID SPECIES

ROCK SCRAPERS

FIN CHOPPER

EYE BITER

LEAF CHOPPER

DIGGERS

SCALE EATER

FIG. 101. Part of the adaptive radiation of cichlid fishes in Lake Malawi, Africa. The species are: (1) *Haplochromis cyaneus*; (2) *Cynotilapia afra*; (3) *Haplochromis paradalis*; (4) *Ramphochromis macrophthalmus*; (5) *Haplochromis livingstonii*; (6) *Haplochromis rostratus*; (7) *Haplochromis placodon*; (8) *Haplochromis compressiceps*; (9) *Genyochromis mento*; (10) *Lethrinops brevis*; (11) *Haplochromis euchilus*; (12) *Haplochromis similis*; (13) *Pseudotropheus zebra*; (14) *Petrotilapia tridentiger*; (15) *Labeotropheus fuelleborni*; (16) *Aulonocara nyassae*; (17) *Labidochromis vellicans*. Redrawn from Fryer and Iles (1972).

readily interpretable in adaptive terms, some clusters of species seem to be differentiated relatively little except in male color (Greenwood 1981a, 1984), which raises the possibility that sexual selection has been a major driving force in the radiation (Dominey 1984), as apparently it has been in the evolutionary diversification of many groups of insects (West-Eberhard 1983).

Although each radiation took place in a single lake basin, there is enough spatial complexity to allow sexual selection to act in small and isolated groups. Furthermore, the fish disperse relatively little. They have complex (polygamous) mating systems, and male parental care is lacking. Hybridization is common among many freshwater fish species (Smith and Todd 1984) but is almost lacking in this group, which suggests that specific mate-recognition systems diverged in allopatry and served to isolate the new species reproductively at times of secondary contact (Mayr 1984). Founder events may therefore have triggered genetic reorganizations in sexually selected traits in these fish, and perhaps played an important role in the radiation of the group, although once again the question must be raised whether founder events were necessarily involved, since sexual selection can lead to reproductive isolation without any major genetic reorganizations taking place (Lande 1981). The interplay of adaptive processes (natural selection) and non-adaptive processes (e.g. sexual selection) is also not clear. Adaptive differentiation in allopatry cannot be ruled out just because species appear to be ecologically similar. This warning comes from the work of Witte (1984) who has shown that morphologically similar species differ more in feeding niches than was previously supposed.

To summarize, speciation occurs in a variety of ways depending on the particular genetic, ecological, and behavioral characteristics of organisms and on their environments. Genetic changes taking place in speciation include profound chromosomal rearrangement and subtle alteration of polygenic complexes. They are governed by natural selection, sexual selection, and an interplay of selection and drift. No single, and simple, theory of speciation encompasses such diversity. Nevertheless, according to our current state of knowledge, the allopatric model of gradual genetic change links Darwin's Finch evolution with the radiations of many other groups of organisms as the most probable, general, mode of speciation.

Having attempted to place the evolution of Darwin's Finches in a broad context, I will conclude by stressing that an understanding of an evolutionary radiation, adaptive or otherwise, requires attention to four basic aspects of the group under consideration. First, the morphology of the members of the group: the patterns of adult structure, the functions of each structure, and the way they arise in development (ontogeny) and in history (phylogeny). Second, their behavior: the way they feed, avoid being eaten, choose mates, and breed. Third, their ecology: the structure of populations imposed by age, sex, and genetic relatedness, and the dynamics of those populations. Fourth, their genetics: the genetic control of phenotypes, the frequencies of genes, and their exposure to random processes and to the forces of selection.

Knowledge of these four aspects of Darwin's Finches is not uniformly good. I am impressed in the accounts and interpretations of other evolutionary radiations, and in particular in discussions of speciation, by a similar unevenness in our knowledge of other species-rich groups as well. Correspondingly, those aspects that are best known because they are most suitable for study (e.g. chromosomes, ontogeny, male courtship, feeding ecology) tend to be given the greatest prominence in the explanations for the evolution of the groups. I hope this book shows, by the strengths in our knowledge of Darwin's Finch evolution, and by the shortcomings, what needs to be known for us to be able to declare that we approach a full understanding of the evolutionary diversification of a group of organisms.

Appendix

Spanish and English names of the seventeen major Galápagos islands in Table 2, listed in decreasing order of size.

Spanish	English
Isabela	Albemarle
Santa Cruz	Indefatigable
Fernandina	Narborough
Santiago (San Salvador)	James
San Cristóbal	Chatham
Floreana (Santa María)	Charles
Marchena	Bindloe
Pinta	Abingdon
Española	Hood
Baltra	South Seymour
Santa Fe	Barrington
Pinzón	Duncan
Genovesa	Tower
Rábida	Jervis
Wolf	Wenman
Darwin	Culpepper
Seymour	North Seymour

References

Abbott, I. 1972. The ecology and evolution of passerine birds on islands. Ph.D. thesis, Monash University.

Abbott, I. 1973. Birds of Bass Strait. Evolution and ecology of the avifaunas of some Bass Strait Islands, and comparisons with those of Tasmania and Victoria. *Proceedings of the Royal Society of Victoria* 85:197–224.

Abbott, I. 1974a. The avifauna of Kangaroo Island and causes of its impoverishment. *Emu* 74:124–134.

Abbott, I. 1974b. Morphological changes in isolated populations of some passerine bird species in Australia. *Biological Journal of the Linnean Society* 6:153–168.

Abbott, I. 1977. The role of competition in determining differences between Victorian and Tasmanian passerine birds. *Australian Journal of Zoology* 25:429–447.

Abbott, I. 1980. Theories dealing with the ecology of landbirds on islands. *Advances in Ecological Research* 11:329–371.

Abbott, I., and L. K. Abbott. 1978. Multivariate study of morphological variation in Galápagos and Ecuadorean mockingbirds. *Condor* 80:302–308.

Abbott, I., P. R. Grant, and L. K. Abbott. 1975. Seed selection and handling ability of four species of Darwin's Finches. *Condor* 77:332–335.

Abbott, I., L. K. Abbott, and P. R. Grant. 1977. Comparative ecology of Galápagos Ground Finches (*Geospiza* Gould): evaluation of the importance of floristic diversity and interspecific competition. *Ecological Monographs* 47:151–184.

Abs, M., E. Curio, P. Kramer, and J. Niethammer. 1965. Zur Ehrnahrungsweise der Eulen auf Galápagos. Ergebnisse der Deutschen Galápagos-Expedition 1962/63. IX. *Journal für Ornithologie* 106:49-57.

Adsersen, H. 1976. A botanist's notes on Pinta. *Noticias de Galápagos* No. 24:26–28.

Alatalo, R. V. 1982. Bird species distributions in the Galápagos and other archipelagos: competition or chance? *Ecology* 63:881-887.

Alatalo, R. V., L. Gustafsson, and A. Lundberg. 1982. Hybridization and breeding success of Collared and Pied Flycatchers on the island of Gotland. *Auk* 99:285–291.

Alberch, P., S. J. Gould, G. F. Oster, and D. B. Wake. 1979. Size and shape in ontogeny and phylogeny. *Paleobiology* 5:296–317.

Alerstam, T., B. Ebenman, M. Sylvén, S. Tamm and S. Ulfstrand. 1978. Hybridization as an agent of competition between two bird allospecies: *Ficedula albicollis* and *F. hypoleuca* on the island of Gotland in the Baltic. *Oikos* 31:326–331.

Alexander, R. D. 1968. Life cycle, origins, speciation and related phenomena in crickets (Orthoptera, Gryllidae). *Quarterly Review of Biology* 43:1–41.

Allan, J. D., L. W. Barnthouse, R. A. Prestbye, and D. R. Strong. 1973. On foliage arthropod communities of Puerto Rico second growth vegetation. *Ecology* 54: 628–632.

Allen, G. M. 1937. *Birds and Their Attributes*. Marshall Jones, Boston.

Alpert, L. 1961. The climate of the Galápagos islands. *Occasional Papers of the California Academy of Sciences*. 44:21–44.

Amadon, D. 1950. The Hawaiian Honeycreepers (Aves, Drepaniidae). *Bulletin of the American Museum of Natural History* 95:151–262.

Andrewartha, H. G., and L. C. Birch. 1954. *The Distribution and Abundance of Animals*. University of Chicago Press, Chicago.

Arnold, S. J. 1983. Morphology, performance and fitness. *American Zoologist* 23:347-361.

Arthur, W. 1982. The evolutionary consequences of interspecific competition. *Advances in Ecological Research* 12:127–187.

Baldwin, P. H. 1953. Annual cycle, environment and evolution in the Hawaiian honeycreepers (Aves: Drepaniidae). *University of California Publications in Zoology* 52:285–398.

Barigozzi, C. (ed.). 1982. *Mechanisms of Speciation*, A. R. Liss Inc., New York.

Barrowclough, G. F. 1983. Biochemical studies of microevolutionary processes. *In* A. H. Brush and G. A. Clark (eds.), *Perspectives in Ornithology*, pp. 223–261. Cambridge University Press, Cambridge.

Barton, N. H., and B. Charlesworth. 1984. Genetic revolutions, founder effects, and speciation. *Annual Reviews of Ecology and Systematics* 15:133–164.

Bateson, P.P.G. (ed.). 1983. *Mate Choice*. Cambridge University Press, Cambridge.

Beebe, W. 1924. *Galápagos: World's End*. Putnams, New York.

Berry, R. J. (ed.). 1985. Evolution in the Galápagos Islands. Academic Press, London. (Reprinted from the *Biological Journal of the Linnean Society*, Vol. 21, Nos. 1 and 2, 1984.)

Beverley, S. M., and A. C. Wilson. 1985. Ancient origin for Hawaiian Drosophilinae inferred from protein comparisons. *Proceedings of the National Academy of Sciences USA* 82: 4753–4757.

Boag, P. T. 1983. The heritability of external morphology in Darwin's Ground Finches (*Geospiza*) on Isla Daphne Major, Galápagos. *Evolution* 37:877-894.

Boag, P. T. 1984. Growth and allometry of external morphology in Darwin's finches (*Geospiza*) on Isla Daphne Major, Galápagos. *Journal of Zoology*, London 204:413–441.

Boag, P. T., and P. R. Grant. 1978. Heritability of external morphology in Darwin's Finches. *Nature* 274:793–794.

Boag, P. T., and P. R. Grant. 1981. Intense natural selection in a population of Darwin's Finches (Geospizinae) in the Galápagos. *Science* 214:82–85.

Boag, P. T., and P. R. Grant. 1984a. Darwin's Finches (*Geospiza*) on Isla Daphne Major, Galápagos: breeding and feeding ecology in a climatically variable environment. *Ecological Monographs* 54:463–489.

Boag, P. T., and P. R. Grant. 1984b. The classical case of character release: Darwin's Finches (*Geospiza*) on Isla Daphne Major, Galápagos. *Biological Journal of the Linnean Society* 22:243–287.

Bock, W. J. 1963. Morphological differentiation and adaptation in the Galápagos finches. *Auk* 80:202–207.

Bock, W. J. 1970. Microevolutionary sequences as a fundamental concept in macroevolutionary models. *Evolution* 24:704–722.

Boecklen, W. J., and C. NeSmith. 1985. Hutchinsonian ratios and log-normal distributions. *Evolution* 39:695–698.

Bond, J. 1948. Origin of the bird fauna of the West Indies. *Wilson Bulletin* 60:207–229.

Bowman, R. I. 1961. Morphological differentiation and adaptation in the Galápagos finches. *University of California Publications in Zoology* 58:1–302.

Bowman, R. I. 1963. Evolutionary patterns in Darwin's finches. *Occasional Papers of the California Academy of Sciences* 44:107–140.

Bowman, R. I. 1979. Adaptive morphology of song dialects in Darwin's Finches. *Journal für Ornithologie* 120:353–389.

Bowman, R. I. 1983. The evolution of song in Darwin's Finches. *In* R. I. Bowman, M. Berson, and A. E. Leviton (eds.), *Patterns of Evolution in Galápagos Organisms*, pp. 237–537. American Association for the Advancement of Science, Pacific Division, San Francisco, California.

Bowman, R. I., M. Berson, and A. E. Leviton (eds.). 1983. *Patterns of Evolution in Galápagos Organisms*. American Association for the Advancement of Science, Pacific Division, San Francisco, California.

Bowman, R. I., and S. I. Billeb. 1965. Blood-eating in a Galápagos Finch. *Living Bird* 4:29–44.

Bowman, R. I., and A. Carter. 1971. Egg-pecking behavior in Galápagos mockingbirds. *Living Bird* 9:243–270.

Boyce, M. S. 1979. Seasonality and patterns of natural selection for life histories. *American Naturalist* 114:569–583.

Bronowski, J. 1977. *A Sense of the Future*. M.I.T. Press, Cambridge, Mass.

Brown, J. H., O. J. Reichman, and D. W. Davidson. 1979. Granivory in desert ecosystems. *Annual Reviews of Ecology and Systematics* 10:201–227.

Brown, W. 1983. Evolution of animal mitochondrial DNA. *In* M. Nei and R. K. Koehn (eds.), *Evolution of Genes and Proteins*, pp. 62–88. Sinauer, Sunderland, Mass.

Brown, W. L., Jr., and E. O. Wilson. 1956. Character displacement. *Systematic Zoology* 5:49–64.

Burtt, E. H., Jr. 1979. Tips on wings and other things. *In* E. H. Burtt, Jr. (ed.), *The Behavioral Significance of Color*. STPM Press, Garland, New York.

Bush, G. L. 1975. Modes of animal speciation. *Annual Reviews of Ecology and Systematics* 6:339–364.

Cain, A. J. 1983. Ecology and ecogenetics of terrestrial molluscan populations. *In* W. D. Russell-Hunter (ed.), *The Mollusca*, Vol. 6, pp. 597–647. Academic Press, New York.

Calder, W. A., III. 1974. Consequences of body size for avian energetics. *In* R. A.

Paynter, Jr. (ed.), *Avian Energetics*. Publications of the Nuttall Ornithological Club 15:86–151.

Calder, W. A., III. 1984. *Size, Function, and Life History*. Harvard University Press, Cambridge, Mass.

Cane, M. A. 1983. Oceanographic events during El Niño. *Science* 222:1189–1195.

Cane, M. A., and S. E. Zebiak. 1985. A theory for El Niño and the southern oscillation. *Science* 228:1085–1087.

Carlquist, S. J. 1970. *Hawaii: A Natural History*. Doubleday, Natural History Press, Garden City, New York.

Carson, H. L. 1968. The population flush and its genetic consequences. *In* R. C. Lewontin (ed.), *Population Biology and Evolution*, pp. 123–137. Syracuse Unversity Press, Syracuse, New York.

Carson, H. L. 1978. Chromosomes and species formation. *Evolution* 32:925–927.

Carson, H. L. 1982a. Evolution of *Drosophila* on the newer Hawaiian volcanoes. *Heredity* 48:3–25.

Carson, H. L. 1982b. Speciation as a major reorganization of polygenic balances. *In* C. Barigozzi (ed.), *Mechanisms of Speciation*, pp. 411–433. A. R. Liss Inc., New York.

Carson, H. L. 1983. Chromosomal sequences and interisland colonization in Hawaiian *Drosophila*. *Genetics* 103:465–482.

Carson, H. L., and A. R. Templeton. 1984. Genetic revolutions in relation to speciation phenomena: the founding of new populations. *Annual Reviews of Ecology and Systematics* 15:97–131.

Case, T. J. 1979. Character displacement and coevolution in some *Cnemidophorus* lizards. *Fortschritte Zoologie* 25:235–282.

Case, T., and R. Sidell. 1983. Pattern and chance in the structure of model and natural communities. *Evolution* 37:832–849.

Casey, T.L.C., and J. D. Jacobi. 1974. A new genus and species of bird from the island of Maui, Hawaii (Passeriformes: Drepanididae). *Bernice P. Bishop Museum Occasional Papers* 24: 215–226.

Charlesworth. B., R. Lande, and M. Slatkin. 1982. A neo–Darwinian commentary on macroevolution. *Evolution* 36:474–498.

Cheverud, J. M. 1984. Quantitative genetics and developmental constraints on evolution by selection. *Journal of Theoretical Biology* 110:155–171.

Chitty, D. 1967. What regulates bird populations? *Ecology* 48:698–701.

Clark, D. A. 1984. Native land mammals. *In* R. Perry (ed.), *Galápagos*, pp. 225–231. Pergamon Press, Oxford.

Clarke, B. C., and J. J. Murray. 1969. Ecological genetics and speciation in land snails of the genus *Partula*. *Biological Journal of the Linnean Society* 1:31–42.

Cock, A. G. 1963. Genetical studies on growth and form in the fowl. I. Phenotypic variation in the relative growth pattern of shank length and body weight. *Genetic Research, Cambridge* 4:167–192.

Colinvaux, P. A. 1972. Climate and the Galápagos Islands. *Nature* 240:17–20.

Colinvaux, P. A. 1984. The Galápagos climate: present and past. *In* R. Perry (ed.), *Galápagos*, pp. 55–69. Pergamon Press, Oxford.

Colinvaux, P. A., and E. K. Schofield. 1976a. Historical ecology in the Galápagos

Islands. I. A Holocene pollen record from El Junco lake, Isla San Cristóbal. *Journal of Ecology* 64:989–1012.

Colinvaux, P. A. and E. K. Schofield. 1976b. Historical ecology in the Galápagos islands. II. A Holocene spore record from El Junco lake, Isla San Cristóbal. *Journal of Ecology* 64:1013–1028.

Colnett, J. 1798. *A Voyage to the South Atlantic and round Cape Horn into the Pacific Ocean.* Bennett, n.p.

Colwell, R. K. 1974. Predictability, constancy, and contingency of periodic phenomena. *Ecology* 55:1148–1153.

Colwell, R. K., and D. Winkler. 1984. A null model for null models in evolutionary ecology. *In* D. R. Strong, Jr., D. Simberloff, L. G. Abele, and A. B. Thistle (eds.), *Ecological Communities: Conceptual Issues and the Evidence*, pp. 344–359. Princeton University Press, Princeton, N.J.

Connor, E. F., and D. S. Simberloff. 1978. Species number and compositional similarity of the Galápagos flora and avifauna. *Ecological Monographs* 48:219–248.

Connor, E. F., and D. Simberloff. 1983. Interspecific competition and species co-occurrence patterns on islands: null models and the evaluation of evidence. *Oikos* 41:455–465.

Cox, A. 1983. Ages of the Galápagos Islands. *In* R. I. Bowman, M. Berson, and A. E. Leviton (eds.), *Patterns of Evolution in Galápagos Organisms*, pp. 11–23. American Association for the Advancement of Science, Pacific Division, San Francisco, California.

Curio, E. 1965. Zur geographischen Variation des Feinderkennens einiger Darwinfinken (Geospizidae). *Zoologische Anzeiger*, suppl. 28:466–492.

Curio. E. 1969. Funktionsweise und Stammesgeschichte des Flugfeinderkennens einiger Darwinfinken (Geospizinae). *Zeitschrift für Tierpsychologie* 26:394–487.

Curio, E., and P. Kramer. 1964. Vom Mangrovefinken (*Cactospiza heliobates*) *Zeitschrift für Tierpsychologie* 21:223–234.

Curio, E., and P. Kramer. 1965a. *Geospiza conirostris* auf Abingdon und Wenman entdeckt. *Journal für Ornithologie* 106: 355–357.

Curio, E., and P. Kramer. 1965b. On plumage variation in male Darwin's Finches. *Bird-Banding* 36:27–44.

Cutler, B. D. 1970. Anatomical studies of the syrinx of Darwin's finches. M.A. thesis, San Franciso State University, San Francisco.

Dalrymple, G. B., and A. Cox. 1968. Paleomagnetism, potassium-argon ages and petrology of some volcanic rocks. *Nature* 217:323–326.

Darwin, C. R. 1839. *Journal of Researches into the Geology and Natural History of the various countries visited during the Voyage of H.M.S. 'Beagle', under the command of Captain FitzRoy, R. N. from 1832 to 1836.* Henry Colborn, London.

Darwin, C. R. (ed.). 1841. *The Zoology of the Voyage of H.M.S. Beagle, under the command of Captain FitzRoy, R.N., during the Years 1832–1836. Part III: Birds.* Smith Elder, London.

Darwin, C. R. 1842. *Journal of Researches into the Geology and Natural History of the various countries visited during the Voyage of H.M.S.* 'Beagle', *under the command of Captain FitzRoy, R. N. from 1832 to 1836.* Henry Colborn, London (reissue).

Darwin, C. R. 1845. *Journal of Researches into the Geology and Natural History of the various countries visited during the voyage of H.M.S. Beagle, under the command of Captain FitzRoy, R. N.* 2nd ed. John Murray, London.

Darwin, C. R. 1859. *On the Origin of Species by Means of Natural Selection.* John Murray, London.

Davidson, D. W., D. A. Samson, and R. S. Inouye. 1985. Granivory in the Chihuahuan desert: interactions within and between trophic levels. *Ecology* 66:486–502.

de Vries, Tj. 1975. The breeding biology of the Galápagos Hawk, *Buteo galapagoensis. Le Gerfaut* 65:29–58.

de Vries, Tj. 1976. Prey selection and hunting methods of the Galápagos Hawk, *Buteo galapagoensis. Le Gerfaut* 66:3–42.

Dhondt, A. A. 1982. Heritability of blue tit tarsus length from normal and cross-fostered broods. *Evolution* 36:418–419.

Dickey, D. R., and A. J. van Rossem. 1938. *The Birds of El Salvador.* Field Museum of Natural History, Zoological Series 23: 1–609.

Dickinson, H., and J. Antonovics. 1973. Theoretical considerations of sympatric divergence. *American Naturalist* 107:256–274.

Dobzhansky, T. 1937. *Genetics and the Origin of Species.* Columbia University Press, New York.

Dobzhansky, T. 1940. Speciation as a stage in evolutionary divergence. *American Naturalist* 74:312–321.

Dominey, W. J. 1984. Effects of sexual selection and life history on speciation: species flocks in African cichlids and Hawaiian *Drosophila. In* A. A. Echelle and I. Kornfield (eds.), *Evolution of Fish Species Flocks*, pp. 231–249. University of Maine at Orono Press.

Downhower, J. F. 1976. Darwin's Finches and the evolution of sexual dimorphism in body size. *Nature* 263:558–563.

Downhower, J. F. 1978. Observations on the nesting of the small ground finch *Geospiza fuliginosa* and the large cactus ground finch *G. conirostris* on Española, Galápagos. *Ibis* 120:340–346.

Downhower, J. F., and C. H. Racine. 1976. Darwin's Finches and *Croton scouleri*: an analysis of the consequences of seed predation. *Biotropica* 8:66–70.

Dunn, E. K. 1976. Laying dates of four species of tits in Wytham Wood, Oxfordshire. *British Birds* 69:45–50.

Eibl-Eibesfeldt, I. 1984. The large iguanas of the Galápagos islands. *In* R. Perry (ed.), *Galápagos*, pp. 157–173. Pergamon Press, Oxford.

Eliasson, U. 1984. Native climax forests. *In* R. Perry (ed.), *Galápagos*, pp. 101–114. Pergamon Press, Oxford.

Endler, J. A. 1977. *Geographic Variation, Speciation, and Clines.* Princeton University Press, Princeton, N.J.

Endler, J. A. 1984. Progressive background matching in moths and a quantitative measure of crypsis. *Biological Journal of the Linnean Society* 22:187–231.

Endler, J. A. 1986. *Natural Selection in the Wild*. Princeton University Press, Princeton, N.J.

Emlen, S. T. 1971. The role of song in individual recognition in the Indigo bunting. *Zeitschrift für Tierpsychologie* 28:241–246.

Falconer, D. S. 1981. *Introduction to Quantitative Genetics*. 2nd ed. Longman, London.

Felsenstein, J. 1981. Skepticism towards Santa Rosalia, or why are there so few kinds of animals? *Evolution* 35:124–138.

Fisher, R. A. 1937. The relation between variability and abundance shown by the measurements of the eggs of British nesting birds. *Proceedings of the Royal Society of London B* 122:1–26.

Ford, H. A., A. W. Ewing, and D. T. Parkin. 1974. Blood proteins in Darwin's Finches. *Comparative Biochemistry and Physiology* 47B:369–375.

Ford, H. A., D. T. Parkin, and A. W. Ewing. 1973. Divergence and evolution in Darwin's Finches. *Biological Journal of the Linnean Society* 5:289–295.

Fryer, G., and T. D. Iles. 1972. *The Cichlid Fishes of the Great Lakes of Africa*. Oliver and Boyd, Edinburgh.

Futuyma, D. J. 1979. *Evolutionary Biology*. Sinauer, Sunderland, Mass.

Futuyma, D. J., and G. C. Mayer. 1980. Non-allopatric speciation in animals. *Systematic Zoology* 29:254–271.

Futuyma, D. J., and M. Slatkin (eds.). 1983. *Coevolution*. Sinauer, Sunderland, Mass.

Gause, G. F. 1934. *The Struggle for Existence*. Williams and Wilkins, Baltimore.

Gause, G. F. 1939. Discussion *in* Analytical population studies in relation to general ecology, by T. Park. *American Midland Naturalist* 21:235–255.

Gibbs, H. L., P. R. Grant, and J. Weiland. 1984. Breeding of Darwin's Finches at an unusually early age in an El Niño year. *Auk* 101:872–874.

Gifford, E. W. 1919. Field notes on the land birds of the Galápagos Islands and of Cocos Island, Costa Rica. *Proceedings of the California Academy of Sciences*, ser. 4, 2:189–258.

Gilpin, M. E., and J. M. Diamond. 1984. Are species co-occurrences on islands non-random, and are null hypotheses useful in community ecology? *In* D. R. Strong, Jr., D. Simberloff, L. G. Abele, and A. B. Thistle (eds.), *Ecological Communities: Conceptual Issues and the Evidence*, pp. 297–315. Princeton University Press, Princeton, N.J.

Gould, J. 1837. Description of new species of finches collected by Darwin in the Galápagos. *Proceedings of the Zoological Society of London* 5:4–7.

Gould, S. J. 1966. Allometry and size in ontogeny and phylogeny. *Biological Reviews* 41:587–640.

Gould, S. J. 1977. *Ontogeny and Phylogeny*. Harvard University Press, Cambridge, Mass.

Gould, S. J. 1980. Is a new and general theory of evolution emerging? *Paleobiology* 6:119–130.

Gould, S. J. 1982. Darwinism and the expansion of evolutionary theory. *Science* 216:380–387.

Gould, S. J. 1984. Covariance sets and ordered geographic variation in *Cerion* from

Aruba, Bonaire and Curaçao: a way of studying nonadaptation. *Systematic Zoology* 33:217–237.

Gould, S. J., and R. C. Lewontin. 1979. The spandrels of San Marco and the Panglossian paradigm: a critique of the adaptationist program. *Proceedings of the Royal Society of London B* 205:581–598.

Grant, B. R. 1984. The significance of song variation in a population of Darwin's Finches. *Behaviour* 89:90–116.

Grant, B. R. 1985. Selection on bill characters in a population of Darwin's Finches: *Geospiza conirostris* on Isla Genovesa, Galápagos. *Evolution* 39:523–532.

Grant, B. R., and P. R. Grant. 1979. Darwin's Finches: population variation and sympatric speciation. *Proceedings of the National Academy of Sciences USA* 76:2359–2363.

Grant, B. R., and P. R. Grant. 1981. Exploitation of *Opuntia* cactus by birds on the Galápagos. *Oecologia* 49:179–187.

Grant, B. R., and P. R. Grant. 1982. Niche shifts and competition in Darwin's Finches: *Geospiza conirostris* and congeners. *Evolution* 36:637–657.

Grant, B. R., and P. R. Grant. 1983. Fission and fusion in a population of Darwin's Finches: an example of the value of studying individuals in ecology. *Oikos* 41:530–547.

Grant, P. R. 1965. The adaptive significance of some size trends in island birds. *Evolution* 19:355–367.

Grant, P. R. 1966. Ecological compatibility of bird species on islands. *American Naturalist* 100:451–462.

Grant, P. R. 1967. Bill length variability in birds of the Tres Marías Islands, Mexico. *Canadian Journal of Zoology* 45:805–815.

Grant, P. R. 1968. Bill size, body size, and the ecological adaptations of bird species to competitive situations on islands. *Systematic Zoology* 17:319–333.

Grant, P. R. 1969. Colonization of islands by ecologically dissimilar species of birds. *Canadian Journal of Zoology* 47:41–43.

Grant, P. R. 1972a. Interspecific competition among rodents. *Annual Reviews of Ecology and Systematics* 3:79–106.

Grant, P. R. 1972b. Convergent and divergent character displacement. *Biological Journal of the Linnean Society* 4:39–68.

Grant, P. R. 1972c. Bill dimensions of the three species of *Zosterops* on Norfolk Island. *Systematic Zoology* 21:289–291.

Grant, P. R. 1975a. The classical case of character displacement. *Evolutionary Biology* 8:237–337.

Grant, P. R. 1975b. Four Galápagos Islands. *Geographical Journal* 141:76–87.

Grant, P. R. 1977. Review of D. Lack, 1976, *Island Biology, Illustrated by the Land Birds of Jamaica*. *Bird-Banding* 48:296–300.

Grant, P. R. 1979a. Ecological and morphological variation of Canary Island blue tits, *Parus caeruleus* (Aves: Paridae). *Biological Journal of the Linnean Society* 11:103–129.

Grant, P. R. 1979b. Evolution of the chaffinch, *Fringilla coelebs*, on the Atlantic Islands. *Biological Journal of the Linnean Society* 11:301–332.

Grant, P. R. 1980. Colonization of Atlantic islands by chaffinches (*Fringilla* spp.). *Bonner zoologische Beiträge* 31:311–317.

Grant, P. R. 1981a. Patterns of growth in Darwin's Finches. *Proceedings of the Royal Society of London B* 212:403–432.

Grant, P. R. 1981b. The feeding of Darwin's Finches on *Tribulus cistoides* (L.) seeds. *Animal Behaviour* 29:785–793.

Grant, P. R. 1981c. Speciation and the adaptive radiation of Darwin's Finches. *American Scientist* 69:653–663.

Grant, P. R. 1982. Variation in the size and shape of Darwin's Finch eggs. *Auk* 99:15–23.

Grant, P. R. 1983a. Inheritance of size and shape in a population of Darwin's Finches, *Geospiza conirostris*. *Proceedings of the Royal Society of London B* 220:219–236.

Grant, P. R. 1983b. The relative size of Darwin's Finch eggs. *Auk* 100:228–230.

Grant, P. R. 1983c. The role of interspecific competition in the adaptive radiation of Darwin's Finches. *In* R. I. Bowman, M. Berson, and A. E. Leviton (eds.), *Patterns of Evolution in Galápagos Organisms*, pp. 187–199. American Association for the Advancement of Science, Pacific Division, San Francisco, California.

Grant, P. R. 1984a. Extraordinary rainfall during the El Niño event of 1982–83. *Noticias de Galápagos* No. 39:10–12.

Grant, P. R. 1984b. The endemic land birds. *In* R. Perry (ed.), *Galápagos*, pp. 175–189. Pergamon Press, Oxford.

Grant, P. R. 1984c. Recent research on the evolution of land birds on the Galápagos. *Biological Journal of the Linnean Society* 21:113–136.

Grant, P. R. 1985a. Climatic fluctuations on the Galápagos Islands and their influence on Darwin's Finches. *In* P. A. Buckley, M. S. Foster, E. S. Morton, R. S. Ridgely, and F. G. Buckley (eds.), *Neotropical Ornithology*, pp. 471–483. Ornithological Monographs of the American Ornithologists' Union, No. 36.

Grant, P. R. 1985b. Interspecific competition in fluctuating environments. *In* J. M. Diamond and T. J. Case (eds.), *Community Ecology*, pp. 173–191. Harper and Row, New York.

Grant, P. R., and I. Abbott. 1980. Interspecific competition, null hypotheses and island biogeography. *Evolution* 34:332–341.

Grant, P. R., I. Abbott, D. Schluter, R. L. Curry, and L. K. Abbott. 1985. Variation in the size and shape of Darwin's Finches. *Biological Journal of the Linnean Society* 25:1–39.

Grant, P. R., and P. T. Boag. 1980. Rainfall on the Galápagos and the demography of Darwin's Finches. *Auk* 97:227–244.

Grant, P. R., and B. R. Grant. 1980a. The breeding and feeding characteristics of Darwin's Finches on Isla Genovesa, Galápagos. *Ecological Monographs* 50:381–410.

Grant, P. R., and B. R. Grant. 1980b. Annual variation in finch numbers, foraging and food supply on Isla Daphne Major, Galápagos. *Oecologia* 46:55–62.

Grant, P. R., and B. R. Grant. 1985. Responses of Darwin's Finches to unusual rainfall. *In* G. Robinson and E. del Pino (eds.), *El Niño in the Galápagos Islands: the 1982–1983 event*, pp. 417–447. ISALPRO, Quito.

Grant, P. R., B. R. Grant, J.N.M. Smith, I. Abbott, and L. K. Abbott. 1976. Darwin's Finches: population variation and natural selection. *Proceedings of the National Academy of Sciences USA* 73:257–261.

Grant, P. R., and K. T. Grant. 1979. Breeding and feeding ecology of the Galápagos Dove. *Condor* 81:397–403.

Grant, P. R., and N. Grant. 1979. Breeding and feeding of Galápagos mockingbirds, *Nesomimus parvulus*. *Auk* 96:723–736.

Grant, P. R., and N. Grant. 1983. The origin of a species. *Natural History* 76:76–80.

Grant, P. R., and T. D. Price. 1981. Population variation in continuously varying traits as an ecological genetics problem. *American Zoologist* 21:795–811.

Grant, P. R., T. D. Price, and H. Snell. 1980. The exploration of Isla Daphne Minor. *Noticias de Galápagos* No. 31:22–27.

Grant, P. R., and D. Schluter. 1984. Interspecific competition inferred from patterns of guild structure. *In* D. R. Strong, Jr., D. Simberloff, L. G. Abele, and A. B. Thistle (eds.), *Ecological Communities: Conceptual Issues and the Evidence*, pp. 201–233. Princeton University Press, Princeton, N.J.

Grant, P. R., D. Schluter, and P. T. Boag. 1979. A bill color polymorphism in young Darwin's Finches. *Auk* 96:800–802.

Grant, P. R., J.N.M. Smith, B. R. Grant, I. Abbott, and L. K. Abbott. 1975. Finch numbers, owl predation and plant dispersal on Isla Daphne Major, Galápagos. *Oecologia* 19:239–257.

Greenwood, P. H. 1981a. *The Haplochromine Fishes of the East African Lakes*. Cornell University Press, Ithaca, New York.

Greenwood, P. H. 1981b. Species flocks and explosive evolution. *In* P. L. Forey (ed.), *Chance, Change, and Challenge: The Evolving Biosphere*, pp. 61–74. Cambridge University Press, Cambridge.

Greenwood, P. H. 1984. African cichlids and evolutionary theories. *In* A. A. Echelle and I. Kornfield (eds.), *Evolution of Fish Species Flocks*, pp. 141–154. University of Maine at Orono Press.

Gulick, A. 1932. Biological peculiarities of oceanic islands. *Quarterly Review of Biology* 7:405–427.

Haldane, J.B.S. 1937. *Adventures of a Biologist*. Harper and Bros., New York.

Haldane, J.B.S. 1954. The measurement of natural selection. *Proceedings of the 9th International Congress of Genetics* 1:480–487.

Hall, M. L., P. Ramon, and H. Yepes. 1983. Origin of Española Island and the age of terrestrial life on the Galápagos Islands. *Science* 221:545–547.

Halpern, D., S. P. Hayes, A. Leetmaa, D. V. Hansen, and S.G.H. Philander. 1983. Oceanographic observations of the 1982 warming of the tropical eastern Pacific. *Science* 221:1173–1175.

Hamann, O. 1981. Plant communities of the Galápagos Islands. *Dansk Botanisk Arkiv* 34:1–163.

Hamann, O. 1984. Changes and threats to the vegetation. *In* R. Perry (ed.), *Galápagos*, pp. 115–131. Pergamon Press, Oxford.

Hamilton, T. H., and I. Rubinoff. 1963. Isolation, endemism, and multiplication of species in the Darwin Finches. *Evolution* 17:388–403.

Hamilton, T. H., and I. Rubinoff. 1964. On models predicting abundance of species and endemics for the Darwin Finches in the Galápagos archipelago. *Evolution* 18:339–342.

Hamilton, T. H., and I. Rubinoff. 1967. On predicting insular variation in endemism and sympatry for the Darwin Finches in the Galápagos archipelago. *American Naturalist* 101:161–171.

Hamilton, T. H., I. Rubinoff, R. H. Barth, Jr., and G. L. Bush. 1963. Species abundance: natural regulation of insular variation. *Science* 142:1575–1577.

Hamilton, W. D., and M. Zuk. 1982. Heritable true fitness and bright plumage in birds: a role for parasites? *Science* 218:384–387.

Hamilton, W. J., III. 1973. *Life's Color Code*. McGraw-Hill, New York.

Hamilton, W. J., III, and F. Heppner. 1967. Radiant solar energy and the function of black homoeotherm pigmentation: an hypothesis. *Science* 155:196–197.

Harris, M. P. 1972. *Coereba flaveola* and the Geospizinae. *Bulletin of the British Ornithological Club* 92:164–168.

Harris, M. P. 1973. The Galápagos avifauna. *Condor* 75:265–278.

Harris, M. P. 1974. *A Field Guide to Birds of Galápagos*. Collins, London.

Harvey, P. H., and G. M. Mace. 1982. Comparisons between taxa and adaptive trends: problems of methodology. *In* King's College Sociobiology Group (eds.), *Current Problems in Sociobiology*, pp. 343–361. Cambridge University Press, Cambridge.

Hendrickson, J. A., Jr. 1981. Community wide character displacement reexamined. *Evolution* 35:794–809.

Hey, R. 1977. Tectonic evolution of the Cocos-Nazca spreading center. *Geological Society of America Bulletin* 88:1404–1420.

Hickman, C. S., and J. H. Lipps. 1985. Geologic youth of Galápagos Islands confirmed by marine stratigraphy and paleontology. *Science* 227:1578–1580.

Hooker, J. D. 1847. On the vegetation of the Galápagos Archipelago, as compared with that of some other tropical islands and of the continent of America. *Transactions of the Linnean Society* 20:235–262.

Houvenaghel, G. T. 1984. Oceanographic setting of the Galápagos islands. *In* R. Perry (ed.), *Galápagos*, pp. 43–54. Pergamon Press, Oxford.

Howarth, F. G. 1980. Cavernicoles in lava tubes on the island of Hawaii. *Science* 175:325–326.

Hutchinson, G. E. 1965. *The Ecological Theater and the Evolutionary Play*. Yale University Press, New Haven, Conn.

Hutt, F. B. 1964. *Genetics of the Fowl*. 2nd edn. McGraw-Hill, New York.

Huxley, J. S. (ed.). *The New Systematics*. Clarendon Press, Oxford.

Huxley, J. S. 1942. *Evolution: the Modern Synthesis*. Allen and Unwin, London.

Huxley, J. S. 1955. Morphism in birds. *Proceedings of the 11th International Ornithological Congress (Basel, 1954)*, pp. 309–328.

Itow, S. 1975. A study of vegetation in Isla Santa Cruz, Galápagos Islands. *Noticias de Galápagos* No. 17:10–13.

Janzen, D. H. 1973. Sweep samples of tropical foliage insects: effects of seasons, vegetation types, elevation, time of day, and insularity. *Ecology* 54:687–708.

Jo. N. 1983. Karyotypic analysis of Darwin's finches. *In* R. I. Bowman, M. Berson, and A. E. Leviton (eds.), *Patterns of Evolution in Galápagos Organisms*, pp. 201–217. American Association for the Advancement of Science, Pacific Division, San Francisco, California.

Johnson, M. P., and P. H. Raven, 1973. Species number and endemism: the Galápagos archipelago revisited. *Science* 179:893–895.

Jones, P. J. 1973. Some aspects of the feeding ecology of the Great Tit *Parus major* L. D. Phil. thesis, Oxford University.

Kaneshiro, K. Y. 1976. Ethological isolation and phylogeny in the *planitibia* subgroup of Hawaiian *Drosophila*. *Evolution* 30:740–745.

Kluge, A. G., and J. S. Farris. 1969. Quantitative phyletics and the evolution of anurans. *Systematic Zoology* 18:1–32.

Lack, D. 1940a. Evolution of the Galápagos finches. *Nature* 146:324–327.

Lack, D. 1940b. The Galápagos finches. *Bulletin of the British Ornithological Club* 60:46–50.

Lack, D. 1945. The Galápagos finches (*Geospizinae*): a study in variation. *Occasional Papers of the California Academy of Sciences* 21:1–159.

Lack, D. 1947. *Darwin's Finches*. Cambridge University Press, Cambridge.

Lack, D. 1950. Breeding seasons in the Galápagos. *Ibis* 92:268–278.

Lack, D. 1968. *Ecological Adaptations for Breeding in Birds*. Methuen, London.

Lack, D. 1969. Subspecies and sympatry in Darwin's Finches. *Evolution* 23:252–263.

Lack, D. 1971. *Ecological Isolation in Birds*. Methuen, London.

Lack. D. 1973. My life as an amateur ornithologist. *Ibis* 115:421–431.

Lande, R. 1976a. The maintenance of genetic variability by mutation in a polygenic character with linked loci. *Genetical Research* 26:221–235.

Lande, R. 1976b. Natural selection and random genetic drift in phenotypic evolution. *Evolution* 30:314–334.

Lande, R. 1979. Quantitative genetic analysis of multivariate evolution, applied to brain:body size allometry. *Evolution* 33:402–416.

Lande, R. 1980a. Sexual dimorphism, sexual selection, and adaptation in polygenic characters. *Evolution* 34:292–305.

Lande, R. 1980b. Genetic variation and phenotypic evolution during allopatric speciation. *American Naturalist* 116:463–479.

Lande, R. 1981. Models of speciation by sexual selection on polygenic traits. *Proceedings of the National Academy of Sciences USA* 78:3721–3725.

Lande, R. 1982. Rapid origin of sexual isolation and character divergence in a cline. *Evolution* 36:213–223.

Lande, R., and S. J. Arnold. 1983. The measurement of selection on correlated characters. *Evolution* 37:1210–1226.

Lasiewski, R. C., and W. R. Dawson. 1967. A reexamination of the relation between standard metabolic rate and body weights in birds. *Condor* 69:13–23.

Lederer, R. J. 1975. Bill size, food size, and jaw forces of insectivorous birds. *Auk* 92:385–387.

Lewin, R. 1983. Finches show competition in ecology. *Science* 219:1411–1412.

Lewontin, R. C. 1978. Adaptation. *Scientific American* 239:213–230.

Lowe, P. R. 1936. The finches of the Galápagos in relation to Darwin's conception of species. *Ibis*, ser. 13, 6:310–321.

MacArthur, R. H. 1972. *Geographical Ecology*. Harper and Row, New York. (Reprinted in paperback, Princeton University Press, 1984.)

MacArthur, R. H., and E. O. Wilson. 1967. *The Theory of Island Biogeography*. Princeton University Press, Princeton, N.J.

MacFarland, C., and W. Reeder. 1974. Cleaning symbiosis involving Galápagos tortoises and two species of Darwin's Finches. *Zeitschrift für Tierpsychologie* 34:464–483.

Manwell, R. D. 1955. The blood protozoa of seventeen species of sparrows and other Fringillidae. *Journal of Protozoology* 2:21–27.

Maynard Smith, J. 1966. Sympatric speciation. *American Naturalist* 100:637–650.

Maynard Smith, J. 1984. The population as a unit of selection. *In* B. Shorrocks (ed.), *Evolutionary Ecology*, pp. 195–202. Blackwell, Oxford.

Maynard Smith, J., R. Burian, S. Kauffman, P. Alberch, J. Campbell, B. Goodwin, R. Lande, D. Raup, and L. Wolpert. 1985. Developmental constraints and evolution. *Quarterly Review of Biology* 60: 265–287.

Mayr, E. 1942. *Systematics and the Origin of Species*. Columbia University Press, New York.

Mayr, E. 1954. Change of genetic environment and evolution. *In* J. S. Huxley, A. Hardy, and E. B. Ford (eds.), *Evolution as a Process*, pp. 157–180. Allen and Unwin, London.

Mayr, E. 1961. Cause and effect in biology. *Science* 134:1501–1506.

Mayr, E. 1963. *Animal Species and Evolution*. Belknap Press, Cambridge, Mass.

Mayr, E. 1982. Processes of speciation in animals. *In* C. Barigozzi (ed.), *Mechanisms of Speciation*, pp. 1–19. A. R. Liss Inc., New York.

Mayr, E. 1983. How to carry out the adaptationist program? *American Naturalist* 121:324–334.

Mayr, E. 1984. Evolution of fish species flocks: a commentary. *In* A. A. Echelle and I. Kornfield (eds.), *Evolution of Fish Species Flocks*, pp. 3–11. University of Maine at Orono Press.

Marchant, S. 1958. The birds of the Santa Elena peninsula, S. W. Ecuador. *Ibis* 100:349–387.

McBirney, A. R., and H. Williams. 1969. Geology and petrology of the Galápagos Islands. *Geological Society of America Memoirs* 118:1–197.

Milkman, R. (ed.). 1981. *Perspectives on Evolution*. Sinauer, Sunderland, Mass.

Millikan, G. C., and R. I. Bowman. 1967. Observations on Galápagos tool-using finches in captivity. *Living Bird* 6:23–41.

Millington, S. J., and P. R. Grant. 1983. Feeding ecology and territoriality of the Cactus Finch *Geospiza scandens* on Isla Daphne Major, Galápagos. *Oecologia* 58:76–83.

Millington, S. J., and P. R. Grant. 1984. The breeding ecology of the Cactus Finch *Geospiza scandens* on Isla Daphne Major, Galápagos. *Ardea* 72:177–188.

Millington, S. J., and T. D. Price, 1985. Song inheritance and mating patterns in Darwin's Finches. *Auk* 102:342–346.

Montgomery, S. L. 1982. Biogeography of the moth genus *Eupethicia* in Oceania and the evolution of ambush predation in Hawaiian caterpillars (Lepidoptera: Geometridae). *Entomologia Generalis* 8:27–34.

Moore, J. A. 1957. An embryologist's view of the species concept. *In* E. Mayr (ed.), *The Species Problem*, pp. 325–338. A.A.A.S., Washington, D.C.

Morgan, W. J. 1971. Convection plumes in the lower mantle. *Nature* 230:42–43.

Mossiman, J. E., and F. C. James. 1979. New statistical methods for allometry with application to Florida red-winged blackbirds. *Evolution* 33:444–459.

Muller, H. J. 1940. Bearings of the *Drosophila* work on systematics. *In* J. S. Huxley (ed.), *The New Systematics*, pp. 185–268. Clarendon Press, Oxford.

Murray, J. J., and B. C. Clarke. 1980. The genus *Partula* on Moorea: speciation in progress. *Proceedings of the Royal Society of London B*, 211:83–117.

Newton, I. 1967. The adaptive radiation and feeding ecology of some British finches. *Ibis* 109:33–98.

O'Donald, P. 1970. Change of fitness by selection for a quantitative character. *Theoretical Population Biology* 1:219–232.

Olson, S. L., and H. F. James. 1982. Fossil birds from the Hawaiian Islands: evidence for wholesale extinction by man before western contact. *Science* 217:633–635.

Orr, R. T. 1945. A study of captive Galápagos finches of the genus *Geospiza*. Condor 47:177–201.

Paterson, H. E. 1980. A comment on "mate recognition systems." *Evolution* 34:330–331.

Patton, J. L., and S. W. Sherwood. 1983. Chromosome evolution and speciation in rodents. *Annual Reviews of Ecology and Systematics* 14:139–158.

Paulay, G. 1985. Adaptive radiation on an isolated oceanic island: The Cryptorhynchinae (Curculionidae) of Rapa revisited. *Biological Journal of the Linnean Society* 26:95–187.

Payne, R. B. 1982. Ecological consequences of song matching: breeding success and intraspecific song mimicry in Indigo Buntings. *Ecology* 63:401–411.

Paynter, R. A., Jr. 1970. Subfamily Emberizinae. *In* R. A. Paynter, Jr. (ed.), *Checklist of Birds of the World*, Vol. 13, pp. 3–214. Museum of Comparative Zoology, Harvard University, Cambridge, Mass.

Perry, R. (ed.) 1984. *Galápagos*, Pergamon Press, Oxford.

Philander, S.G.H. 1983. El Niño Southern Oscillation phenomena. *Nature* 302:295–301.

Pimm, S. L. 1979. Sympatric speciation: a simulation model. *Biological Journal of the Linnean Society* 11:131–140.

Polans, N. 1983. Enzyme polymorphisms in Galápagos finches. *In* R. I. Bowman, M. Berson, and A. E. Leviton (eds.), *Patterns of Evolution in Galápagos Organisms*, pp. 219–235. American Association for the Advancement of Science, Pacific Division, San Francisco, California.

Porter, D. M. 1976. Geography and dispersal of Galápagos Islands vascular plants. *Nature* 264:745–746.

Porter, D. M. 1983. Vascular plants of the Galápagos: origins and dispersal. *In* R. I. Bowman, M. Berson, and A. E. Leviton (eds.), *Patterns of Evolution in Galápagos Organisms*, pp. 33–96. American Association for the Advancement of Science, Pacific Division, San Francisco, California.

Porter, D. M. 1984. Endemism and evolution in terrestrial plants. *In* R. Perry (ed.), *Galápagos*, pp. 85–99. Pergamon Press, Oxford.

Power, D. M. 1975. Similarity among avifaunas of the Galápagos islands. *Ecology* 56:616–626.

Price, T. D. 1984a. Sexual selection and body size, plumage and territory variables in a population of Darwin's Finches. *Evolution* 38:327–341.

Price, T. D. 1984b. The evolution of sexual size dimorphism in a population of Darwin's Finches. *American Naturalist* 123:500–518.

Price, T. D. 1985. Reproductive responses to varying food supply in a population of Darwin's Finches: clutch size, growth rates and hatching synchrony. *Oecologia* 66:411–416.

Price, T. D., and P. R. Grant. 1984. Life history traits and natural selection for small body size in a population of Darwin's Finches. *Evolution* 38:483–494.

Price, T. D., and P. R. Grant. 1985. The evolution of ontogeny in Darwin's Finches: a quantitative genetic approach. *American Naturalist* 125:169–188.

Price, T. D., P. R. Grant, H. L. Gibbs, and P. T. Boag. 1984. Recurrent patterns of natural selection in a population of Darwin's finches. *Nature* 309:787–789.

Price, T. D., P. R. Grant, and P. T. Boag. 1984. Genetic changes in the morphological differentiation of Darwin's Ground Finches. *In* K. Wöhrmann and V. Loeschcke (eds.), *Population Biology and Evolution*, pp. 49–66. Springer, Berlin.

Price, T., S. Millington, and P. Grant. 1983. Helping at the nest in Darwin's Finches as misdirected parental care. *Auk* 100:192–194.

Proctor-Gray, E., and R. T. Holmes, 1981. Adaptive significance of delayed attainment of plumage in male American Redstarts: a test of two hypotheses. *Evolution* 35:742–751.

Pulliam, H. R. 1980. Do Chipping Sparrows forage optimally? *Ardea* 68:75–82.

Quinn, W. H. 1971. Late quaternary meteorological and oceanographic developments in the equatorial Pacific. *Nature* 229:330–331.

Racine, C. H., and J. F. Downhower. 1974. Vegetative and reproductive strategies of *Opuntia* (Cactaceae) in the Galápagos Islands. *Biotropica* 6:175–186.

Raikow, R. J. 1977. The origin and evolution of the Hawaiian Honeycreepers (Drepanididae). *Living Bird* 15:95–117.

Rasmusson, E. M. 1985. El Niño and variations in climate. *American Scientist* 73:168–177.

Ratcliffe, L. M. 1981. Species recognition in Darwin's Ground Finches (*Geospiza* Gould). Ph.D. thesis, McGill University, Montreal.

Ratcliffe, L. M., and P. T. Boag. 1983. Foreword to *Darwin's Finches*, by David Lack, 2nd ed. Cambridge University Press, Cambridge.

Ratcliffe, L. M., and P. R. Grant. 1983a. Species recognition in Darwin's Finches

(*Geospiza*, Gould). I. Discrimination by morphological cues. *Animal Behaviour* 31:1139–1153.

Ratcliffe, L. M., and P. R. Grant. 1983b. Species recognition in Darwin's Finches (*Geospiza*, Gould). II. Geographic variation in mate preference. *Animal Behaviour* 31:1154–1165.

Ratcliffe, L. M., and P. R. Grant. 1985. Species recognition in Darwin's Finches (*Geospiza*, Gould). III. Male responses to playback of different song types, dialects and heterospecific songs. *Animal Behaviour* 33:290–307.

Reeder, W. G., and S. E. Riechert. 1975. Vegetation change along an altitudinal gradient, Santa Cruz Island, Galápagos. *Biotropica* 7:162–175.

Rensch, B. 1933. Zoologische Systematik und Artbildungsproblem. *Zoologische Anzeiger*, suppl. 6:19–83.

Richards, O. W. 1948. Species formation on islands. Review of D. Lack (1947), *Darwin's Finches*, Cambridge University Press. *Journal of Animal Ecology* 17:83–84.

Ricklefs, R. E. 1976. Growth rates of birds in the humid New World tropics. *Ibis* 118:179–207.

Ridgway, R. 1890. Scientific results of explorations by the U.S. Fish Commission Steamer *Albatross* . . . No. 1. Birds collected on the Galápagos Islands in 1888. *Proceedings of the U.S. National Museum* 12:101–128.

Rothschild, W., and E. Hartert. 1899. A review of the ornithology of the Galápagos islands. With notes on the Webster-Harris expedition. *Novitates Zoologicae* 6:85–205.

Rothschild, W., and E. Hartert. 1902. Further notes on the fauna of the Galápagos Islands. *Novitates Zoologicae* 9:381–418.

Rothstein, S. I. 1973. The niche-variation model—is it valid? *American Naturalist* 107:598–620.

Roughgarden, J. 1972. Evolution of niche width. *American Naturalist* 106:683–718.

Rowher, S., S. D. Fretwell, and D. M. Niles. 1980. Delayed maturation in passerine plumages and the deceptive acquisition of resources. *American Naturalist* 115:400–437.

Salt, G. W. (ed.). 1984. *Ecology and Evolutionary Biology. A Round Table on Research*. University of Chicago Press, Chicago.

Salvin, O. 1876. On the avifauna of the Galápagos archipelago. *Transactions of the Zoological Society of London* 9:447–510.

Sarich, V. M. 1977. Rates, sample sizes and the neutrality hypothesis for electrophoresis in evolutionary studies. *Nature* 265:24–25.

Schluter, D. 1982a. Seed and patch selection by Galápagos ground finches: relation to foraging efficiency and food supply. *Ecology* 63:1106–1120.

Schluter, D. 1982b. Distributions of Galápagos ground finches along an altitudinal gradient: the importance of food supply. *Ecology* 63:1504–1517.

Schluter, D. 1984a. Feeding correlates of breeding and social organization in two Galápagos Finches. *Auk* 101:59–68.

Schluter, D. 1984b. Morphological and phylogenetic relations among the Darwin's Finches. *Evolution* 38:921–930.

Schluter, D. 1986. Character displacement between distantly-related taxa? Finches and bees in the Galápagos. *American Naturalist* 127:95–102.

Schluter, D., and P. R. Grant. 1982. The distribution of *Geospiza difficilis* in relation to *G. fuliginosa* in the Galápagos islands: tests of three hypotheses. *Evolution* 36:1213–1226.

Schluter, D., and P. R. Grant. 1984a. Ecological correlates of morphological evolution in a Darwin's Finch species. *Evolution* 38:856–869.

Schluter, D., and P. R. Grant. 1984b. Determinants of morphological patterns in communities of Darwin's Finches. *American Naturalist* 123:175–196.

Schluter, D., T. D. Price, and P. R. Grant. 1985. Ecological character displacement in Darwin's finches. *Science* 227:1056–1059.

Schoener, T. W. 1984. Size differences among sympatric, bird-eating hawks: a worldwide survey. *In* D. R. Strong, Jr., D. Simberloff, L. G. Abele, and A. B. Thistle (eds.), *Ecological Communities: Conceptual Issues and the Evidence*, pp. 254–281. Princeton University Press, Princeton, N. J.

Searcy, W. A., P. Marler, and S. S. Peters. 1982. Species song discrimination in adult female song and swamp sparrows. *Animal Behaviour* 29:997–1003.

Selander, R. K. 1962. Feeding adaptations in Darwin's Finches. *Evolution* 16:391–393.

Sherry, T. W. 1985. Adaptation to a novel environment: food, foraging, and morphology of the Cocos Island Flycatcher. *In* P. A. Buckley, M. S. Foster, E. S. Morton, R. S. Ridgely, and F. G. Buckley (eds.), *Neotropical Ornithology*, pp. 908–920. Ornithological Monographs of the American Ornithologists' Union, No. 36.

Sibley, C. G., and J. E. Ahlquist. 1982. The relationships of the Hawaiian honeycreepers (Drepaninae) as indicated by DNA–DNA hybridization. *Auk* 99:130–140.

Sibley, C. G., and J. Ahlquist. 1984. The phylogeny and classification of the passerine birds, based on comparisons of the genetic material, DNA. *Proceedings of the 18th International Ornithological Congress (Moscow, 1982)*.

Simberloff, D. 1983a. Biogeography: the unification and maturation of a science. *In* A. H. Brush and G. A. Clark, Jr. (eds.), *Perspectives in Ornithology*, pp. 411–455. Cambridge University Press, Cambridge.

Simberloff, D. 1983b. Sizes of coexisting species. *In* D. J. Futuyma and M. Slatkin (eds.), *Coevolution*, pp. 404–430. Sinauer, Sunderland, Mass.

Simberloff, D. 1984. Morphological and taxonomic similarity and combinations of coexisting birds in two archipelagos. *In* D. R. Strong, Jr., D. Simberloff, L. G. Abele, and A. B. Thistle (eds.), *Ecological Communities: Conceptual Issues and the Evidence*, pp. 234–253. Princeton University Press, Princeton, N. J.

Simberloff, D., and W. Boecklin. 1981. Santa Rosalia reconsidered: size ratios and competition. *Evolution* 35:1206–1228.

Simberloff, D., and E. F. Connor. 1981. Missing species combinations. *American Naturalist* 118:215–239.

Simkin, T. 1984. Geology of Galápagos islands. *In* R. Perry (ed.), *Galápagos*, pp. 15–41. Pergamon Press, Oxford.

Simpson, B. B. 1974. Glacial migrations of plants: island biogeographical evidence. *Science* 185:698–700.

Slatkin, M. 1984. Ecological causes of sexual dimorphism. *Evolution* 38:622–630.

Slud. P. 1967. The birds of Cocos Island (Costa Rica). *Bulletin of the American Museum of Natural History* 134:263–295.

Smith, G. R., and T. N. Todd. 1984. Evolution of species flocks of north temperate lacustrine fishes. *In* A. A. Echelle and I. Kornfield (eds.), *Evolution of Fish Species Flocks*, pp. 47–68. University of Maine at Orono Press.

Smith, J.N.M., and A. A. Dhondt. 1980. Experimental confirmation of heritable morphological variation in a natural population of song sparrows. *Evolution* 34:1155–1160.

Smith, J.N.M., P. R. Grant, B. R. Grant, I. Abbott, and L. K. Abbott. 1978. Seasonal variation in feeding habits of Darwin's Ground Finches. *Ecology* 59:1137–1150.

Smith, J.N.M., and H.P.A. Sweatman. 1976. Feeding habits and morphological variation in Cocos Finches. *Condor* 78:244–248.

Smith, J.N.M., and R. Zach. 1979. Heritability of some morphological characters in a song sparrow population. *Evolution* 33:460–467.

Snell, H. L., H. M. Snell, and C. R. Tracy. 1984. Variation among populations of Galápagos land iguanas (*Conolophus*): contrasts of phylogeny and ecology. *Biological Journal of the Linnean Society* 21:185–207.

Snodgrass, R. E. 1902. The relation of the food to the size and shape of the bill in the Galápagos genus *Geospiza*. *Auk* 19:367–381.

Snodgrass, R. E. 1903. Notes on the anatomy of *Geospiza, Cocornis* and *Certhidia*. *Auk* 20:402–417.

Snodgrass, R. E., and E. Heller. 1901. A new species of *Geospiza* collected by the Hopkins-Stanford expedition to the Galápagos Islands. *Condor* 3:96.

Snodgrass, R. E., and E. Heller. 1904. Papers from the Hopkins-Stanford Galápagos expedition, 1898–99. XVI. Birds. *Proceedings of the Washington Academy of Science* 5:231–372.

Snow, D. 1966. Moult and the breeding cycle in Darwin's Finches. *Journal für Ornithologie* 107:283–291.

Soulé, M., and B. R. Stewart. 1970. The "niche-variation" hypothesis: a test and alternatives. *American Naturalist* 104:85–97.

Steadman, D. 1981. Vertebrate fossils in lava tubes in the Galápagos Islands. *Proceedings of the 8th International Congress in Speleology*, pp. 549–550.

Steadman, D. 1982. The origin of Darwin's Finches (Fringillidae, Passeriformes). *Transactions of the San Diego Society of Natural History* 19:279–296.

Steadman, D. 1984. The status of *Geospiza magnirostris* on Isla Floreana, Galápagos. *Bulletin of the British Ornithological Club* 104:99–102.

Steadman, D. 1985. Holocene vertebrate fossils from Isla Floreana, Galápagos. *Smithsonian Contributions to Zoology*, No. 413.

Stewart, A. 1911. Expedition of the California Academy of Sciences to the Galápagos Islands, 1905-1906. II. A botanical survey of the Galápagos Islands. *Proceedings of the California Academy of Sciences*, ser. 4, 1:7–288.

Stewart, A. 1915. Some observations concerning the botanical conditions on the Galápagos islands. *Transactions of the Wisconsin Academy of Science, Arts and Letters* 18:272–340.

Stresemann, E. 1931. Aves, in *Handbuch der Zoologie* 7:645.

Stresemann, E. 1936. Zur Frage der Artbildung in der Gattung *Geospiza*. *Orgaan der Club Van Nederlandische Vogelkunde* 9:13–21.

Strong, D. R., Jr., D. Simberloff, L. G. Abele, and A. B. Thistle (eds.). 1984. *Ecological Communities: Conceptual Issues and the Evidence*. Princeton University Press, Princeton, N. J.

Strong, D. R., Jr., L. Szyska, and D. Simberloff. 1979. Tests of community-wide character displacement against null hypotheses. *Evolution* 35:897–913.

Swarth, H. S. 1931. The avifauna of the Galápagos Islands. *Occasional Papers of the California Academy of Sciences* 18:1–299.

Swarth, H. S. 1934. The bird fauna of the Galápagos Islands in relation to species formation. *Biological Reviews* 9:213–234.

Sulloway, F. J. 1982a. The *Beagle* collections of Darwin's Finches (Geospizinae). *Bulletin of the British Museum (Natural History), Zoology series* 43:49–94.

Sulloway, F. J. 1982b. Darwin and his Finches: the evolution of a legend. *Journal of the History of Biology* 15:1–53.

Sulloway, F. J. 1984. Darwin and the Galápagos. *Biological Journal of the Linnean Society* 21:29–59.

Sushkin, P. P. 1929. On some peculiarities of adaptive radiation presented by insular faunae. *Proceedings of the 6th International Ornithological Congress (Moscow, 1926)*, pp. 375–378.

Tauber, C. A., and M. J. Tauber, 1978. Sympatric speciation based on allelic changes at three loci: evidence from natural populations in two habitats. *Science* 197:1298–1299.

Templeton, A. R. 1980. The theory of speciation via the founder principle. *Genetics* 94:1011–1038.

Templeton, A. R. 1981. Mechanisms of speciation—a population genetic approach. *Annual Reviews of Ecology and Systematics* 12:23–48.

Thorpe, R. S. 1982. The molecular clock hypothesis: biochemical evolution, genetic differentiation and systematics. *Annual Reviews of Ecology and Systematics* 13:139–168.

Tordoff, H. B. 1954. A systematic study of the avian family Fringillidae based on the structure of the skull. *Miscellaneous Publications of the Museum of Zoology, University of Michigan*, No. 81.

Turelli, M. 1984. Heritable genetic variation via mutation-selection balance; Lerch's zeta meets the abdominal bristle. *Theoretical Population Biology* 25: 138–193.

Turelli, M. 1985. Effects of pleiotropy on predictions concerning mutation-selection balance for polygenic traits. *Genetics* 111:165–195.

Turner, J.R.G. 1983. Mimetic butterflies and punctuated equilibria: some old light on a new paradigm. *Biological Journal of the Linnean Society* 20:277–300.

Van Riper, C., III. 1980. Observations on the breeding of the Palila *Psittarostra bailleui* of Hawaii. *Ibis* 122:462–475.

Van Valen, L. 1965. Morphological variation and the width of the ecological niche. *American Naturalist* 99:377–390.

Warner, R. E. 1968. The role of introduced diseases in the extinction of the endemic Hawaiian avifauna. *Condor* 70:101–120.

West-Eberhard, M. J. 1983. Sexual selection, social competition, and evolution. *Quarterly Review of Biology* 58:155–183.

White, M.J.D. 1978. *Modes of Speciation*. W. H. Freeman and Co., San Francisco.

Wiens, J. A. 1984. On understanding a non-equilibrium world: myth and reality in community patterns and processes. *In* D. R. Strong, Jr., D. Simberloff, L. G. Abele, and A. B. Thistle (eds.), *Ecological Communities: Conceptual Issues and the Evidence*, pp. 439–457. Princeton University Press, Princeton, N. J.

Wiggins, I. L., and D. M. Porter. 1971. *Flora of the Galápagos Islands*. Stanford University Press, Stanford, California.

Williams, G. C. 1966. *Adaptation and Natural Selection*. Princeton University Press, Princeton, N. J.

Williamson, M. 1981. *Island Populations*. Oxford University Press, Oxford.

Willson, M. F. 1971. Seed selection in some North American finches. *Condor* 73:415–429.

Wilson, D. S., and A. Hedrick. 1982. Speciation and the economics of mate choice. *Evolutionary Theory* 6:15–24.

Witte, F. 1984. Ecological differentiation in Lake Victoria haplochromines: comparison of cichlid species flocks in African lakes. *In* A. A. Echelle and I. Kornfield (eds.), *Evolution of Fish Species Flocks*, pp. 155–168. University of Maine at Orono Press.

Woodruff, R. C., and J. N. Thompson, Jr. 1980. Hybrid release of mutator activity and the genetic structure of natural populations. *Evolutionary Biology* 13:129–162.

Wright, J. W. 1983. The evolution and biogeography of the lizards of the Galápagos Archipelago: evolutionary genetics of *Phyllodactylus* and *Tropidurus* populations. *In* R. I. Bowman, M. Berson, and A. E. Leviton (eds.), *Patterns of Evolution in Galápagos Organisms*, pp. 123–155. American Association for the Advancement of Science, Pacific Division, San Francisco, California.

Wright, S. 1931. Evolution in Mendelian populations. *Genetics* 10:97–159.

Wright, S. 1977. *Evolution and the Genetics of Populations*, Vol. 3. University of Chicago Press, Chicago.

Wright, S. 1980. Genic and organismic selection. *Evolution* 34:825–843.

Wright, S. 1982. Character change, speciation, and the higher taxa. *Evolution* 36:427–433.

Wyles, J. S. and V. Sarich. 1983. Are the Galápagos iguanas older than the Galápagos? Molecular evolution and colonization models for the archipelago. *In* R. I. Bowman, M. Berson, and A. E. Leviton (eds.), *Patterns of Evolution in Galápagos Organisms*, pp. 177–186. American Association for the Advancement of Science, Pacific Division, San Francisco, California.

Wynne-Edwards, V. C. 1947. Review of D. Lack, *Darwin's Finches*. *Ibis* 89:685–687.

Yang, S. Y., and J. L. Patton. 1981. Genic variability and differentiation in Galápagos finches. *Auk* 98:230–242.

Zimmerman, E. C. 1938. Cryptorhynchinae of Rapa. *Bernice P. Bishop Museum Bulletin* 151:1–75.

Author Index

Abbott, I. (1972) 288; (1973) 330; (1974a) 330; (1974b) 330; (1977) 408; (1980) 273; and L. K. Abbott (1978) 272; P. R. Grant, and L. K. Abbott (1975) 62, 133, 134; L. K. Abbott, and P. R. Grant (1977) 7, 64, 79, 80, 118, 119, 120, 122, 123, 125, 158, 160, 180, 208, 223, 273, 275, 288, 291, 293, 316, 318, 322, 330, 333

Abs, M., E. Curio, P. Kramer, and J. Niethammer (1965) 64, 256, 365

Adsersen, H. (1976) 21

Alatalo, R. V. (1982) 317, 320, 328, 330; L. Gustafsson, and A. Lundberg (1982) 355

Alerstam, T., B. Ebenman, M. Sylvén, S. Tamm, and S. Ulfstrand (1978) 355

Alexander, R. D. (1968) 405

Alberch, P., S. J. Gould, G. F. Oster, and D. B. Wake (1979) 100

Allan, J. D., L. W. Barnthouse, R. A. Prestbye, and D. R. Strong (1973) 408

Allen, G. M. (1937) 117

Alpert, L. (1961) 27

Amadon, D. (1950) 404

Andrewartha, H. G., and L. C. Birch (1954) 273

Arnold, S. J. (1983) 357

Arthur, W. (1982) 273

Baldwin, P. H. (1953) 404

Barigozzi, C. (1982) 405

Barrowclough, G. F. (1983) 283, 386

Barton, N. H., and B. Charlesworth (1984) 405, 407, 408

Bateson, P.P.G. (1983) 249

Beebe, W. (1924) 6, 60, 117, 288, 358, 359

Berry, R. J. (1985) 31

Beverley, S. M., and A. C. Wilson (1985) 403

Boag, P. T. (1983) 80, 89, 181-183, 209, 250, 377, 378, 382-384

Boag, P. T. (1984) 80, 89, 90, 102, 103, 105, 109; and P. R. Grant (1978) 183, 250; and P. R. Grant (1981) 167, 184, 187, 189, 196; and P. R. Grant (1984a) 64, 80, 119, 123, 125, 148, 150, 154, 159, 166, 171, 173, 196, 208, 214, 218, 223, 240, 365, 371; and P. R. Grant (1984b) 122, 125, 136, 138, 159, 168, 174, 199, 204-206, 208, 209, 213, 341, 345

Bock, W. J. (1963) 116; (1970) 404

Boecklen, W. J., and C. NeSmith (1985) 323

Bond, J. (1948) 254

Bowman, R. I. (1961) xi, 5, 6, 29, 51, 55, 62, 80, 86, 111, 115-118, 131, 132, 176, 199, 208, 254, 259, 276, 286, 288, 289, 291, 293, 294, 300, 301, 310, 341, 364-368, 373; (1963) 113, 114; (1979) 230, 231, 235, 280, 372; (1983) 47, 80, 199, 230-232, 234, 235, 237, 241, 244, 247-249, 254, 280, 353, 372; M. Berson, and A. E. Leviton (1983) 31; and S. I. Billeb (1965) 3, 129, 373; and A. Carter (1971) 268

Boyce, M. S. (1979) 362

Bronowski, J. (1977) 401

Brown, J. H., O. J. Reichman, and D. W. Davidson (1979) 148

Brown, W. (1983) 258

Brown, W. L., Jr., and E. O. Wilson (1956) xii, 273, 314

Burtt, E. H., Jr., (1979) 371

Bush, G. L. (1975) 274, 405

Cain, A. J. (1983) 382

Calder, W. A. III. (1974) 362; (1984) 82

Cane, M. A. (1983) 27; and S. E. Zebiak (1985) 25

Carlquist, S. J. (1970) 404

Carson, H. L. (1968) 407; (1978) 406; (1982a) 403, 406, 407; (1982b) 406, 407;

Carson, H. L. (*cont.*)
(1983) 266, 403; and A. R. Templeton (1984) 404, 405, 407, 408
Case, T. J. (1979) 339; and R. Sidell (1983) 315, 320, 328
Casey, T.L.C., and J. D. Jacobi (1974) 404
Charlesworth, B., R. Lande, and M. Slatkin (1982) 404, 407
Cheverud, J. M. (1984) 407
Chitty, D. (1967) 330
Clark, D. A. (1984) 148
Clarke, B. C., and J. Murray (1969) 404
Cock, A. G. (1963) 103
Colinvaux, P. A. (1972) 30; (1984) 21, 30; and E. K. Schofield (1976a) 30; and E. K. Schofield (1976b) 30
Colnett, J. (1798) 8, 45
Colwell, R. K. (1974) 25; and D. Winkler (1984) 317, 323, 328, 329
Connor, E. F., and D. S. Simberloff (1978) 273, 288, 311, 330; (1983) 11
Cox, A. (1983) 19, 20, 261
Curio, E. (1965) 365; (1969) 365; and P. Kramer (1964) 3; and P. Kramer (1965a) 60; and P. Kramer (1965b) 64, 368
Cutler, B. D. (1970) 254

Dalrymple, G. B., and A. Cox (1968) 31
Darwin, C. R. (1839) 8, 27; (1841) 8; (1842) 6, 8, 12, 286, 311; (1845) 8; (1859) 9, 285, 286, 311, 359, 375
Davidson, D. W., D. A. Samson, and R. S. Inouye (1985) 166
de Vries, T. J. (1975) 64, 365; (1976) 64, 365
Dhondt, A. A. (1982) 183
Dickey, D. R., and A. J. van Rossem (1938) 255
Dickinson, H., and J. Antonovics (1973) 275
Dobzhansky, T. (1937) 273; (1940) 285
Dominey, W. J. (1984) 408, 410
Downhower, J. F. (1976) 128, 196, 197, 362, 363; (1978) 244, 363; and C. H. Racine (1976) 122
Dunn, E. K. (1976) 363

Eibl-Eibesfeldt, I. (1984) 3
Eliasson, V. (1984) 28

Endler, J. A. (1977) 275, 405; (1984) 366; (1986) 184, 357, 359
Emlen, S. T. (1971) 241

Falconer, D. S. (1981) 180, 184
Felsenstein, J. (1981) 275
Fisher, R. A. (1937) 207
Ford, H. A., A. W. Ewing, and D. T. Parkin (1974) 258
Ford, H. A., D. T. Parkin, and A. W. Ewing (1973) 208, 275
Fryer, G., and T. D. Iles (1972) 408, 409
Futuyma, D. J. (1979) 353, 357; and G. C. Mayer (1980) 275, 280; and M. Slatkin (1983) 339

Gause, G. F. (1934) 286; (1939) 285
Gibbs, H. L., P. R. Grant, and J. Weiland (1984) 154
Gifford, E. W. (1919) 3, 129, 199, 274, 298, 304
Gilpin, M. E., and J. M. Diamond (1984) 317, 330
Gould, J. (1837) 8, 47, 51
Gould, S. J. (1966) 103; (1977) 381; (1980) 404; (1982) 404; (1984) 382; and R. C. Lewontin (1979) 357, 368
Grant, B. R. (1984) 234, 235, 237, 242, 243, 245, 246, 249-251, 278, 279; (1985) 180, 191, 214, 217, 218, 279, 369; and P. R. Grant (1979) 180, 218, 235, 245, 249, 275, 276, 278; and P. R. Grant (1981) 86, 125, 131, 150, 310, 380; and P. R. Grant (1982) 86, 199, 214, 215, 280, 294, 299-304, 380; and P. R. Grant (1983) 214, 218, 234, 235, 238, 240, 245, 246, 249, 279, 369
Grant, P. R. (1965) 330, 408; (1966) 330, 408; (1967) 175, 207; (1968) 323, 330, 408; (1969) 315; (1972a) xii; (1972b) xii, 273, 314, 330, 341; (1972c) 323; (1975a) xii; (1975b) 340; (1977) 330; (1979a) xii; (1979b) xii, 175; (1980) 268; (1981a) 102, 104, 109-111, 250, 378, 386; (1981b) 121, 122, 136, 137, 191; (1981c) 264, 321, 325, 328; (1982) 101, 102; (1983a) 80, 106, 181, 183, 214, 250, 277, 279, 376-378, 382, 383, 386; (1983b) 101; (1983c) 54, 328, 329, 331; (1984a) 27; (1984b) 47, 62, 65, 268, 314;

(1984c) 65, 256, 261, 262, 264, 268, 269; (1985a) 25; (1985b) 153, 154, 158, 161, 165, 166, 168-170, 173, 174, 314; and I. Abbott (1980) 317, 323, 326-329; I. Abbott, D. Schluter, R. L. Curry, and L. K. Abbott (1985) 47, 52, 55, 57, 79, 80, 83, 86, 88, 91-94, 131, 175, 176, 197, 199, 207, 307, 384; and P. T. Boag (1980) 24-26, 170; and B. R. Grant (1980a) 64, 117, 123, 125, 128, 129, 131, 148, 150, 153, 159, 171, 196, 223, 234, 244, 250, 298, 300, 363, 365, 371, 372; and B. R. Grant (1980b) 125, 166, 173, 333; and B. R. Grant (1985) 27; B. R. Grant, J.N.M. Smith, I. Abbott, and L. K. Abbott (1976) 132, 133, 178, 196, 208, 209; and K. T. Grant (1979) 148; and N. Grant (1979) 148; and N. Grant (1983) 289; and T. D. Price (1981) 178, 199, 204, 209; T. D. Price, and H. Snell (1980) 185; and D. Schluter (1984) 52, 53, 55, 56, 273, 306, 307, 309, 316-319, 324, 326, 327; D. Schluter, and P. T. Boag (1979) 277; J.N.M. Smith, B. R. Grant, I. Abbott, and L. K. Abbott (1975) 60, 64, 164, 199, 303, 365

Greenwood, P. H. (1981a) 408, 410; (1981b) 408; (1984) 408, 410

Gulick, A. (1932) 272

Haldane, J.B.S. (1937) 331; (1954) 184

Hall, M. L., P. Ramon, and H. Yepes (1983) 19, 366

Halpern, D., S. P. Hayes, A. Leetmaa, D. V. Hansen, and S.G.H. Philander (1983) 25

Hamann, O. (1981) 22, 28-30, 288; (1984) 28-31, 311

Hamilton, T. H., and I. Rubinoff (1963) 62, 272; (1964) 62, 272; (1967) 62, 272, 330, 352

Hamilton, T. H., I. Rubinoff, R. H. Barth, Jr., and G. L. Bush (1963) 273, 311

Hamilton, W. D., and M. Zuk (1982) 370

Hamilton, W. J., III (1973) 370; and F. Heppner (1967) 371

Harris, M. P. (1972) 254; (1973) 53, 62, 64, 65, 304; (1974) 8, 24, 51, 65

Harvey, P. H., and G. M. Mace (1982) 86

Hendrickson, J. A., Jr. (1981) 323, 330

Hey, R. (1977) 19, 20

Hickman, C. S., and J. H. Lipps (1985) 19

Hooker, J. D. (1847) 27

Houvenaghel, G. T. (1984) 21

Howarth, F. G. (1980) 404

Hutchinson, G. E. (1965) 404

Hutt, F. B. (1964) 278

Huxley, J. S. (1940) 10; (1942) 10, 285, 286; (1955) 367, 368

Itow, S. (1975) 29

Janzen, D. H. (1973) 408

Jo, N. (1983) 281

Johnson, M. P., and P. H. Raven (1973) 30, 273, 311

Jones, P. J. (1973) 363

Kaneshiro, K. Y. (1976) 403

Kluge, A. G., and J. S. Farris (1969) 384

Lack, D. (1940a) 222, 265, 285; (1940b) 222, 265, 285; (1945) 9, 45-47, 52, 55, 60, 62, 79, 92, 94, 117, 131, 201, 222-225, 227, 230, 231, 234, 237, 241, 254, 257, 260, 263, 265, 268, 272, 276, 280, 282, 285-287, 300, 304, 340, 348, 352, 353, 359, 362, 367, 369, 388; (1947) xi, 3, 6, 8-12, 46, 47, 51, 55, 60, 62-64, 79, 92, 95, 111, 131, 150, 152, 176, 177, 199, 201, 207, 224, 230, 234, 257-259, 263, 265-267, 272, 274-276, 282, 285-288, 291, 294, 295, 298-300, 302, 304, 307, 310, 320, 340, 341, 353, 359, 364-366, 369, 375, 385, 387, 401; (1950) 152, 367; (1968) 101; (1969) 47, 51, 60, 62, 131, 268, 294, 300, 302, 304; (1971) 294; (1973) 286

Lande, R. (1976a) 177, 206; (1976b) 381; (1979) 376, 382; (1980a) 197; (1980b) 287, 407, 408; (1981) 354, 410; (1982) 275, 405; and S. J. Arnold (1983) 189

Lasiewski, R. C., and W. R. Dawson (1967) 161

Lederer, R. J. (1975) 116

Lewin, R. (1983) 321

Lewontin, R. C. (1978) 357

Lowe, P. R. (1936) 6, 222, 281, 359

MacArthur, R. H. (1972) 321; and E. O. Wilson (1967) 263

MacFarland, C., and W. Reeder (1974) 3, 373

Manwell, R. D. (1955) 171

Maynard Smith, J. (1966) 275, 405; (1984) 408; R. Burian, S. Kauffman, P. Alberch, J. Campbell, B. Goodwin, R. Lande, D. Raup, and L. Wolpert (1985) 407

Mayr, E. (1942) 10, 285, 408; (1954) 407, 408; (1961) 11; (1963) xi, 183, 184, 287, 353, 405; (1982) 408; (1983) 357; (1984) 405, 407, 408, 410

Marchant, S. (1958) 332

McBirney, A. R., and H. Williams (1969) 19

Milkman, R. (1981) 258

Millikan, G. C., and R. I. Bowman (1967) 372

Millington, S. J., and P. R. Grant (1983) 209; (1984) 128, 168, 194, 371

Millington, S. J., and T. D. Price (1985) 237, 242, 249

Montgomery, S. L. (1982) 404

Moore, J. A. (1957) 354

Morgan, W. J. (1971) 19

Mossiman, J. E., and F. C. James (1979) 80

Muller, H. J. (1940) 273, 287

Murray, J. J., and B. C. Clarke (1980) 404

Newton, I. (1967) 360

O'Donald, P. (1970) 184

Olson, S. L., and H. F. James (1982) 404

Orr, R. T. (1945) 103, 199, 222-225, 230, 242, 367

Paterson, H. E. (1980) 251

Patton, J. L., and S. W. Sherwood (1983) 406

Paulay, G. (1985) 404

Payne, R. B. (1982) 370

Paynter, R. A., Jr. (1970) 254

Perry, R. (1984) 31

Philander, S.G.H. (1983) 25

Pimm, S. L. (1979) 275

Polans, N. (1983) 258, 268

Porter, D. M. (1976) 27; (1983) 27; (1984) 27, 28

Power, D. M. (1975) 64, 273, 288

Price, T. D. (1984a) 46, 193, 197, 250, 251, 369, 370; (1984b) 193, 195-197, 209; (1985) 103; and P. R. Grant (1984) 106, 168, 194, 250, 381; and P. R. Grant (1985) 109, 381-383; P. R. Grant, H. L. Gibbs, and P. T. Boag (1984) 170, 190, 192; P. R. Grant, and P. T. Boag (1984) 181, 209, 377-380; S. Millington, and P. Grant (1983) 272

Proctor-Gray, E., and R. T. Holmes (1981) 370

Pulliam, H. R. (1980) 134

Quinn, W. H. (1971) 30

Racine, C. H., and J. F. Downhower (1974) 28

Raikow, R. J. (1977) 401, 403

Rasmusson, E. M. (1985) 25

Ratcliffe, L. M. (1981) 223, 225, 227, 233-237, 247, 280, 352; and P. T. Boag (1983) xi; and P. R. Grant (1983a) 223, 225-227, 230, 349; and P. R. Grant (1983b) 349-351; and P. R. Grant (1985) 238-240, 242, 244, 248

Reeder, W. G., and S. E. Reichert (1975) 28

Rensch, B. (1933) 285

Richards, O. W. (1948) xi, 301

Ricklefs, R. E. (1976) 103

Ridgway, R. (1890) 9

Rothschild, W., and E. Hartert (1899) 51, 55, 62, 64, 272, 273, 289, 364, 369; (1902) 45, 51, 55

Rothstein, S. I. (1973) 196

Roughgarden, J. (1972) 209

Rowher, S., S. D. Fretwell, and D. M. Niles (1980) 370

Salt, G. W. (1984) 328

Salvin, O. (1876) 6, 51, 65, 222, 272, 358, 369

Sarich, V. M. (1977) 260

Schluter, D. (1982a) 86, 119, 122-125, 134, 135, 159, 161, 304; (1982b) 119, 123, 150, 159, 304; (1984a) 150, 362; (1984b) 164, 223, 384-387; (1986) 361; and P. R. Grant (1982) 111, 120, 159, 167, 294, 304-306, 308, 320; and P. R. Grant (1984a) 3, 52, 54, 80, 111, 116, 120, 129-131, 159, 304, 307, 310, 317; and

P. R. Grant (1984b) 167, 332, 334-340;
T. D. Price, and P. R. Grant (1985) 196,
342-345
Schoener, T. W. (1984) 321, 323, 328
Searcy, W. A., P. Marier, and S. S. Peters
(1982) 241
Selander, R. K. (1962) 368
Sherry, T. W. (1985) 31, 65
Sibley, C. G., and J. E. Alquist (1982) 266,
401; (1984) 256
Simberloff, D. (1983a) 273, 286; (1983b)
273, 286, 323, 330; (1984) 273, 286,
331; and W. Boecklin (1981) 316, 323,
330; and E. F. Connor (1981) 330
Simkin, T. (1984) 19, 20
Simpson, B. B. (1974) 21
Slatkin, M. (1984) 196
Slud, P. (1967) 219
Smith, G. R., and T. N. Todd (1984) 405,
410
Smith, J.N.M., and A. A. Dhondt (1980)
183
Smith, J.N.M., P. R. Grant, B. R. Grant,
I. Abbott, and L. K. Abbott (1978) 123,
125-127, 159
Smith, J.N.M., and H.P.A. Sweatman
(1976) 219
Smith, J.N.M., and R. Zach (1979) 183
Snell, H. L., H. M. Snell, and C. R. Tracey
(1984) 22
Snodgrass, R. E. (1902) 51, 117, 125, 131,
224, 293, 294, 358, 359; (1903) 45, 103,
110, 373, 387; and E. Heller (1901) 9;
and E. Heller (1904) 46, 110, 125, 257,
258, 304, 366
Snow, D. (1966) 150, 152, 208, 276
Soulé, M., and B. R. Stewart (1970) 196
Steadman, D. (1981) 256; (1982) 10, 47,
51, 254, 255, 387; (1984) 52, 54; (1985)
47, 54, 256, 268, 272
Stewart, A. (1911) 28, 288; (1915) 288
Stresemann, E. (1931) 289; (1936) 9, 51,
199, 222, 263, 265, 272, 282, 285, 287,
289, 352
Strong, D. R., Jr., D. Simberloff, L. G.

Abele, and A. B. Thistle (1984) 328
Strong, D. R., Jr., L. Szyska, and D. Simberloff (1979) 286, 327-330
Swarth, H. S. (1931) 5, 9, 51, 52, 55, 60;
(1934) 6, 62, 94, 222, 257, 272, 282,
355, 358, 359, 364, 367
Sulloway, F. J. (1982a) 8, 9, 47, 52, 54, 63,
64; (1982b) 8, 47, 54; (1984) 8
Sushkin, P. P. (1929) 272

Tauber, C. A., and M. J. Tauber (1978) 405
Templeton, A. R. (1980) 280, 281, 287,
406, 407; (1981) 280, 281, 353, 405-408
Thorpe, R. S. (1982) 260
Tordoff, H. B. (1954) 254
Turelli, M. (1984) 206; (1985) 206
Turner, J.R.G. (1983) 406

Van Riper, C., III (1980) 360
Van Valen, L. (1965) 178, 179, 219

Warner, R. E. (1968) 171
West-Eberhard, M. J. (1983) 410
White, M.J.D. (1978) 405, 406
Wiens, J. A. (1984) 11
Wiggins, I. L., and D. M. Porter (1971) 28,
159, 288
Williams, G. C. (1966) 290
Williamson, M. (1981) 263
Willson, M. F. (1971) 133, 134
Wilson, D. S., and A. Hedrick (1982) 251
Witte, F. (1984) 410
Woodruff, R. C., and J. N. Thompson, Jr.
(1980) 206
Wright, J. W. (1983) 261
Wright, S. (1931) 40; (1977) 406; (1980)
406, 407; (1982) 404-406
Wyles, J. S., and V. Sarich (1983) 261
Wynne-Edwards, V. C. (1947) xi

Yang, S. Y., and J. L. Patton (1981) 51, 52,
258-260, 262, 263, 386, 387

Zimmerman, E. C. (1938) 404

Subject Index

The main entry (with page numbers) for a species will be found under the name of the genus (e.g. *Butorides sundevalli*). A listing (without page numbers) is also given under the specific name (*sundevalli, Butorides*) and under the English vernacular name (lava heron) as a means of cross-reference to the main entry.

achatinellid snail, 404
adaptation, adaptive, xi, xii, 3, 10, 82, 95, 132, 259, 284, 286-290, 293, 294, 307, 310, 312, 341, 357-374, 381, 386, 398-400, 405, 408, 410; adaptive habitat selection, 132, 133, 139; adaptive radiation, xi, 3, 5, 8, 9, 12, 77, 253, 260, 261, 266, 272, 285, 286, 289, 290, 300, 312, 313, 359, 380, 399-401, 403-405, 409, 410; adaptive landscape, 282, 335, 381; adaptive peak, 335, 336, 338, 340-345, 347, 380, 381, 396-398, 407; adaptive trough or valley, 282, 381; adaptive variation model, 178, 179, 196, 219; adaptive zone, 399. *See also* nonadaptive traits
aegyptica, Merremia
affinis, Camarhynchus psittacula
afra, Cyanotilapia
African great lakes, 408, 409
age of islands, 19, 20, 31, 32, 260, 261
aggression and dominance, 191, 208, 218, 224, 227, 229, 238, 241, 242, 252, 290, 370
'ākepa, *see Loxops coccinea*
'akialoa, *see Hemignathus procerus*
'akiapola'au, *see Hemignathus wilsoni*
alba, Tyto
albus, Casmerodius
allelic variation and change, alleles, 177, 199, 259, 278, 280, 283, 287, 386, 406; neutral, 259-261, 280, 369, 386
allometry, 52, 82, 86, 87, 89, 95, 106, 107, 110-112, 308, 326, 327, 336, 361, 381; dynamic, 82, 107-110; interpopulation, 82, 87, 325, 326, 361; interspecific, 82, 87, 108; static (intrapopulation), 52, 82, 87, 89, 107

allopatry, 9, 11, 51, 57, 66, 73, 206, 263-266, 268, 269, 283-289, 291, 294, 299, 302, 307, 313, 314, 320, 326, 328, 340-350, 352, 389, 396-399, 405, 410
altitude, *see* islands, elevation
'amakihi, *see Loxops virens*
Amblyrhynchus cristatus (marine iguana), 3, 260, 364, 399, Pl. 6
America, Central, 9, 27, 31, 32, 253-255, 283, 389; South, 9, 19, 27, 32, 101, 253-256, 283, 331, 389
amplexicaulis, Chamaesyce
anagenesis, 405
anatomy, 9, 82, 360, 381, 401
ancestral species, stock, 6, 10, 11, 253-255, 257, 260, 263, 265, 266, 268, 282-284, 323, 371, 378, 387, 389, 399-401, 405
angiospermum, Heliotropium
angustissima, Sarcostemma
ani, Crotophaga
anna, Ciridops
anthers, 149
ants, 148
'apapane, *see Himatione sanguinea*
Ardea herodias (great blue heron), 65
arils, 212, 217, 276, 307, Pl. 10
arthropods, 6, 48, 116-118, 120, 126, 128, 130, 139, 147, 150, 152, 158, 161, 171, 196, 215, 217, 219, 300, 305, 307, 311, 332, 372, 374, 393, 399, 400, 403, Pl. 3
Asio flammeus (short-eared owl), 64, 65, 76, 171, 364-366, 371, 391, 399
Aulonocara nyassae, 409
Australia, 404
Azores islands, xii

Bahía Academía (Santa Cruz), 72, 126, 127,

132, 152, 162, 163, 178, 208, 275, 296
Bahía Borrero (Santa Cruz), 132, 145, 152,
    161-163, 180, 204, 205, 208
Bahía Darwin (Genovesa), 245
Bahía Elizabeth (Isabela), 68
bailleui, Loxioides
Bainbridge, 4, 35, 61
Baltra (South Seymour), 4, 53, 63, 148, 413
barbadense, Gossypium
bark of trees, 128, 219, 372, 393, 400, 401
barn owl, *see Tyto alba*
Bartolomé, 4, 61
bauri, Oryzomys
Beagle, H.M.S., 6
Beagle (Isla), 61
beak, *see* bill
beetle larvae, 167, 212; pupae, 167, 212
behavior, xii, 3, 45, 128, 132, 134, 138,
    139, 178, 192, 193, 208, 219, 221, 224,
    225, 227, 238, 242, 247, 250, 255, 306,
    350, 371-374, 401, 410
Bella Vista (Santa Cruz), 24
bill, xi, xii, 3, 6, 8, 12, 56, 89, 95, 125;
    color, 277-280, Pl. 4, 7; color poly-
    morphism, 277-280, Pl. 7; curvature,
    116, 131, 138, 360, 393, 403; depth, 56,
    77-81, 85, 97, 102, 105, 107, 109-112,
    116, 118-120, 132, 136-138, 175, 176,
    178, 180, 181, 189-194, 204, 205, 209,
    213, 214, 217-220, 275, 277-295, 298,
    299, 318, 321-325, 327, 332-334, 336-
    342, 347, 377-379, 381, 383, 392, 393,
    395, 399; differences between species,
    54, 66, 77, 80, 87, 91, 104, 111, 117-
    120, 138, 139, 147, 208, 286, 296, 297,
    320, 323-330, 341, 346, 348-353, 375,
    377, 381, 385, 387, 390, 392, 393, 396;
    length, 77, 79, 81, 82, 84, 87-91, 95,
    102, 105-107, 109-112, 116, 120, 125,
    129, 131, 132, 138, 181, 189-191, 195,
    203, 208, 209, 217, 218, 230, 266, 276-
    279, 295, 298, 310, 320-325, 327, 362,
    377-380, 383, 392, 393, 399; mechanical
    properties, 113, 115, 116, 120, 393; size
    (dimensions), 45, 55, 57, 66, 71, 72, 80,
    103, 104, 106, 107, 109, 113, 116-120,
    123, 125, 128, 130-136, 138, 139, 147,
    175, 183, 185, 189, 191, 193-195, 206,
    209, 217, 218, 220, 221, 224, 225, 229,
    230, 241, 249, 250, 252, 257, 285, 286,

294, 295, 298, 299, 301, 308, 310, 312,
    315, 321, 323, 326, 331, 333-336, 338,
    339, 341, 343-347, 353, 354, 357-362,
    364, 373, 374, 376, 378, 384, 392-394,
    396, 408; shape (proportions), 45, 55-57,
    66, 70-72, 75, 80, 86, 87, 89, 91, 92, 95,
    100, 104, 109, 111-114, 116, 117, 120,
    123, 129, 131, 139, 147, 206, 208, 214,
    217, 218, 224, 225, 230, 257, 276, 277,
    280, 295, 298, 301, 302, 307, 308, 320,
    358, 360, 363, 381, 392-395, 399; to-
    mium, 45; width, 77, 79, 84, 87-90, 102,
    107, 109, 110, 112, 116, 118, 120, 136,
    137, 181, 189-194, 209, 323, 325, 377,
    379, 392. *See also* variation, morphologi-
    cal
bimodal frequency distribution, 55, 275,
    276
biochemical, biochemistry, 12, 51, 52, 55,
    256, 258, 260-262, 268, 272, 276, 283
    386-388, 401
biogeography, 263. *See also* geographical
    distributions
biomass of finches, correlated with seed bio-
    mass, 154, 158, 160-167, 173, 333, 391
birds, North American, 225, 241, 254. *See*
    also finches, North American
blue-footed booby, *see Sula nebouxii*
blue tit, *see Parus caeruleus*
body size, 45, 52, 55, 56, 60, 66, 75, 80,
    82, 86, 87, 89, 91, 94-96, 100-103, 107,
    109, 111, 112, 161, 176, 182-185, 187,
    189, 190, 192-196, 204, 207, 209, 214,
    220, 222, 224, 229, 236, 250, 252, 257,
    272, 279, 282, 286, 295, 301, 303, 307,
    308, 315, 349, 352, 358-364, 372-374,
    376, 384, 386, 388, 390, 392, 394, 395,
    397, 398, 400, 408; proportions, 52, 55,
    82, 89, 91, 100, 106, 109, 257, 272, 282,
    361, 376, 378, 381, 386, 388, 392, 399,
    408. *See also* weight
bones, 86, 115
boobies, *see Sula*
breeding: habits, behavior, and success, 10,
    12, 148, 150, 152, 154, 157, 171, 184,
    185, 218, 223, 255, 303, 310, 363, 367,
    369, 373, 374, 391, 395; between spe-
    cies (*see* interbreeding); distributions,
    304; season (and nonbreeding season),
    62, 125, 148, 150, 160, 168, 171, 193,

breeding (*cont.*)
  196, 199, 202, 208, 245, 353, 365, 368,
  391, 395, 405; status, 53, 60; with kin
  (*see* kin recognition); at an unusually
  early age, 154, 221; age at first reproduction, 154, 194, 195
brevis, Lethrinops
Bubulcus ibis (cattle egret), 65
bud, 6, 86, 98, 117, 360, 393
budgerigar, 257
bullfinch, *see Pyrrhula pyrrhula*
Bursera, 28, 118, 131, 135, 160, 209, 214,
  215, 300, 302; *graveolens*, 33, 37, 39-41,
  128, 145, 147, 216, Pl. 10; *malacophylla*,
  132
Buteo galapagoensis (Galapagos hawk), 60,
  64, 65, 76, 364, 365, 391, 399
Butorides striatus (green-backed heron), 65
Butorides sundevalli (lava heron), 64, 65,
  171, 364, 366
butterflies, 3, 406

Cacabus miersii, Pl. 9
Cactornis, 8
Cactospiza, 118, 215, 231, 274, 298, 320,
  372, 374, 389
Cactospiza heliobates, 3, 5, 9, 11, 46, 47,
  51, 53, 68, 91-93, 151, 176, 201, 262,
  274, 284, 371, 372, 391, 393, 394
Cactospiza pallida, 3, 5, 10, 11, 15, 46, 47,
  51, 53, 68, 83, 91-93, 116, 138, 152,
  176, 199, 201, 258, 266, 267, 274, 282,
  284, 298, 300, 371, 372, 391, 393, 394,
  399, 401, 403
cactus: nectaries, extrafloral (*see* extrafloral
  nectaries); rotting pads, 167, 208, 212,
  217, 276, 300, 393; spines, 3, 208, 372,
  374, 393, 403. *See also Opuntia* and *Jasminocereus*
cactus finch, *see Geospiza scandens*
caeruleus, Parus
Caldwell, 4, 60, 61
California Academy of Sciences, expedition
  and collections, 9, 60, 61
caltrop, *see Tribulus cistoides*
Camarhynchus, 8, 46, 51, 116, 118, 151,
  215, 231, 267, 295, 298, 320, 360, 389;
  Camarhynchus parvulus (small tree finch),
  5, 11, 47, 53, 58, 60, 80, 91-93, 94, 117,

  118, 176, 198, 199, 201, 247, 248, 257,
  268, 285, 298, 308, 391; *C. parvulus salvini*, 367; *C. pauper* (medium tree finch),
  5, 9, 11, 47, 53, 58, 59, 62, 66, 67, 91-
  93, 176, 198, 201, 257, 266-268, 300,
  390-391; *C. psittacula* (large tree finch),
  5, 11, 47, 50, 53, 58, 61, 62, 66, 67, 83,
  91-94, 117, 118, 176, 198, 201, 231,
  257, 262, 266-268, 282, 285, 298, 300,
  372, 384, 391, 403; *C. psittacula affinis*,
  267, 268; *C. psittacula habeli*, 267;
  C. p. psittacula, 267
Canary islands, xii
cantans, Telespiza
cardinal, *see Richmondena cardinalis*
cardinalis, Richmondena
cardueline finch, 401
Caribbean islands, 382
Carnegie ridge, 20
carpenter bee, 399
carrying capacity, 154, 158, 162-164, 168,
  171, 264, 291, 407
Casmerodius albus (great egret), 64, 65,
  171, 366
Castela galapageia, 28, 33, 126, 127
caterpillars, 118, 125, 128, 139, 143, 147,
  148, 150, 152, 168, 196, 391
cattle, 30
cattle egret, *see Bubulcus ibis*
cats, 30, 365
cave cricket, 404
Cepaea, snail, 3
Cerro Ballena (Isabela), 33, 162
Certhidea olivacea (warbler finch), 10, 14,
  45, 47, 51, 53, 61, 62-66, 69, 80, 83, 92,
  94, 101, 102, 106, 109, 110, 116, 117,
  138, 150, 175, 199, 218, 222, 234, 247,
  248, 257-261, 272, 282, 283, 287, 298,
  362-366, 374, 385, 387, 389-391, 393,
  397, 399-401
chaffinch, *see Fringilla coelebs*
Chamaesyce, 147, 166, 185; *amplexicaulis*,
  69, 133, 146
Champion, 4, 61, 73, 268, 270, 337, 366,
  400
character displacement, xii, 314, 315, 323,
  328, 330, 332, 341, 343, 345, 346, 353,
  396, 399; ecological, 353-355, 396; reproductive, 353-355
character release, xii, 296, 340, 341, 345

Charles Darwin Research Station, xiii, 12, 22, 23, 26, 40, 276
Charles mockingbird, *see Nesomimus trifasciatus*
Chatham mockingbird, *see Nesomimus melanotis*
chickens, 257, 278
chromosomes, 197, 198, 280, 281, 283, 403, 406, 410, 411
cichlid fish, *see* haplochromine fish
cilianensis, Eragrostis
Ciridops anna (ula-ai-hawana), 403
cistoides, Tribulus
cladogenesis, *see* speciation
classification, 8, 9, 28, 55, 66, 204, 222, 256, 295
climate, 21, 29, 30, 46, 257, 263, 286, 287; garúa, 22, 391; rain (precipitation), 21-28, 30, 31, 64, 145, 147, 148, 150, 152, 154, 155, 165, 168, 174, 184, 196, 218, 223, 391; temperature, 21-24, 31
clutch of eggs, 148, 195, 208, 218, 363, 371, 374, 391
coadapted gene complexes, 406, 407
coccinea, Loxops
coccinea, Vestiaria
Coccyzus melacorhyphus (dark-billed cuckoo), 64, 65
Cocos finch, *see Pinaroloxias inornata*
Cocos island, 3, 4, 9, 10, 31, 32, 44, 63, 65, 218, 253, 255, 264, 283, 357, 371, 389, 391, 397, 400
Cocos plate, 19, 20
Cocos ridge, 20
Coereba flaveola, 254
coevolution, 339, 380
coexistence of closely related species, 267, 294, 300, 304, 312, 321, 323, 326, 328, 329, 331, 332, 338, 339, 346, 355, 396, 398
collared flycatcher, *see Ficedula collaris*
collaris, Ficedula
colonization of islands from continent, 10, 255, 258, 260, 264, 265, 283, 287, 289, 323, 357, 371, 397, 400
colonization of islands from other islands, 62, 266, 267, 280, 287-289, 300, 303, 304, 310, 311, 313-315, 320, 326, 332, 336, 338-340, 380, 388, 399
combinations of species occurring together,

315-320, 326, 327, 331, 339, 346, 396
Commicarpus, 125, 126
communities of finches, 314, 338-340, 396; structure, xi, 294, 315, 331, 347; models of assembly, 321, 332, 336, 338-340, 346, 347; randomness, 315-317, 320, 321, 326, 327, 330, 336, 338, 339, 346, 396, 397
competition, 64, 148, 288-291, 293, 294, 299, 300, 302, 303, 305, 307, 310-347, 354-356, 364, 391, 396, 398, 399, 408; intraspecific, 196, 219, 221, 265, 284, 290; interspecific, xii, 10, 173, 174, 267, 285, 286, 288, 290, 291, 293, 298-301, 304, 310-347, 354-356, 391, 396, 398, 399
competitive displacement, 295, 299, 304, 387; exclusion, 298, 300, 303, 304, 312, 315, 317, 320, 329, 340, 346-354, 355, 396, 397, 399; release, 295, 299, 301, 312, 341
Compositae, 27
compressiceps, Haplochromis
conirostris, Geospiza
Conolophus subcristatus (land iguana), 3, 16, 260, 399
constraints: developmental, 407; genetic, 283, 376, 381, 386-388, 399, 407; morphological, 381; physiological, 86
continuous morphism, 368
Convolvulaceae, 148, 333
copulation, 223, 225-227, 229, 350
Cordia leucophlyctis, 40
Cordia lutea, 28, 40, 118, 120, 127, 128, 147, 148, 291, 300, 303, Pl. 10
correlation, 79, 82, 89, 92, 94, 101, 103, 120, 132, 133, 160-162, 164-168, 173, 189, 190, 193, 195, 207, 218, 250, 251, 273, 293, 323, 325, 326, 333, 339, 340, 354, 355, 357, 358, 361, 362, 369, 373, 374, 376, 377, 379, 381-383, 388, 391, 392; genetic, 197, 283, 360, 376-378, 383, 385, 386, 388, 407; genotype-environment, 182, 183; phenotypic, 80
courtship, 150, 222-224, 228-230, 241, 251, 252, 348, 350, 352, 355, 369, 371, 374, 392, 403, 408, 411
covariance, 130; genetic, 375-379, 382, 384, 387, 407; phenotypic, 384
Cowley, 4, 61

crassirostris, Platyspiza
crested honeyeater, *see Palmeria dolei*
cricket, 405. *See also* cave cricket
cristatus, Amblyrhynchus
Crossmans, *see* Hermanos
Croton scouleri, 18, 28, 36, 37, 40, 122, 127, 143, 147, 151, 300
Crotophaga ani (smooth-billed ani), 65
Crotophaga sulcirostris (groove-billed ani), 65
cuckoo, on Cocos island, 65
culmen curvature, *see* bill, curvature
cyaneus, Haplochromis
Cyanotilapia afra, 409

dactylatra, Sula
Daphne Major, 4, 35, 60, 61, 72-79, 94-98, 102, 105-106, 119-125, 131-139, 146-150, 152-159, 162-175, 178-199, 202-208, 213-230, 234, 237-242, 246-250, 278-282, 291-296, 299-302, 321, 322, 340-356, 361-369, 372-378, 381, 388, 391-397, Pl. 8, 9
Daphne Minor, 4, 61, 185
dark-billed cuckoo, *see Coccyzus melacor-hyphus*
Darwin, Charles, 6, 8-10, 12, 52, 287, 358
Darwin (Culpepper), 3, 4, 21, 30, 37, 52, 53, 55-57, 63, 129-131, 176, 258, 282, 295, 310, 352, 373, 390, 393, 413
darwini, Geospiza
debilirostris, Geospiza difficilis
delanotis, Tropidurus
demography, 64
Dendroica petechia (yellow warbler), 64, 65, 69, 272, 400
diet, 12, 86, 113, 117-119, 123, 126, 127, 129, 132, 136, 154, 158-161, 171, 214, 279, 285, 293, 294, 301, 305, 307, 312, 317, 332, 346, 359, 373, 393, 396, 398; convergence, 125; development through learning, 128, 210, 219, 399; determined by availability of seeds, 123, 125-127, 131, 138, 219, 293, 306, 396; determined by profitability of seeds, 123, 124, 138, 307, 333; differences between species, 3, 6, 10, 113, 116-118, 120, 123, 125-127, 130, 133, 138, 139, 147, 208, 285, 286, 290, 301, 302, 304-308, 310, 317, 318, 321, 326, 346, 348, 358, 393; diver-

gence, 123, 139, 173, 174, 391; generalized (broad), 120, 125, 136, 174, 178, 179, 208, 209, 215, 219, 260, 266; inferred from stomach contents, 117, 118, 132, 294; overlap, 173, 174, 317, 318, 391; seasonal changes, 123-127; specialized (narrow), 125, 131, 136, 174, 178-180, 208, 209, 217, 219-221, 265, 266, 288, 408
differential colonization, 315, 317, 332, 346, 396, 397
differentiation (diversification), 3, 11, 51, 92, 139, 263-267, 270, 272, 275, 280, 281, 283, 284, 287, 289, 291, 293, 294, 310, 311, 347, 358, 388, 389, 398, 400, 404, 405, 408, 410, 411
difficilis, Geospiza
dinosaurs, 404
Dirección General de Desarrollo Forestal, Quito, xiii
disease, 64, 171, 173, 174
dispersal between islands, 150, 263-265, 272, 283, 289, 291, 397-399, 403, 410
dispersal within islands, 150, 362
display nests, 148, 223, 363
displays, reproductive, 148, 193, 223, 224, 226, 227, 255, 369
distributions: frequency, *see* frequency distributions; geographical, *see* geographical, distributions
diversity, 3, 6, 272, 275, 287, 288, 293, 403
DNA: mitochondrial, 258; nuclear, 256, 258; hybridization, 256
dogs, 30
dolei, Palmeria
donkeys, 30
dove, *see Zenaida galapagoensis*
Drepanis pacifica (mamo), 403
Dromicus (snake), 366
Drosophila, flies, 3, 206, 403, 406, 408
drought, 97, 152, 157, 165-170, 173, 174, 183-185, 188, 193, 196, 209, 214, 217, 220, 250, 279, 326, 344, 345, 347, 354, 357, 367, 369, 379, 381, 391, 395, 398, Pl. 9

echios, Opuntia
ecological distinctiveness of islands, 273, 293, 390, 400, 406

ecological incompatibility of species, 300
ecological isolation, *see* speciation, factors
of importance
ecological restriction, 10, 62, 265, 266,
272-274, 294, 295, 298, 300-307, 310,
312, 313, 341, 343-347, 396-397
Ecuador, xiii, 12, 19, 25
Eden, 4, 60, 61
eggs, 100-103, 106, 112, 148, 150, 154,
171, 183, 195, 202, 207, 247, 362, 363,
366, 371, 372, 374, 391, 392. *See also*
clutch of eggs
electrophoresis, 52, 258, 260, 283, 389
El Junco lake, 29
El Niño, 23, 25, 27, 31, 32, 125, 154-156,
174, 367, 391, Pl. 9
emberizine finch, 253, 254, 256, 389
emigration, 185, 209
endemic species and subspecies: finches, 62,
272, 273, 390; flies, 403; plants, *see*
plants, endemic
Enderby, 4, 61
energy: balance, 150; costs, 370, 371;
needs, 122, 161, 194, 221, 303, 362; rewards, 121-123, 128, 135, 138, 303, 371;
storage, 361, 362; use, 360-362
epiphytes, 29, 31, 41; bromeliads, 44; moss,
29, 116, Pl. 3
equilibrial number of species, 262, 263
Eragrostis cilianensis, 155
erythrops, Neocrex
Española (Hood), 4, 19, 31, 37, 53, 55, 57,
61-63, 70, 86, 130, 131, 140, 159, 161-
163, 214, 217, 233, 234, 247, 268, 271,
280, 295, 298-304, 312, 335, 350, 366,
413
euchilus, Haplochromis
Europe, 360, 362
evolution, xi, 3, 6, 8, 158, 253-284, 376,
389, 411; convergent, 69, 111, 254, 256,
307, 378, 386; divergent, 257, 258, 261,
263, 266, 276, 280, 283, 285, 286, 312,
315, 346, 352, 354, 355, 375, 386, 387,
396, 398, 406, 407; parallel, 255; multivariate model, 376, 377, 382. *See also*
shifting balance theory
evolutionary plasticity, 100, 376
evolutionary response to selection, 180,
184, 192, 195, 282, 293, 314, 375-377,
380, 386-388, 395, 398

expected population density, 332, 333, 335-
345, 347, 396. *See also* adaptation, adaptive peak
experiments, 6, 12; clutch exchange, 183;
cryptic avoidance of predators, 365; natural, 6, 355; predator recognition by
finches, 365; song learning, 244, 246;
species discrimination, 224-230, 238-
242, 249, 252, 348-352, 354, 392
exposed rocks on shore, feeding site, 6, 393
extinction, 52-54, 61, 64, 66, 93, 148, 160,
170, 253, 256, 261-263, 268, 280, 283,
304, 305, 310, 315, 316, 337, 339, 388,
390, 399, 400, 404
extrafloral nectaries, 167, 208

fagara, Zanthoxylum
feeding behavior, habits, and skills (*see also*
diet), 3, 6, 10, 113, 117, 118, 128, 132,
139, 175, 190, 194, 214, 217, 219, 266,
298, 300, 360, 372, 389, 393, 398-400,
411; bark stripping, 3, 18, 48, 128, 214,
216, 217, 300, 372, 393; base crushing,
116, 138, 360, 393; blood drinking, 3,
129, 131, 266, 310, 373, 374, 393, Pl. 4;
cactus pad ripping, 6, 167, 208, 212,
217, 276, 300, 393; caterpillar gleaning,
160; egg breaking, 3, 131, 310, 393; leaf
eating, 6, 266; parasitic, 138; probing, 6,
116, 131, 138, 149, 208, 308, 380, 373;
scratching in leaf litter, 6, 86, 96, 99,
116, 131, 307, 393; selective, 135, 136,
303; stone kicking, 6, 373; tick feeding,
3, 16, 373, 399, Pl. 6; tip biting, 116,
138, 360, 393; tool use, 3, 266, 372, 374,
393. *See also* seeds *and* diet
feeding efficiency, in relation to bill size,
118-120, 123, 128, 129, 132-137, 139,
189, 191, 193, 194, 210, 220, 303, 335,
343, 344, 381, 394
Fernandina (Narborough), 4, 21, 30, 36, 52-
54, 60, 63, 130, 148, 159, 161, 304, 413,
Pl. 6
Ficedula collaris (collared flycatcher), 355
Ficedula hypoleuca (pied flycatcher), 355
finches, 8, 101; continental, 101, 103, 171,
253, 254, 256, 258, 260, 283, 387; North
American, 133; West Indian, 253, 256
finch-tanager, 256

fish, 404, 405, 410. *See also* haplochromine cichlid fish
fitness, 191, 193, 209, 213, 251, 335, 345, 347, 353, 358, 370; absolute, 189; relative, 189
FitzRoy, Captain, 8, 52
flammeus, Asio
flaveola, Coereba
fledging, 102, 106, 128, 148, 154, 195, 202, 215, 244, 381
fledgling, 100, 106, 111, 112, 128, 144, 165, 171, 181, 202, 207, 244, 247, 366, 383
flocks, 150, 160, 162, 287, 370
Floreana (Santa María, Charles), 4, 29, 30, 47, 52-54, 58-64, 66, 93, 94, 160, 234, 256, 266, 268, 300, 304, 305, 316, 337, 390, 413
floribunda, Pisonia
flowers, 98, 99, 116, 125, 128, 131, 143, 146, 148, 149, 152, 162-164, 168, 208, 209, 211, 218, 219, 306-308, 310, 361, 363, 373, 380, 393
flycatchers, 65, 272, 355, 400
food limitation and shortage, 125, 152, 154, 158, 160, 162, 164, 166-168, 170, 171, 173, 174, 208, 290, 291, 326, 344, 359, 362, 373, 390, 398
food supply, 10, 12, 125-127, 129, 139, 147, 152, 153, 158-160, 167, 168, 173, 174, 185, 209, 221, 279, 289-291, 293, 294, 306, 308, 312-314, 320, 332, 333, 336, 338-341, 343, 345, 347, 359, 361-363, 372, 386, 390, 391, 396-400
foraging, *see* feeding behavior, habits, and skills
fortis, Geospiza
fossils, 3, 11, 47, 54, 253, 256, 257, 260, 261, 263, 268, 272, 280, 283, 390, 404
founder effects, *see* speciation, factors of importance
frequency distributions, of measurements, 77, 78, 80, 81, 196, 206, 214, 230, 292, 321, 322, 326-328, 333, 336, 340, 347, 396. *See also* bimodal frequency distribution
Fringilla coelebs (chaffinch), xii
fruits, 6, 48, 117, 118, 120, 123, 125-127, 138, 142, 143, 146-148, 152, 158, 159, 211, 212, 217-220, 290-292, 300, 312, 313, 360, 393

fuelleborni, Labeotropheus
fuliginosa, Geospiza

galapageia, Castela
galapageia, Paspalum
galapageum, Psidium
galapagoensis, Buteo
galapagoensis, Zenaida
Galápagos, *see* islands, physical features
Galápagos dove, *see Zenaida galapagoensis*
Galápagos hawk, *see Buteo galapagoensis*
Galápagos martin, *see Progne modesta*
Galápagos mockingbird, *see Nesomimus parvulus*
Galápagos rail, *see Laterallus spilonotus*
Gardner by Española, 4, 55, 61, 86, 217, 268, 280
Gardner by Floreana, 4, 61, 400
garúa, *see* climate, garúa
gecko, *see Phyllodactylus*
genera, difficulties of classifying, 51
generalized forms and generalists, 136, 178, 218, 257, 265, 300, 332, 347, 387, 395, 396, 398
generation length, 170, 175
genetic characteristics, parameters, and change, xi, 12, 55, 258, 278, 287, 375, 376, 383, 384, 386, 387, 405-407, 410, 411; bottlenecks, 407; introgression and gene flow, 177, 178, 180, 199, 202, 206, 207, 215, 217, 219, 220, 263, 265, 268, 272, 353, 387, 394-396, 398; linkage disequilibrium, 204, 376; mutation, 177, 178, 199, 206, 220, 282, 287-289, 294, 312, 378, 386, 395; number of loci, 204, 278, 406; pleiotropy, 376; relatedness, 256, 372, 411; sex-linked genes, 197. *See also* genetic variance *and* covariance, genetic
genetic (random) drift, 177, 178, 197, 199, 214, 220, 274, 287, 357, 378, 381, 395, 406, 407, 410
genetic transilience, 281, 407, 408
genetic variance and variation, 52, 180, 183, 184, 199, 204, 206, 220, 275, 282, 341, 376-380, 382, 384, 387, 394, 398, 406, 407; additive, 177, 180, 183, 206, 220, 406, 407; dominance, 177, 180; epistasis, 177, 180, 278
genotype, 183, 184, 204
Genovesa (Tower), 4, 30, 37, 40, 52-57,

63, 64, 69-75, 84, 86, 99-112, 123, 128-131, 143, 149-156, 159-162, 165, 171, 180, 183, 200, 211-217, 221, 227, 228, 234-240, 245-250, 270, 276-284, 291-295, 298-313, 340, 347-353, 360-369, 376-380, 390, 394, 395, 401, 413

Genyochromis mento

geographical: contiguity of populations, 274, 284 (*see also* speciation, parapatric); distributions, 9, 10, 12, 60, 62, 64, 66, 257, 259, 266, 269, 274, 293, 306, 315-317, 320, 321, 326; forms, 10, 266, 268, 288, 305, 353 (*see also* subspecies); isolation, 10, 268, 272-274, 288, 293, 294, 301, 313, 390, 396, 407, 408; replacement, 8; restriction, 62, 66, 298, 305, 306, 390; variation in morphology, 71, 92-96, 207, 222, 266, 267, 274, 293, 360, 361. *See also* speciation, factors of importance

Geospiza (ground finches), 6, 8, 45-47, 51, 55, 56, 61-66, 78-83, 86-93, 96, 101-104, 110-118, 122, 125, 129, 130, 138, 148-170, 176, 180, 202, 207, 220-229, 234-240, 252-260, 282-295, 298-301, 313-323, 326, 331-337, 346, 358-360, 364-366, 371, 374, 377, 381, 384-393, 396-403

Geospiza conirostris (large cactus finch), 5-11, 47, 53-57, 60, 61, 66, 70-75, 84, 86, 91-94, 102, 106-111, 120, 128, 129, 140, 150, 153, 171, 176, 180-183, 200, 201, 212-221, 234-250, 276-284, 295, 298-304, 312-320, 332-337, 346, 359, 363-372, 376-377, 380, 384, 390-395

Geospiza darwini, 55, 390

Geospiza difficilis (sharp-beaked ground finch), 3-7, 11, 47, 52-56, 61-66, 70, 75, 80, 83-87, 92, 93, 99-112, 116, 120, 128-131, 150, 153, 161, 167, 176, 180, 199, 201, 215-219, 225-230, 234-241, 247, 256-260, 266, 295, 298-300, 304-321, 332-337, 340, 347-351, 360-367, 373, 374, 384, 389-395, Pl. 3, 4; *debilirostris*, 305; *difficilis*, 305; *nebulosa*, 47, 305; *septentrionalis*, 305

Geospiza fortis (medium ground finch), 3-7, 11, 16, 46, 47, 53-56, 61, 62, 74-83, 87-97, 102-127, 132-139, 152-157, 164-167, 170-209, 213-250, 257-266, 275-278, 284, 285, 293-303, 310, 312, 316-326,

333-350, 355, 356, 360-363, 367-372, 375-384, 387, 388, 394-396, Pl. 6

Geospiza fuliginosa (small ground finch), 3-7, 11, 47, 53-56, 61, 62, 66, 77-83, 89-96, 99, 102, 109-112, 115-135, 150, 151, 164-167, 171-176, 199-207, 213, 219-231, 234-242, 247, 257-261, 285, 295-301, 304-310, 313, 316, 319-326, 332-351, 354, 356, 361-365, 370, 378, 379, 383, 387, 390, 394, 396

Geospiza magnirostris (large ground finch), 5-7, 11, 14, 45, 47, 52-66, 75-80, 83-96, 102-129, 134-138, 151, 153, 164-167, 171, 175, 176, 199-202, 207, 208, 212-215, 218, 222-225, 231, 234, 241, 247, 256-258, 276, 285, 295-304, 312, 316, 319, 320, 333-337, 345, 349, 358-361, 367, 378-383, 388, 390, 394, 397-400, Pl. 3, 6

Geospiza nebulosa, 47

Geospiza scandens (cactus finch), 5-7, 11, 17, 47, 53-56, 60-66, 73, 80-94, 97-109, 116, 125-131, 144, 150, 154, 157, 164-167, 170-176, 181, 185, 194, 199-210, 214-234, 238-242, 247-249, 260, 266, 278-281, 295, 299-303, 310, 312, 316-326, 332, 345, 350, 359, 363-372, 377-384, 393-395, Pl. 4

G. conirostris × *G. difficilis*, 215

G. conirostris × *G. magnirostris*, 215

G. fortis × *G. fuliginosa*, 204

G. fortis × *G. scandens*, 200, 249

Geospizinae, xi, 286, 388

geospizine, 234, 241

gizzard, 86, 87, 96, 360, 373, 374, 394

goats, 30, 31, 64, 304

gonys, *see* mandible, lower

Gossypium barbadense, 145

Gotland (island, Swedish), 355

Gould, John, 8, 9

granivore, 332, 333, 337, 340, 347, 387, 396, 399, 403

graveolens, Bursera

great blue heron, *see Ardea herodias*

great egret, *see Casmerodius albus*

great tit, *see Parus major*

green-backed heron, *see Butorides striatus*

groove-billed ani, *see Crotophaga sulcirostris*

ground finches, *see Geospiza*

growth (ontogeny), 12, 82, 100-112, 182,

growth (ontogeny) (*cont.*)
194, 250, 379, 381-383, 388, 392, 410;
start (onset), 100, 106, 110-112; end (offset), 100, 106, 110-112; duration, 100,
106, 112, 194; rate, 100, 106; absolute,
100, 102, 103, 106; relative, 100, 106-
112, 392; curves (trajectories), 100, 105,
107, 109-112, 194, 381, 382; stages, 382,
383, 388; embryonic, 100, 101, 109, 110,
112; nestling, 100-104, 107-110, 112;
post-fledging, 111, 112, 383. *See also* allometry, dynamic
guilds (feeding), 299, 300, 313, 314, 329,
331, 332, 346
guyot (submerged island), 19, 261

habeli, Camarhynchus psittacula
habeliana, Ipomoea
habitat selection, *see* adaptation, adaptive
habitat selection
habitats, 10, 94, 117, 139, 152, 218, 256,
261, 274, 275, 284, 287, 288, 332, 367,
404; arid, 3, 28-30, 32, 39, 40, 62, 145,
147, 148, 255, 263, 272, 298, 362, 364,
368, 390, 397; fern-sedge-grass, 28, 29,
31, 42; littoral, 29, 43, 287; mesic, 263,
362, 390, 397; moist (humid) forest, 28-
31, 62, 150, 272, 298, 371, 390; rain-
forest, *see* humid tropics; transition forest, 28, 29, 31, 41, 62, 150, 272, 368,
390; parkland, 132, 145; woodland, 3,
132; differences between species in occupancy, 10, 117
hallux (hind toe plus claw), 56, 86, 96, 99,
307, 394
haplochromine cichlid fish, 408
Haplochromis compressiceps, 409
H. cyaneus, 409
H. euchilus, 409
H. livingstonii, 409
H. pardalis, 409
H. placodon, 409
H. rostratus, 409
H. similis, 409
hatching, 100-104, 106, 108-112, 128, 148,
277, 381, 392
Hawaiian islands, 46, 266, 360, 401, 403,
404, 408
hawk, *see Buteo galapagoensis*
heart weight, 86, 96, 360, 373, 374

Heliconius butterflies, 3
heliobates, Cactospiza
Heliotropium, 168, 185; *angiospermum*,
133, 155, 165, Pl. 9
helleri, Opuntia
helping at the nest, 372
Hemignathus procerus ('akialoa), 403
Hemignathus wilsoni ('akiapola'au), 403
heritability, 180-183, 192, 197, 209, 214,
220, 279, 376, 382, 394, 395. *See also*
genetic variance, inheritance
Hermanos, Los (Crossmans), 4, 33, 34, 61,
77, 78, 282, 295, 297, 340-343, 352, 396
herodias, Ardea
heterochrony, 381
Himatione sanguinea ('apapane), 403
hippoboscid flies, 373
Hood mockingbird, *see Nesomimus macdonaldi*
honeycreeper finches, 401, 403, 404
horses, 30
hotspot theory of island origin, 19, 31
howellii, Portulaca
human activity, 64, 66, 390, 404, 405; habitat destruction, 61; introduction of species, 65
humid tropics, 27, 32, 219, 255
hybridization, *see* interbreeding
hybrids, *see* interbreeding
hypoleuca, Ficedula

ibis, Bubulcus
identification of specimens, difficulties, 6,
8, 111, 222, 340
iguana, land, *see Conolophus subcristatus*
iguana, marine, *see Amblyrhynchus cristatus*
i'iwi, *see Vestiaria coccinea*
immatures, 60, 62, 303, 322
immigration and immigrants, 52, 61, 62,
64, 75, 164, 165, 171, 172, 199, 202,
205, 207, 208, 213, 217, 220, 263, 264,
267, 272, 275, 280, 285, 303, 310, 311,
314, 348-356, 378, 394, 397, 398
imprinting, 242, 244, 247, 249, 252, 354,
392. *See also* misimprinting
inbreeding, 197, 250, 251
incubation, 101, 148, 150, 363, 371, 372,
374
inheritance, 278, 375

inornata, Pinaroloxias
insect larvae, 3, 208, 212, 216, 276, 372, 393, 399, 401; pupae, 3, 209, 212, 216
interbreeding of species (hybridization), 12, 51, 52, 55, 199, 201, 202, 204-206, 208, 215, 218, 220-223, 247, 248, 251, 261, 263, 276, 280-283, 340, 341, 349-356, 387, 390, 392, 394, 395, 397, 398, 410; hybrids formed, 10, 51, 199-205, 207, 215, 220, 282, 341, 353; backcrossing of hybrids, 206, 207, 215, 220, 394; fertility of hybrids, 206, 207, 353, 394, 398; intergeneric hybrids, 199, 282, 288, 353; intersterility, 10, 265, 288, 353, 355; premating barrier, 353, 354, 398; postmating disadvantage, 206, 353; selective disadvantage, 10, 353, 355; viability of hybrids, 204, 206, 220, 288, 353, 355
intergradation of species in size, 6-8, 199, 201, 202, 204, 205, 222, 355, 358
intertropical convergence zone, 21
intestine, length, 86, 360, 373, 374, 394
introgression of genes, *see* genetic characteristics
invertebrates, 131, 305-307, 309
Ipomoea habeliana, 156
Ipomoea linearifolia, 148
Isabela (Albermarle), 4, 21, 22, 29, 30, 33, 43, 53, 60-63, 67, 79, 94, 148, 151, 162, 176, 222, 235, 267, 268, 274, 304, 316, 340, 341, 413
islands, physical features: area, 21, 31, 62, 94, 207, 273, 293, 311, 362, 390, 397; elevation, 21, 24, 29, 31, 62, 79, 94, 152, 159, 164, 165, 167, 274, 287, 298, 304, 306, 311, 362, 364, 366, 367, 390, 391, 397; isolation, 21, 31, 62, 94, 268, 273, 293, 310, 311, 328, 352, 390, 397, 400; lava flows, 21, 22, 33, 36, 43, 366, 403; lava tubes, 256; sea level, changes, 30; volcanic, 19, 21, 31
isometry, 82, 86, 89, 95, 106, 361. *See also* allometry

jacarina, Volatinia
Jasminocereus thouarsii, 28, 40
Java sparrow, 45

kin recognition, 250, 251
kona, Loxioides

Labeotropheus fuelleborni, 409
Labidochromis vellicans, 409
Lack, David, 10-12, 286, 287, 312, 314
lace-wing flies, 405
Lake Malawi, 408, 409
Lake Tanganyika, 408
Lake Victoria, 408
land bridge, 10
land iguana, *see Conolophus subcristatus*
Lantana peduncularis, 147
large-billed flycatcher, *see Myiarchus magnirostris*
large cactus finch, *see Geospiza conirostris*
large ground finch, *see Geospiza magnirostris*
large tree finch, *see Camarhynchus psittacula*
lava gull, 364
lava heron, *see Butorides sundevalli*
lava lizard, *see Tropidurus*
Laterallus spilonotus (Galápagos rail), 65
leaf litter, 6
leaf petiole (used as a tool), 3, 372, 393
leaves, 6, 86, 117, 128, 147, 148, 219, 393, 400
Lethrinops brevis, 409
leucophlyctis, Cordia
limiting similarity, theory, 321, 323, 331
lineage, 3, 283, 284, 399
linearifolia, Ipomoea
livingstonii, Haplochromis
Loxioides balleui (palila), 360, 403
Loxioides kona (grosbeak finch), 403
Loxioides psittacea (o'u), 403
Loxops coccinea ('ākepa), 403
Loxops virens ('amakihi), 403
lutea, Cordia

macdonaldi, Nesomimus
macroevolutionary patterns, 404
"macro"-*Geospiza*, 400
macrophthalmus, Ramphochromis
maculata, Paroreomyza
magnirostris, Geospiza
magnirostris, Myiarchus
major, Parus
malacophylla, Bursera
malaria, *see* parasites, avian malaria
mammals, 64, 106; introduced, 30, 65
mamo, *see Drepanis pacifica*

mandibles, 45, 117; lower, 55, 80, 116, 122, 138, 308, 401; upper, 45, 78, 80, 81, 116, 122, 138, 401, 403

mangrove, 43, 274

mangrove finch, *see Cactospiza heliobates*

Marchena (Bindloe), 4, 21, 22, 28, 31, 37, 50, 53, 55, 62, 63, 94, 102, 112, 123, 159, 161, 162, 280, 281, 306-310, 340, 347, 361, 366, 413

marine iguana, *see Amblyrhynchus cristatus*

marsupials, 404

masked booby, *see Sula dactylatra*

mass, *see* weight

mate choice, 12, 193, 222, 224, 244, 249, 251, 252, 352-354, 392, 410; by females, 193, 195, 220, 244, 245, 250, 251, 277, 354, 370, 392

mate recognition system, 251, 408, 410

mating behavior (*see also* pair formation), 222, 223, 249, 352, 369, 370; random, 192, 249, 250, 279; nonrandom (assortative), 192, 193, 241, 249, 250, 276, 278, 279

Maytenus octagona, 151

means, of morphological traits, 77, 79-81, 83, 87-89, 92, 94, 95, 101, 105, 111, 175, 185, 189, 196, 197, 204, 209, 215, 217, 218, 220, 250, 272, 299, 301, 317, 324, 340-343, 360, 361, 376, 382, 384, 395

medium ground finch, *see Geospiza fortis*

medium tree finch, *see Camarhynchus pauper*

megasperma, Opuntia

melacorhyphus, Coccyzus

Melamprosops phaeosoma, 403

melanin, 277, 370

Melanospiza richardsonii, 254, 255

melanotis, Nesomimus

melodia, Melospiza

Melospiza melodia (song sparrow), 183

mento, Genyochromis

mericarp (of *Tribulus cistoides*), 121,122, 135-137, 142, 190

Merremia aegyptica, 148

metabolic, 86, 122, 161, 190, 193, 307, 361, 373

mice, 30

Miconia robinsonia, 29

microevolutionary processes, 404

"micro"-*Geospiza*, 332, 339, 400

microhabitat, 118, 219, 221

Microcryptorhynchus beetles, *see Miocalles miersii, Cacabus*

millet, 133, 134

misimprinting, 246-249, 252, 392

Miocalles (Microcryptorhynchus) beetles, 404

mockingbirds (*Nesomimus*), 8, 64, 148, 171, 268-270, 272, 314, 363, 366

models: of community assembly, *see* communities of finches, models; of feeding efficiency, *see* feeding efficiency in relation to bill size; of the maintenance of variation, *see* variation, model; of resource exploitation, *see* adaptation, adaptive variation model; of speciation, *see* speciation, models; null, *see* null model

modern synthesis of evolutionary thought, 10, 285

modesta, Progne

molt, 148, 150, 254-256, 367

monogamy, 193, 223

monophyly, 389, 390

Moorea, 404

morabine grasshoppers, 406

morphological complementarity between ground finches and tree finches, 92, 95

mortality and survival, 150, 152, 154, 157, 160, 163-165, 168-171, 173, 174, 184, 185, 189-192, 194-196, 199, 202, 204, 209, 213, 214, 217, 220, 300, 343-345, 353, 361, 362, 368, 372, 395

moths, 366, 404

Muller's theory of differential mutation, 287, 288, 294, 312

muscles, *see* skull, muscles

M. adductor mandibulae externus superficialis, 113, 115

M. depressor mandibulae, 115

M. pterygoideus ventralis lateralis, 115

museum specimens, xi, 6, 9, 77-81, 125, 199, 222, 268, 369

mutation, *see* genetic characteristics, *and* Muller's theory of differential mutation

Myiarchus magnirostris (large-billed flycatcher), 65

names of Darwin's Finches, 47

narboroughi, Nesoryzomys

native rat, *see Oryzomys and Nesoryzomys*
natural selection, xi, xii, 9, 175, 177, 178, 180, 183, 184, 190-194, 196, 197, 199, 206, 209, 217, 219-221, 250, 258, 265, 268, 272, 281, 283, 284, 286-288, 290, 294, 307, 311-313, 335, 341, 345-348, 353-355, 357-362, 368, 373, 375, 376, 378, 381-383, 395, 398, 399, 406, 407, 410; antagonistic (countervailing or opposing), 193-196, 209, 214, 218, 220, 221, 360, 378, 387, 395; differential, 180, 189, 190, 192, 214; directional, 184, 185, 189, 192, 193, 195, 196, 209, 214, 217, 218, 220, 221, 282, 293, 344, 354, 361, 375-377, 380, 381, 383, 394, 395, 398, 407; directional but oscillating, 184, 193, 195, 196, 218, 220; disruptive, 196, 208, 217, 218, 221, 275, 279, 284 344, 347, 395, 398; distance, 383-386, 388; divergent, 265, 288; forces or pressures, xii, 178, 183, 193, 195, 197, 214, 217, 220, 259, 274, 282, 284, 287, 288, 293, 360, 361, 370, 371, 375, 377-379, 383, 386-388, 394, 396-398, 405-407, 411; gradient, 189-192, 377, 379, 380, 382-384; intensity, magnitude, or strength, 184, 189, 192, 195, 196, 209, 220, 221, 274, 361, 375, 376, 381, 387, 388; reproductive, 195-197, 214; stabilizing, 178, 184, 185, 193, 196, 197, 209, 214, 217-221, 360, 381, 386, 394-396; survival, 193, 195, 214, 361; target, 184, 189-192, 195, 375, 379, 383, 386. *See also* evolutionary response to selection
Nazca plate, 19, 20
nebouxii, Sula
nebulosa, Geospiza difficilis
nectar 6, 99, 116-118, 125, 131, 148-150, 156, 161-165, 196, 208, 218, 219, 305-307, 361, 363, 380, 393, 403. *See also* extrafloral nectaries
Neocrex erythrops (paint-billed crake), 65
Nesomimus macdonaldi (Hood mockingbird), 65, 268, 271
N. melanotis (Chatham mockingbird), 65, 268, 270
N. parvulus (Galápagos mockingbird), 65, 270, 272
N. trifasciatus (Charles mockingbird), 65, 268, 271

Nesoryzomys narboroughi (native rat), 148
nestling, (chick), 100, 102, 103, 106, 111, 112, 128, 129, 139, 148, 171, 181, 202, 207, 244, 247, 277-280, 363, 365, 366, 371, 372, 374, 391, 392
nests, 102, 111, 112, 150, 151, 154, 183, 225, 230, 247, 255, 256, 363, 371
nest site, 171, 234
niche (ecological), xii, 173, 209, 218, 219, 275, 284, 285, 294, 295, 298, 300-303, 305, 308, 310, 312, 338, 339, 341, 345, 360, 386, 387, 393, 396, 399, 400, 410; included, 173, 208; width (breadth), 209, 219, 265, 301, 302
Nihoa finch, *see Telespiza cantans*
nonadaptive traits, and changes, 259, 287, 294, 359, 365, 369, 381, 382, 406, 410
null model, 321, 323, 328, 330, 346
number of Darwin's Finch species on islands, 62, 262, 272, 293, 331, 338, 340, 346, 347, 396
nyassae, Aulonacara
Nyctanassa violacea (yellow-crowned night heron), 64, 65, 171, 366

Oceanodroma (storm petrel), 171
octagona, Maytenus
Öland (island, Swedish), 355
olivacea, Certhidea
ontogeny, *see* growth
Opuntia (cactus), 3, 28, 64, 125, 128, 130, 131, 135, 148, 150-152, 162-165, 168, 171, 185, 191, 208, 214, 218, 225, 240, 276, 300, 302, 304, 310, 363, 366, 380; *O. echios*, 38-40, 98, 126, 127, 132, 145, 151, 194, 210, 291, 344, 345, 380; *O. helleri*, 37, 149, 156, 211, 212, 217, 291, Pl. 10; *O. megasperma*, 37, 380, 400
origin of species, 10, 253, 254, 275, 282, 283, 352
oriole, 272
Oryzomys bauri (native rat), 148
o'u, *see Loxioides psittacea*
ovata, Waltheria
owls, *see Asio flammeus* and *Tylo alba*

pacifica, Drepanis
paint-billed crake, *see Neocrex erythrops*

pair formation, 192, 223, 249, 250, 265, 279, 354
palate, horny, 115, 138
pallida, Cactospiza
palila, *see Loxioides bailleui*
Palmeria dolei (crested honeycreeper), 403
parasites, 65, 171; avian malaria (*Plasmodium*), 171; avian pox, 65, 171, 172; worms, 65
pardalis, Haplochromis
parental care, 128, 150, 223, 251, 410; misdirected, 372
Paroreomyza Maculata (creeper), 403
parrotbill, *see Pseudonestor xanthophrys*
Partula, snails, 404
Parus caeruleus (blue tit), xii, 183
Parus major (great tit), 363
parvulus, Camarhynchus
parvulus, Nesomimus
Paspalum galapageia, 135
passerine, 46, 79, 103, 148, 170, 175, 408
Passiflora, 125-127
pauciflora, Scutia
pauper, Camarhynchus
peduncularis, Lantana
peduncularis, Scalesia
perching ability, 99, 307
Peru, 21, 25-27, 254, 255
petechia, Dendroica
Petrotilapia tridentiger, 409
phaeosoma, Melamprosops
phenology: finches, 147, 148, 363, 390; plants, *see* plants, phenology
phenotypic variation, *see* variation, morphological
Phyllodactylus (gecko), 260
phylogeny, 375, 381, 383, 384, 410; reconstruction, 375-388; relationships (affinities), 6, 10, 254, 255, 257-260, 268, 283, 304, 384, 385, 387; tree (evolutionary), 10, 11, 257, 266, 384, 385, 387
physiological, 82, 86, 360, 362, 364, 371, 373
pied flycatcher, *see Ficedula hypoleuca*
pigs, 30
Pinaroloxias inornata (Cocos finch), 5, 11, 45-49, 51, 65, 66, 91, 116, 150, 175, 176, 218, 221, 231, 254, 255, 258, 277, 371, 387, 389, 391, 393, 395, 400, 401, 403

Pinta (Abingdon), 4, 21, 22, 29, 31, 41, 52, 53, 60, 63, 70, 94, 99, 119, 123, 130-132, 134, 135, 159, 161, 162, 164, 165, 167, 173, 223, 224, 226-230, 235, 238-240, 247, 249, 298, 304, 306-310, 312, 316, 349, 351, 352, 366, 391, 413, Pl.3
Pinzón (Duncan), 4, 28, 53, 62-64, 66, 160, 413
Piscidia, 29
Pisonia floribunda, 29, 41
placodon, Haplochromis
plants: angiosperm, 293; colonization, 21, 27, 311; dry-deciduous, 28, 29, 31, 147; endemic, 27-30; evergreen, 28, 44; human introductions, 27; indigenous, 27; number of species, 27, 94, 273, 293, 311; origins, 27, 28; patchy distribution, 21; phenology, 147, 148; weeds, 27. *See also* habitats
Plasmodium, malaria, *see* parasites, avian malaria
Platyspiza crassirostris (vegetarian finch), 3-5, 11, 18, 46, 47, 50, 51, 53, 61, 62, 66, 83, 86, 87, 91-95, 116, 117, 122, 176, 198, 201, 231, 234, 266, 320, 360, 361, 373, 389, 394, Pl.4
Plazas, 4, 21, 61
Plaza Sur, 31, 144, 223, 224, 227, 229, 233, 238, 239, 349, 350, 352
plumage, 66, 92, 94, 188, 224, 227, 252, 254, 255, 257-259, 272, 358, 364-371, 373, 384, 387, 391, 392, 395; adult black, 45, 46, 50, 62, 64, 193, 250, 254, 255, 258, 260, 358, 364-371, 391, 399; cryptic, 364, 365, 368, 370, 374, 399; geographical variation, 94, 259, 366, 369, 374; heat absorption, 370, 371; immature, 45, 62, 365, 369; mosaic molt, 46, 254, 391; protection function, 364, 370, 371; signaling function, 364, 369-371; throat patch, 45. *See also* sexual dichromatism
pollen, 6, 98, 118, 125, 128, 148-150, 156, 161-165, 168, 196, 208, 217, 218, 307, 332, 363, 391, 393
polygamy, 223, 410
polymorphism: bill color, *see* bill, color polymorphism; protein, *see* protein
Polynesian islands, 404
population: age structure, 154, 157, 411;

ecological subdivision, 136, 178-180, 196, 219, 275, 276, 279, 280, 367, 368; numbers, size, 147, 152, 154, 159-169, 171, 173, 174, 184, 187, 199, 202, 207, 214, 279, 287, 291, 298, 309, 333, 357, 391, 395; reproductive subdivision, 249, 276, 278-280, 284

Portulaca, 127, 185; *P. howellii*, 156

precipitation, *see* climate

predators and predation, 64, 171, 173, 174, 357, 363-366, 368, 370, 372, 374, 391, 399

prevailing winds, 24, 29

primitive traits, 253, 258, 259

procerus, Hemignathus

production, finches, 148, 157

Progne modesta (Galápagos martin), 65

protein, 258-260, 262, 389

Pseudonestor xanthrophrys (parrotbill), 403

Pseudotropheus zebra, 409

Psidium galapageum, 29

psilostachya, Tournefortia

psittacea, Loxioides

psittacula, Camarhynchus

Puerto Bacquerizo (San Cristóbal), 24, 26, 27

Punta Tortuga, Isabela, 151

Pyrocephalus rubinus (vermilion flycatcher), 65

Pyrrhula pyrrhula (bullfinch), 360

pyrrhula, Pyrrhula

Rábida (Jervis), 4, 31, 53, 63, 222, 337, 413

rain, *see* climate

rainshadow, 24, 29

Rapa, 60-62, 64

rarity, 207, 208, 268

Ramphochromis macrophthalmus, 409

rat (*Rattus rattus*)

Rattus rattus (rat), 30, 148, 256

rattus, Rattus

red-footed booby, *see Sula sula*

reinforcement of bill differences between species, 348, 353, 355

relative, closest to Darwin's Finches, 254-256, 258

repeatability and error of measurement, 89, 181, 209, 214, 376

reproduction, *see* breeding

reproductive absorption into populations, 355, 356, 397; confusion, 265, 348-350, 353, 355; isolation, *see* speciation, factors of importance

reptiles, 3

residents, 61, 75, 97, 125, 165, 169, 171, 202, 206-208, 214, 220, 224, 246, 263, 272, 280, 285, 298, 303, 310, 348-356, 365, 394, 397, 398

resources: partitioning, 197 (*see also* diets, differences between species); spectrum, 178, 179, 219, 220. *See also* adaptation, adaptive models of variation

Rhynchosia, 125-127

richardsonii, Melanospiza

Richmondena cardinalis (cardinal), 133, 134

ritualized behavior, 3

robinsonia, Miconia

rodents, xii, 256

rostratus, Haplochromis

rubinus, Pyrocephalus

saguinea, Himatione

San Cristóbal (Chatham), 4, 24, 26, 27, 29-31, 52-54, 60, 61, 63, 64, 66, 93, 94, 148, 160, 222, 268, 270, 276, 298, 304, 305, 316, 337, 367, 390, 413

San Francisco, 9

Santa Cruz (Indefatigable), xiii, 4, 6, 21-30, 38-42, 53, 54, 63-72, 75-81, 89, 94, 123-127, 132, 139, 145, 148-152, 159-164, 171, 175-180, 199-210, 222-224, 233, 242, 247, 256, 261, 267, 268, 275-284, 296-298, 304, 305, 316, 341-343, 350, 365, 366, 380, 390, 394, 413, Pl. 4, 10

Santa Fe (Barrington), 4, 31, 53, 60, 63, 148, 162-164, 337, 366, 413

Santiago (San Salvador, James), 4, 21, 29-31, 35, 45, 50, 52, 53, 63, 68, 94, 130, 148, 159, 162, 171, 176, 235, 261, 269, 298, 304, 316, 360, 390, 413

Sarcostemma angustissima, 125, 126, Pl. 10

Scalesia, 27-29, 151, 364; *S. pendunculata*, 29, 39, 42

scandens, Geospiza

scouleri, Croton

Scutia pauciflora, 28, 125, 126

Scutia spicata, 290

seabird, 129, 131, 393

seasonality (climatic), 21, 23, 24, 31, 362, 390, 391, 397; dry season, xiii, 24, 25, 39, 60, 80, 118, 123, 125-130, 138, 139, 145, 147, 148, 150, 161-168, 173, 174, 214, 217, 223, 276, 279, 292, 301, 306, 308, 317, 318, 322, 332, 333, 335, 343, 359, 362-364, 366, 368, 370, 374, 391, 393, 396; wet season, xiii, 24, 25, 39, 125-127, 138, 145, 147, 148, 152, 166, 185, 292, 308, 333, 391, Pl. 9

secondary contact of previously allopatric forms, 265, 266, 268, 273, 276, 283-285, 312, 348, 354, 355, 405, 410

seeds, 6, 49, 117, 118, 122, 123, 125, 128, 130-138, 141, 142, 147, 148, 154, 156, 158-163, 173, 185, 190, 194, 196, 208, 209, 214, 217-220, 291-293, 295, 303, 305, 307, 309, 312, 313, 332-335, 345, 358, 360, 380, 391, 393, 398; size (depth) and hardness, 117-120, 122-124, 131, 133, 134, 136-139, 185, 187, 194, 196, 209, 212, 217, 220, 291-293, 295, 303, 309, 321, 333, 334, 336, 343, 344, 358, 380, 393-395, 398, 400; production, 147, 148, 152, 154, 166, 168, 174, 185; supply, 126, 127, 148, 165, 166, 168, 169, 184, 185, 187, 193, 305, 335, see also food supply

seedcracker (to measure the force to crack a seed), 119, 141

selection, see natural selection and sexual selection

selective peaks, multiple, 406

septentrionalis, Geospiza difficilis

Servicio Parque Nacional Galápagos (Galápagos National Parks Service), xiii, 31

Sex chromosomes, see chromosomes

sex ratio, 192, 193, 220, 250, 369, 395

sexual dichromatism, 70, 79, 196, 371, 374

sexual dimorphism, 70, 136, 192, 195-197, 395

sexual selection, 192, 195, 196, 209, 214, 219, 220, 354, 369, 371, 374, 395, 408, 410; differential, 193; directional, 209; stabilizing, 209

Seymour, 4, 53, 413

sharp-beaked ground finch, see Geospiza difficilis

shifting balance theory of evolution, 406, 408

short-eared owl, see Asio flammeus

similis, Haplochromis

size neighbors on a morphological axis, 77, 222, 317-319, 338

skeletons, skeletal traits, 6, 86, 254, 381, 386, 387

skull, 113-115, 128, 138, 139, 367, 373, 374; muscles of adduction, 113, 115-117, 128, 138, 194, 373, 374, 393; pneumatization (ossification), 115, 194, 367, 373, 374; weight, 86, 96; zygomatic bar, 45

small ground finch, see Geospiza fuliginosa

small tree finch, see Camarhynchus parvulus

smooth-billed ani, see Crotophaga ani

snails, 305-307, 382, 404. See also Cepaea and Partula

song, 52, 148, 193, 199, 215, 217, 223, 224, 230-252, 254, 276-281, 340, 352, 363, 372, 374, 392, 401; whistle or hiss, 230, 231; advertising, 230-233, 235, 241, 252, 392; types, 217, 218, 231-238, 240-250, 252, 276, 277, 279-281; type A (derived), 217, 218, 231, 232, 234-236, 238, 240, 243, 245-248, 276, 278, 281; type B (basic), 217, 218, 231, 232, 234-236, 238, 240, 243, 245, 246, 248, 254, 276, 278; subtypes, 250; antiphonal, 234; bilingual singing, 234, 236; characteristics in relation to habitat variation, 372, 374; countersinging, 247; convergence, 234, 254; divergence, 234, 279, 280, 352; heterotypic, 238, 240, 244-246, 277, 279; homotypic, 238, 240, 245, 246, 249, 277, 279; repertoire, 234, 235, 280; species specific, 230, 234, 235, 237, 241, 242, 280, 340, 392; variants, 235, 240. See also imprinting, misimprinting, experiments on species discrimination, and species discrimination by song

song sparrow, see Melospiza melodia

southern oscillation, 25

specialized forms and specialists, 10, 86, 136, 178, 179, 196, 208, 219, 220, 257, 265, 298, 300, 332, 395. See also diet, specialized

speciation, events, 197, 257-284, 288, 289, 312, 347, 371, 399, 403-408, 410

speciation, factors of importance: ecological interactions and isolation, 273, 275, 282,

285, 286, 288, 289, 294, 295, 298-314,
348, 353-355, 388, 399, 401; founder effects (events), 281, 283, 287, 359, 407,
408, 410; genetic reorganization, revolutions, 283, 407, 408, 410 (*see also* genetic transilience); geographical isolation
and replacement, 8, 268, 272, 403, 405-
408; hybridization, 281, 282; population
structure, 405-407; reproductive interactions and isolation, 265, 273, 275, 279,
282, 284, 285, 312, 348-356, 388, 392,
398, 399, 401, 405, 410
speciation, models and theories: allopatric,
9, 263-268, 273, 275, 276, 280, 283-286,
288, 289, 294, 299, 312, 314, 348, 351,
353, 399, 404, 405, 410; complete allopatric, 289, 291, 399; founder-flush, 407;
parapatric, 274, 275, 284, 405; partial allopatric, 289; peripatric, 407, 408; stasi-
patric, 406; sympatric, 274, 275, 279,
284, 405
species discrimination, 222-252, 348-352,
355, 387, 392, 398; by morphology, 224-
230, 241, 242, 248, 249, 251, 252, 348-
352, 355, 392; by song, 224, 230, 237-
242, 247, 248, 251, 252, 352, 392
species-recognition, 222, 224, 230, 241,
242, 247-251, 348, 352
species transitions and transformations, *see*
transformation of species
spicata, Scutia
spiders, 69, 128, 129, 139, 148, 196
spilontus, Laterallus
Sporophila telasco, 332
stasipatry, *see* speciation, models and theories
stigmas, 149, 211
St. Lucia (West Indies), 253
stochastic (random) processes, 406, 407,
411. *See also* communities of finches,
randomness, *and* genetic drift
storm petrel, *see Oceanodroma*
Stresemann, E., 10, 12
striatus, Butorides
stuffed specimens, used as models in experiments, 224-226, 242, 252, 348, 365, 392
style (plant part), 149
subcristatus, Conolophus
subdivided population, *see* population
subspecies, 11, 47, 55, 60, 63, 66, 222,

285, 295, 305, 359, 390
Sula (booby), 3, 374, 393
Sula dactylatra (masked booby), Pl. 4
Sula nebouxii (blue-footed booby), Pl. 9
Sula sula (red-footed booby), Pl. 4
sula, Sula
sulcirostris, Crotophaga
sundevalli, Butorides
supernormal stimulus, 350
survival, *see* mortality
Swarth, H., 10, 12
sympatry, 11, 51, 52, 119, 123, 138, 139,
215, 222, 225, 231, 234, 235, 241, 251,
252, 264, 266, 274-276, 279, 280, 284,
286, 288-290, 299, 302, 307, 313, 314,
318, 321, 323-329, 340-346, 348, 349,
351-354, 393, 396-399
syrinx (voice box), 254

tameness of Darwin's Finches, 118, 140
tanagers, 8, 256
tarsus, length, 52, 56, 57, 80, 83, 85, 86,
89, 95, 102, 103, 105, 107, 109-112,
181, 189, 307, 358, 377, 392
taxonomy, 9, 12, 51, 52, 56, 57, 60, 389,
390
telasco, Sporophila
Telespiza cantans (Nihoa finch), 403
Terminalia catappa, 48
termites, 3, 216, 393
territories, 148, 182, 192, 193, 208, 218,
223, 225, 234, 238, 240, 241, 244-247,
249-251, 277, 279, 370
thouarsii, Jasminocereus
Thraupidae, *see* tanagers
Tiaris, 254, 255
ticks, *see* feeding behavior, tick feeding
tongues, 373, 374
tool use, *see* feeding behavior
tortoise, 3, 8, 399
Tortuga (Brattle), 4, 43, 61, 161, 341
Tournefortia psilostachya, 40, 118, 126,
127
transformation of species, 92, 377-381, 384,
386-388, 405, 406
transmutation of species, Darwin's notebooks, 9
tree finches, 45-47, 50, 51, 62, 65, 66, 80,
82, 83, 86, 87, 91-95, 116, 118, 150,
152, 158, 175, 215, 223, 231, 234, 257-

tree finches (*cont.*)
260, 266, 283, 298-300, 314, 347, 359, 360, 364, 365, 369, 384, 385, 389-391, 393, 399, 401
trial and error learning, 372, 399
Tribulus cistoides (caltrop), 120-122, 135-138, 142, 185, 190, 209, 344, 345
tridentiger, Petrotilapia
trifasciatus, Nesomimus
trophic differential, 191
trophic gradient, 191
Tropidurus (lava lizard), 260; *delanotis*, 366
Tupac Yupanqui, Inca, 6
twigs, 3, 372, 374, 393, 399, 403
Tyto alba (barn owl), 64, 65, 256

ula-ai-hawana, *see Ciridops anna*
uropygial gland, 255

vagrant, 53, 62, 303, 308
variance, 77, 79, 80, 89, 175, 185, 197, 204, 209, 214, 217, 332; phenotypic, 180, 197, 204, 209, 376; genetic, *see* genetic variance
variants, xii, 6, 218
variation, morphological, xi, 6, 8, 10, 12, 55, 60, 71, 77-96, 113, 132-135, 139, 177, 178, 180, 183, 184, 193, 197, 207, 208, 214, 215, 218-221, 224, 253, 314, 354, 357-360, 373, 394-396, 398; in relation to abundance, 207, 208; coefficient of, 175, 176, 197; population (intra-), xii, 72, 132-135, 139, 175, 177, 178, 180, 196, 207, 208, 214, 215, 218-220, 263, 354, 360, 394-396, 398, 401; model of maintenance, 177, 178, 180, 219, 220
vegetarian finch, *see Platyspiza crassirostris*
vellicans, Lapidochromis
vermilion flycatcher, *see Pyrocephalus rubinus*

Vestiaria coccinea (i'iwi), 403
violacea, Nyctanassa
virens, Loxops
Volatinia, iacarina, 254, 255
Volcan Cerro Azul (Isabela), 29
Volcan Darwin (Isabela), 29

Wagner tree, 384, 385
Waltheria ovata, 28, 69, 99, 128, 147, 306, 307, 361, 373
warbler, 8
weight (mass), 52, 80, 82, 83, 86, 87, 89, 90, 95, 102, 103, 105-107, 109-112, 181, 189, 190, 193-195, 336, 361, 362, 377, 379, 390. *See also* body size
West Indies, 27, 253
wilsoni, Hemignathus
wing length, 52, 56, 57, 80, 82, 83, 86, 88-90, 95, 102, 103, 107, 112, 181, 189, 209, 358, 377, 392
Wolf (Wenman), 3, 4, 21, 30, 36, 37, 52, 53, 60-64, 102, 111, 112, 129-131, 199, 295, 298, 300, 310, 350, 365, 373, 393, 413, Pl. 5
woodpecker, 272, 401
woodpecker finch, *see Cactospiza pallida*
wren, 272

xanthrophrys, Pseudonestor

yellow-crowned night heron, *see Nyctanassa violacea*
yellow warbler, *see Dendroica petechia*

Zanthoxylum fagara, 29, 41, 127, Pl. 3
zebra, Pseudotropheus
Zenaida galapagoensis (dove), 65, 148, 314, 400

feeding ecology — p. 147,

p. 171

not seeking to dislodge / engage
with ideas of "ecology" but to
use in general sense of
term — e.g

ecology — detailed wording
see Peter x

natural selection g
intercon between birds
+ ~~good~~ food supply;
other variables, less
signif, predators,
amount of nesting sites etc.
Drought etc. ...
weather, plant, animal,
other life all interconnected.